Computational Analysis and Design of Bridge Structures

Chung C. Fu
Shuqing Wang

CRC Press
Taylor & Francis Group
Boca Raton London New York

CRC Press is an imprint of the
Taylor & Francis Group, an **informa** business

A SPON PRESS BOOK

CRC Press
Taylor & Francis Group
6000 Broken Sound Parkway NW, Suite 300
Boca Raton, FL 33487-2742

First issued in paperback 2017

© 2015 by Taylor & Francis Group, LLC
CRC Press is an imprint of Taylor & Francis Group, an Informa business

No claim to original U.S. Government works

ISBN-13: 978-1-4665-7984-2 (hbk)
ISBN-13: 978-1-138-74837-8 (pbk)

Library of Congress Cataloging-in-Publication Data

Fu, C. C. (Chung C.)
    Computational analysis and design of bridge structures / authors, Chung C. Fu and Shuqing Wang.
        pages cm
    Includes bibliographical references and index.
    ISBN 978-1-4665-7984-2 (hardback)
    1. Bridges--Design and construction--Data processing. I. Wang, Shuqing (Highway engineer) II. Title.

TG300.F785 2014
624.2'5--dc23   2014022853

Visit the Taylor & Francis Web site at
http://www.taylorandfrancis.com

and the CRC Press Web site at
http://www.crcpress.com

This book is dedicated to our wives, Chauling Fu of the first author and Hong Ha of the second author. Without their support, this book would not exist. This book is also dedicated to our family members for their continued support and encouragement.

Contents

14 Stability   435

Preface

Bridges consist of super- and substructures. Superstructures, often called bridge deck structures, are traditionally analyzed by the deck itself for load-distribution behavior. With the invention of computers and the creation of bridge-related software, the approximation can be minimized and tedious processes can be streamlined. It is now possible to change the structural parameters, even structural types, during the design process, because the computer program can now recalculate stresses, deflections, and internal forces in seconds. Through the advances in computer graphic capabilities, meshing in the preprocess and contour displaying on the fly in the postprocess are the norms of almost all bridge analysis and design computer programs. With today's power of both hardware and software, more sophisticated three-dimensional (3D) finite element models have been used in the design of many major structures, in part or all. Based on current availability and future potential, high-performance computer hardware and advanced software technologies can even provide an unprecedented opportunity to develop a new generation of integrated analysis and design systems with roads and bridges to benefit not only new bridge design but also routine load rating and maintenance of existing bridges, which will be discussed more in Chapters 1 and 18.

However, no matter where the computer technology leads, a bridge engineer needs fundamental knowledge of bridge behavior under the combinations of different types of loads during various construction stages. This book serves the role of transferring the fundamental knowledge of bridges to a novel approach of all major bridge types. Several computer programs were used to analyze the illustrated bridge examples throughout this book. We intend to show the principle rather than the capability of each program, so limited details on the data input and the code specifications are provided. The distinctive features are the presentation of a wide range of bridge structural types that are yet fairly code-independent. With this intent, this book is aimed toward students, especially at the master of science (MSc) level, and practicing professionals at bridge design offices and bridge design authorities worldwide.

This book is divided into three parts: Part I covers the general aspects of bridges, Part II covers bridge behavior and modeling of all types of bridges, and Part III covers special topics of bridges. In Part I, Chapter 1 provides an introduction and Chapter 2 covers the methods of computational analysis and design suitable for bridge structures. These methods vary from approximate to refined analyses depending on the size, complexity, and importance of the bridge. With rapidly improving computer technology, the more refined and complex methods of analyses are becoming more and more commonplace. Chapter 3 provides the background and approaches of numerical methods specifically for bridges.

The scope of Part II is to provide information on the methods of analysis and the modeling technique suitable for the design and evaluation of various types of bridges. Chapters include illustrated examples of bridges all over the world, especially in the United States and People's Republic of China. We started from deck-type, especially beam-type, bridges. Chapters 4 through 6 discuss concrete bridges. Chapters 7 and 8 examine steel bridges. The remaining four chapters, 9 through 12, discuss arch bridges, truss bridges, cable-stayed bridges, and suspension bridges, respectively, of which, except for truss bridges, which are mostly built in steel, the other three bridge types can be built in either concrete or steel.

In Part III, for the purpose of analysis, several special topics, such as strut-and-tie modeling (Chapter 13), stability analysis (Chapter 14), redundancy analysis (Chapter 15), integral bridges (Chapter 16), dynamic/earthquake analysis (Chapter 17), and bridge geometry (Chapter 18), are covered to complete the book. In this part, models may include super- and substructures. Some may even need the 3D finite element method of nonlinear analysis. The major issues of recent developments in bridge technology are also discussed in those chapters. The focus is mainly on highway bridges, although some information is also provided for railway bridges.

Overall, this book demonstrates how bridge structures can be analyzed using relatively simple or more sophisticated mathematical models with the physical meanings behind the modeling, so that engineers can gain confidence with their modeling techniques, even for a complicated bridge structure.

Acknowledgments

The authors thank the Department of Civil and Environmental Engineering, University of Maryland, College Park, Maryland, for allowing the use of many examples and photos from many projects that the first author worked on. Also, thanks to the Department of Bridge Engineering, Tongji University, People's Republic of China, for allowing the second author the same privileges for using the materials. The authors thank many people who helped in reviewing and providing critiques of the manuscript. The logistical support of the Bridge Engineering Software and Technology (BEST) Center, Department of Civil and Environmental Engineering, University of Maryland, College Park, is much appreciated.

Finally, the authors express their gratitude to two great institutes, University of Maryland, College Park, Maryland, and Tongji University, People's Republic of China, where the two authors were educated and were/are working, for providing the educational background and working environment for them, and eventually making this book possible.

Authors

Chung C. Fu, PhD, PE, FASCE, is the director/research professor/bridge consultant in the Bridge Engineering Software and Technology (BEST) Center, Department of Civil and Environmental Engineering, University of Maryland, College Park, Maryland. His publications include 50 refereed publications, 20 publications (books, book chapters, etc.), more than 100 presentations and conference proceedings, and 50 public technical reports. Dr. Fu's areas of expertise cover all types of structural engineering, bridge engineering, earthquake engineering, computer application in structures, finite element analysis, ultrahigh-performance concrete (UHPC), steel (fatigue) and composite applications including fiber-reinforced polymer (FRP) and high-performance steel (HPS) for innovative bridge research and construction (IBRC), bridge management, testing (material and structural), and nondestructive evaluation (NDE) application.

Shuqing Wang, PhD, PE, is the senior GIS specialist on contract with Federal Highway Administration; research fellow/bridge consultant, the BEST Center, University of Maryland, College Park, Maryland, in bridge software development and structural analysis; and former director of Bridge CAD Division, Department of Bridge Engineering, Tongji University, People's Republic of China. His areas of expertise span leading edge software technologies to bridge engineering practices, especially in modern bridge modeling and structural analysis system development. Dr. Wang's research interests now focus on visualizing structural behavior in real time and representing bridge geometric and mechanics models in three dimensions.

Part I

General

Chapter I

Introduction

1.1 HISTORY OF BRIDGES

Throughout the late nineteenth and early twentieth centuries, both structural analysis and material science have undergone tremendous progress. Before that time, man-made structures, such as bridges, were designed essentially by art, rather than by science or engineering. Theory of structures did not exist, and structural knowledge was extremely limited. Therefore, bridges designed in that period were based almost entirely on the empirical evidence of what had worked previously. As the principles governing the structural behaviors were better understood, computations of those principles came to serve as a guide to decision making in structural design. Simultaneously, with the progression in production of the main bridge material, concrete and steel, bridge design has become more science than art.

In ancient times, bridges were built from easily accessed natural resources such as wood, stone, and clay with very limited span lengths, until mortar, the early form of Portland cement, was invented. With mortar material and the arch structure shape, Romans were able to build strong and lightweight bridges and even long viaducts, such as the one shown in Figure 1.1, which is built in the first century. In the seventh century, China was able to employ cast iron as dovetails to interlock stone segments during the construction of the Anji Bridge as shown in Figure 9.1, which is still in use after surviving numerous wars, flood, and earthquakes. Techniques did not improve until the eighteenth century when new scientific and engineering knowledge was more widely known. New construction material, iron, especially the cast iron in mass production, enabled the creation of new bridge systems such as trusses. The world's first cast iron truss bridge was built in Coalbrookdale, Telford, England, in 1779, shown in Figure 1.2. This bridge is still in use carrying occasional light transport and pedestrians. Modern bridges are the evolution of the early bridges using modern materials, concrete, and steel. With the aid of modern technology, especially after the invention of the computer and the associated computational

Figure 1.1 Roman viaduct, Pont du Gard, France. (Courtesy of http://en.wikipedia.org/ wiki/File:Pont_du_Gard_BLS.jpg.)

tools, bridges can be built with incredible span lengths. Roman viaducts inspired the building of another incredible Roman viaduct structure, Millau Viaduct (Figure 1.3), a cable-stayed bridge in Southern France. It is the tallest bridge in the world with one of the masts standing at 343 m (1125 ft) above the base of the structure. Currently, the longest span bridge in the world (1991 m or 6532 ft) is the Akashi Kaikyo Bridge, a suspension bridge linking the city of Kobe on the mainland of Honshu to Iwaya on Awaji Island, Japan (Figure 1.4).

Although extra-long span bridges, like cable-stayed and suspension bridges, are the marvels of bridge structures, medium to short span bridges are the norm. In the United States, the most important transportation network is the Interstate Highway System composed of over 44,000 miles (70,800 km) of roadway and around 55,000 bridges. The development of the Interstate Highway System after World War II also propelled the growth of bridge engineering in the last century. The advent of the Interstate Highway System led to the adoption of uniform design standards in the United States and eventually the science of bridge engineering. During this era of the largest public works project, inorder to mass-produce building materials and construct bridges, simplified procedures and simple analysis models were generated and used. The development of the Interstate Highway System

Figure 1.2 The first cast iron truss bridge at Coalbrookdale, Telford, England. (Courtesy of Tata Steel European Limited.)

Figure 1.3 The tallest mast: Millau Viaduct, France.

Figure 1.4 The longest span bridge: Akashi Kaikyo Bridge, Japan. (Courtesy of Yokogawa Bridge Corporation.)

created a workable and efficient method of erecting bridges in a manner that was both consistent and manageable (Tonias 1994). However, with the progress of computational methods and computer tools, more refined and sophisticated methods of analyses have become more common nowadays.

1.2 BRIDGE TYPES AND DESIGN PROCESS

Even though fiber-reinforced polymer (FRP) composites have gradually come to play some roles in civil infrastructures, concrete and steel are still the main materials for bridges. Concrete and steel can form different shapes and build different structural types. According to the U.S. National Bridge Inventory (NBI), as of 2012, the United States has 607,379 highway bridges where 403,072 bridges (72.12%) are slab-, beam-, or frame-type bridges, 10,649 (1.75%) are truss-type bridges, and 7125 (1.17%) are arch-type bridges. Only 45 (0.01%) are stayed-girder bridges and 96 (0.02%) are suspension bridges. Another unique type of bridge popular in the coastal area is the moveable bridge. The moveable bridges are lift-, bascule-, or swing type, and there are 840 (0.14%) of these types of bridges in the United States. The average age of a U.S. highway bridge is about 43 years old, whereas the average age of the 76,000 + U.S. railroad bridges is much older.

For new bridge construction, there are four basic stages for the design process: conceptual design stage, preliminary design stage, detailed design stage, and construction design stage. The conceptual design stage is a process meant to develop a few feasible bridge schemes and decide one or several concepts for further consideration. In the preliminary design stage, the best scheme is selected and cost estimates are conducted. The detailed design stage is a process in which all the details of the bridge structure for construction are finalized. Finally, the construction design stage is the process in which the step-by-step procedures for the building of the bridge are provided. Each of the earlier design stages must carefully consider the requirements of the subsequent stages. For example, the bridge constructability must be considered during the detailed design stage; in addition, costs and construction schedules as well as aesthetics must be considered during the preliminary design stage. An existing bridge in the United States goes through the inspection and load-rating cycles every two years.

Bridge structural analysis, the main subject of this book, is essential for all four stages. Different stages can adopt different modeling techniques, varied from hand calculation to the approximate method and then to the refined method. In this book, constructability, especially constructability of extra-long span bridges, is discussed and demonstrated. Various issues such as deflection, strength of concrete and steel, and stability during critical stages of construction are covered in Chapters 4 through 12, 14, 15, and 17.

In the United States, the load and resistance factor design (LRFD) method is the latest advancement in transportation structures design practice (AASHTO 2013). The combination of the factored loads, termed *limit states* in LRFD, cannot exceed the strength of the material multiplied by a resistance factor less than unity (1.0). Several limit states are included for service, strength, and extreme event considerations. The limit state concept has been universally accepted by many different codes worldwide. A graphical representation of the LRFD process is shown in Figure 1.5a with load (Q) and resistance (R) and later evolved to Figure 1.5b in terms of ($R - Q$). The reliability index β, which shares a similar idea with the safety factor in allowable stress design method, was set at a target of $\beta = 3.5$ in the LRFD code (AASHTO 2013). As can be seen in both figures, the factored safety margin is small, but when the theoretical actual loads and nominal

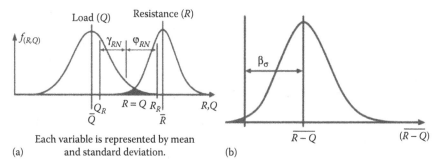

Figure 1.5 Concept of load and resistance factor design. (a) Probability of occurrence based on R and Q. (b) Probability of occurrence based on (R − Q).

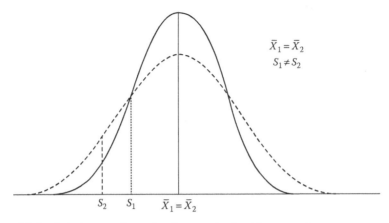

Figure 1.6 Statistical comparison of various methods with exact value.

resistances are observed, the actual safety margin is actually much wider. LRFD also accounts for the different probabilities of occurrence for loads and resistances.

Due to the limitations and assumptions in an analysis, it is not statistically possible to get exact results by any analytical method. In the AASHTO LRFD code, both the approximate method and the refined method, which will be covered in more detail in Chapter 2, are accepted. It is noted that the *bias* values, the difference between the means of the expected result and the exact value ($\bar{X}_1$ and $\bar{X}_2$ in Figure 1.6), of the approximate and refined methods, however, are both close to 1.0. The coefficient of variation (CV), defined as the ratio of the standard deviation σ to the mean μ, is lower with the use of the refined method (shown as the solid line curve versus the approximate method shown as the dotted line curve, respectively, in Figure 1.6). The lowest CV, which means the closest to the exact results, of all methods results from the field load test, which is 4, the least variation. But a field test is costly and time consuming, so oftentimes it is conducted for a few cases and then validated by numerical methods. In these situations, numerical methods are used to simulate all cases. Figure 1.7a shows the side view of a simple-span steel girder bridge on the U.S. Interstate Highway System. Figure 1.7b shows that accelerometers are deployed to detect the modal frequencies (shown in Figure 1.7c) and their associated modes. The results are then compared with the numerical results from the finite element model as shown in Figure 1.8. This process is repeated several times until modal results based on the test and numerical method are close. This technique is called updating. Finite element model updating is the process to ensure that the finite element analysis (FEA) model results better reflect the measured data than the initial models.

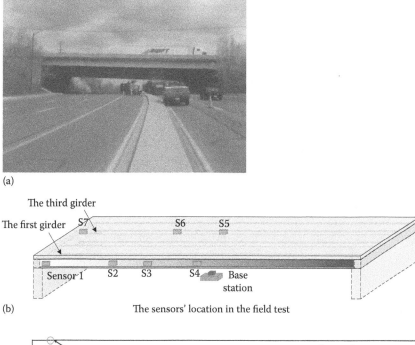

(a)

(b)

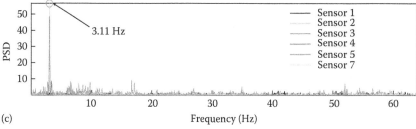

(c)

Figure 1.7 Field test by accelerometers. (a) View of the bridge; (b) placement of the accelerometers; (c) first model frequency from the field test.

1.3 LOADS AND LOAD FACTORS

There are two types of loads on bridges: permanent and transient loads. The most common permanent loads are dead and earth loads. Dead loads include the weight of all components of the structure, appurtenances/utility, wearing surface, future overlays, and future widening. Earth loads include earth pressure, earth surcharge, and down-drag loads.

The most prominent transient load, not necessarily the most damaging one, is the live load: vehicular, rail transit, or pedestrian live loads. Live loads, dynamic impact, centrifugal, braking, and extreme cases such as vehicular

Figure 1.8 The schematic view of a truck on the Middlebrook Bridge finite element model.

collision, have also to be considered in the design process. Definitions of live loads used for bridge designs are different from one specification to another, and usually they are subjected to be amended when traffic demands change years later. For example, HL-93 as shown in Figure 1.9a and b, which is currently used in the U.S. bridge design, specifies two different vehicular loads combined with a lane load, and the extreme values should be taken as the maximum of these two combinations. After 2004, a more simplified live load definition was adopted in China's highway bridge designs, in which one class of live load, for example, class I as shown in Figure 1.10a and b, contains lane load and one single vehicle load, and the extreme values should be taken

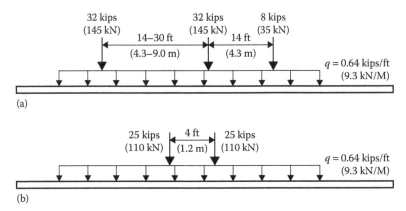

Figure 1.9 U.S. bridge design live loads (US HL-93). (a) Design truck and design lane; (b) design tandem and design lane.

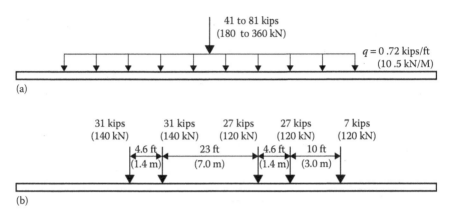

(a)

(b)

Figure 1.10 China (highway class I) bridge design live loads. (a) Design lane; (b) design truck.

as the maximum of these two combinations (JTG D60-2004). In Europe, Eurocode (EN 1991-2) defines traffic load models for road bridges as LM1–4, where load model 1 (LM1 shown in Figure 1.11a) and load model 2 (LM2 shown in Figure 1.11b) are considered normal loads. Ontario highway bridge design code (OHBDC 1991), similar to the AASHTO code, is using the maximum of two loads, truck load (Figure 1.12a) and lane load (Figure 1.12b). In live load applications, various dynamic impact amplification factors, discount due to multiple lanes and load factors, should be employed according to their respective specific design codes. For further study of the live load effect of various codes, refer to *Bridge Loads: An International Perspective* (O'Connor and Shaw 2000). Bridge load rating, other than the bridge design, is a procedure to evaluate the adequacy of various structural components of an existing bridge to carry predetermined live loads (Jaramilla and Huo 2005).

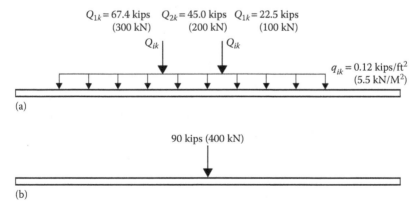

(a)

(b)

Figure 1.11 Eurocode bridge design live loads. (a) Load model 1 (LM1); (b) load model 2 (LM2).

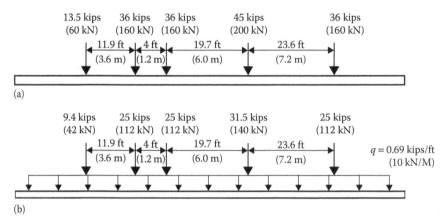

Figure 1.12 Ontario bridge design live loads. (a) Truck load; (b) lane load.

Other types of loads acting on bridges are water loads (buoyancy, stream pressure, and wave), wind loads (on structure and on vehicles as well as aeroelastic effect), ice loads, and earthquake loads. Various codes define load combinations with different load factors. In the design process, it is necessary to go through all load combinations to assume that components and connections of bridges satisfy all the strength, service, and even requirements of extreme events, such as earthquakes, ice load, vehicles and vessel collision. To satisfy specific requirements for bridge design, software that is capable to take care of all load combinations is necessary. In this book, multiple computer software programs, specific or nonspecific for bridges, are used, mainly to demonstrate the modeling technique, not necessarily the capability of the software.

1.4 CURRENT DEVELOPMENT OF ANALYSIS AND DESIGN OF BRIDGES

Structural analysis and computer-aided design (CAD) of bridge structures have long been developed side by side with the development of computer technologies. Many fundamental analysis methods or algorithms were developed based on mainframe and/or minicomputers in the 1970s. When the finite element method was introduced to structural engineering, especially when microcomputer-based FEA packages were available in the 1980s, bridge structural analysis methods and tools advanced a great deal; thus many complicated problems that could not be resolved without taking approximation or simplification assumptions were no longer an obstacle for bridge engineers. After computer graphics and database technologies were widely available in the 1990s, the computer applications in bridge engineering extended even further to computer-aided construction drawing as well, in which a set of detailed drawings could be produced in addition to a set of text reports of structural analysis and design code checking.

In the first decade of the twenty-first century, technologies including computer hardware or software, wide area network (WAN) communication, and parallel computing advanced greatly. As a result, bridge analysis and design tools have advanced from two-dimensional (2D) simplified methods to three-dimensional (3D) detailed methods, from plain console-type operations to intuitive graphical user interfaces (GUIs). Many non-linear problems are now commonly addressed in routine bridge structural analyses. Detailed construction processes are able to be simulated step by step. Many big commercial FEA vendors have had their general-purpose FEA systems expanded to cover special issues found in bridge analysis and design. More sophisticated 3D graphical modeling tools are now available in bridge design firms and institutes.

In recent years, the swift advancement of computer and graphics hardware, such as multiple processing cores, 3D rendering or visualization, fast float-point calculation speed, and vast memory capacity, has drastically increased the potential of computer technology application in bridge engineering. At the same time, fundamental software technologies, including system development and integration, parallel programming, 3D graphics modeling and virtual reality, database and geographic information system (GIS), Internet communication, and cloud computing, have long been ready for a revolution in engineering application development. Although computer applications in bridge analysis and design have greatly progressed, its advancement falls far behind the progress of fundamental technologies and is not in pace with applications in other fields. Current bridge software packages provide engineers a typical process of analysis and design (1) to establish and analyze a bridge's mechanical model, (2) to check design code for each component based on analysis results, and (3) to resize components or adjust structural dimensions and repeat the earlier process if necessary. A new era of computer technology applications is in demand by bridge engineers and transportation administrators.

1.5 OUTLOOK ON ANALYSIS AND DESIGN OF BRIDGES

Based on the current availability and future potential, high-performance computer hardware and advanced software technologies provide an unprecedented opportunity to develop a new generation of analysis and design systems so as to benefit not only new bridge design but also routine load rating and maintenance of existing bridges. There will be several aspects in the analysis and design of bridges that demand great enhancements.

First, the tedious routine work of establishing the mechanic models of a bridge should be completely automated. Bridge engineers should be relieved for more creative works. Taking advantage of modern database and visualization technologies, establishing an engineering model of a true bridge project should be the centerpiece of a bridge software system. It is true that an engineering model is much more complicated than an abstracted mechanical model; however, the goal is achievable when a commonly used bridge type is the focus. Having the engineering model as the core, the engineers' interface will only be editing parameters in a 3D scene that reflects parameter changes in real time as a virtual bridge project. As illustrated in Figure 18.26, engineers should be able to describe a bridge project starting from roadway geometry to girder profiles. Modern visualization technologies should provide engineers instant realization of dimension changes in a virtual project. The design or description process will be interactive and intuitive. For example, an engineer can click a steel plate as shown in Figure 18.26 to pop up a data form that allows verification or changes in its definitions on the fly. When there is a need to perform a certain type of analysis, the questions that need to be asked, such as "what type of analysis model is appropriate" and "how do we establish the required FEA model," will no longer be the direct interest of engineers. The establishment of a required FEA model from the engineering model should be automatic and instant. The analysis result should be directly and instantly represented into the engineering model in terms of engineering meanings, such as color-coded surface rendering that reflects load ratings, rather than ordinary mechanical values. Figure 1.13

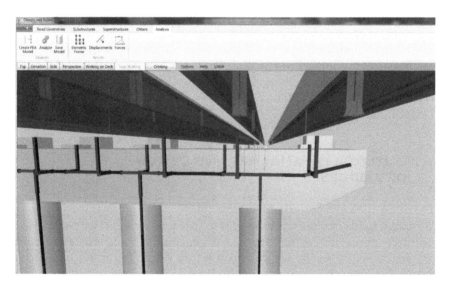

Figure 1.13 A hybrid view of a bridge model and its FEA model.

illustrates the idea to blend one of its FEA model and analysis results into the virtual bridge project scenes.

Second, as a part of critical infrastructures, the whole life cycle management and tracking of each individual component are crucial to a bridge structure. Based on the above-mentioned engineering model, history of dead loading, component geometry and position changes, dimension and deterioration changes, and retrofitting of any component should be established since it was built. Operational live loads should be simulated based on real traffic volume and speed. In addition to the regular extreme values obtained by design live load analyses, each point of interest should have statistical peak values obtained by simulating operational traffic. Local fatigue should be rated by the stress analyses of traffic simulations. Having accumulated the history of a bridge in a certain amount of time, bridge engineers, inspectors, and project managers should be able to obtain a prediction of the imminent actions so as to avoid disastrous failure or high-cost maintenance repairs.

Lastly, as critical points of a national surface transportation network, engineers and/or administrators should be able to overview health conditions of bridges in a large geographic area. Modern GIS technologies including mapping, satellite or aerial imaging (Figure 18.26), spatial data processing, and large area traffic networking should be integrated into a bridge health monitoring system. As each individual bridge structure has had an engineering model associated with real-time history, special queries from an administrative level should be able to be processed, for example, a query for the best routing in terms of structure safety for transporting a special load from place A to B or for the most vulnerable bridges within a certain area of a truck bomb attack. Administrative information including health conditions of bridges, funds allocated for maintenance of bridges, and predictions of future repairs should be able to be displayed on a map overlaid with other traffic volume. Advancement of cloud computing technologies will also greatly impact computer applications in bridge engineering in the near future.

Chapter 2

Approximate and refined analysis methods

2.1 INTRODUCTION

This chapter will serve as the introduction of succeeding chapters, especially chapters in Part II—Bridge Behavior and Modeling. Brief discussion of various bridge structural forms will be made first in this chapter, whereas more details on each bridge type will be covered in their individual chapters (Chapters 4 through 12). Approximate and refined analysis methods with their advantages and disadvantages will then be briefly mentioned (Coletti and Pucket 2012). Although all methods can be categorized as finite element method (FEM), levels of approximation and accuracy are different among various modeling methods. With today's advancing of computer analysis tools, there is a certain advantage to a adopting two-dimensional (2D) model in grillage or three-dimensional (3D) model, as called *refined* analysis models, over one-dimensional (1D) model, as termed *approximate* analysis model. Subsequently, the principle of FEM of all types will be presented in Chapter 3.

2.2 VARIOUS BRIDGE STRUCTURAL FORMS

Bridge systems consist of super- and substructures. The structural model may couple them together where the effect of substructure is essential to the whole analysis, such as earthquake analysis, or have them decoupled at the bearing where the substructure does not affect much on the superstructure behavior, except drastic movement, such as support settlement. This bridge system can be analyzed as a 1D model, which AASHTO termed approximate analysis model, 2D model in grillage, or 3D model, where the latter two can be categorized as refined analysis models. This chapter identifies bridge deck structural forms and basic characteristics of these different types of bridges. More details on all types of bridges and their analyses will be covered in the Chapters 4 through 12 and 14 through 17.

2.2.1 Beam deck type

A bridge can be considered as a beam when the ratio of width to length of the whole bridge is within a certain amount so that the applied loads cause the bridge to bend and twist along its length while the cross sections do not change shape. The most common beam bridges are pedestrian bridges made of either steel, reinforced concrete, or prestressed concrete. Many long-span bridges also behave as beams with dominant concentric loads so that while calculating principal bending stresses the distortion of the cross section under eccentric loads should be considered in the analysis.

Several span arrangements of beam-type bridges and their statical determinacy in bending are shown in Figure 2.1. Continuous bridge with its indeterminacy has many advantages over simple span bridge. Modern steel bridges are usually continuous over the piers and can be considered as continuous for dead and live loads. However, for a precast, prestressed concrete bridge, it is common to be simply supported or cantilevered during construction and then made partially or totally continuous for live loads and long-term movements. The various arrangements of this two-stage analysis are shown in Figure 2.2. With both bending and torsion taken into account, the statical determinacy of the bridge can be determined as shown in Figure 2.3. A beam-type bridge, if there is no skew angle at support, can be simplified to a 1D model with only in-plane shear and bending moment considered.

Frame-type bridges can be regarded as simplified arch structures. The most common types of frame-type bridges are portal frame or slant-leg frame as shown in Figure 2.4. As materials for frame-type bridges are used more efficiently, they can be designed to appear lighter and more slender

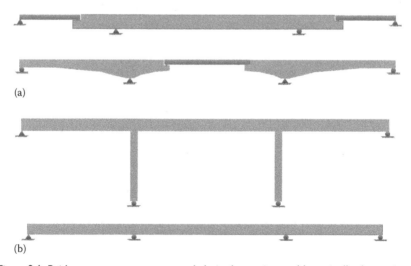

(a)

(b)

Figure 2.1 Bridge span arrangements and their determinacy: (a) statically determinant structure; (b) statically indeterminant structure.

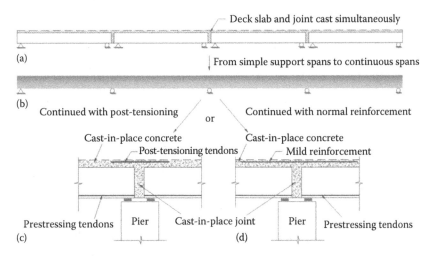

Figure 2.2 Two-stage construction and analysis—from simple support to continuous. (a) Simple support stage; (b) continuous stage; (c) continued with post-tensioning option; and (d) continued with reinforcement option.

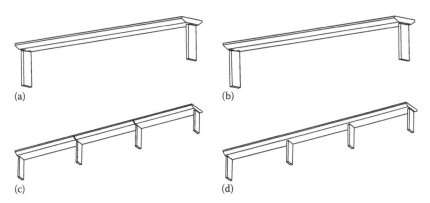

Figure 2.3 Statical determinacy of bridges: (a) simple span with determinate for bending and torsion; (b) simple span with determinate for bending only; (c) multiple simply supported spans with determinate for bending only; (d) continuous spans with indeterminate.

than simply supported bridges, especially when the girder is haunched. A frame-type bridge can be simplified to a 2D model with axial force coupled with in-plane shear and bending moment.

2.2.2 Slab deck type

A slab bridge is usually made of concrete and behaves like a flat plate. The slab is *isotropic* if its stiffness properties are the same in all directions and

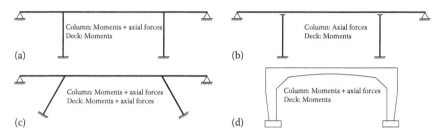

Figure 2.4 Frame-type bridges and internal forces under main loads.

is *orthotropic* if its stiffness properties are different in two perpendicular directions. The slab bridges based on their composition can be divided into the following types:

1. *Solid Slab* (Figure 2.5a). Concrete solid slabs are commonly used where the spans are less than 15 m (50 ft). A solid slab is acting and can be assumed as an isotropic plate, even though the reinforcement may be different in two perpendicular directions.
2. *Void Slab* (Figure 2.5b and c). For spans greater than 15 m (50 ft), the dead load of solid slabs becomes excessive and the structure can be lightened by incorporating cylindrical or rectangular voids. It acts as an orthotropic plate and is treated customarily as one unit. If the void size exceeds 60% of the depth, the deck is generally considered as cellular (box) construction. For the type with large void, the distribution of the loads transversely is by transferring vertical shear through webs; the cross section distorts like a Vierendeel truss.
3. *Corrugated (Coffered) Slab* (Figure 2.5d) or *Precast Beam Slab* (Figure 2.5e). Precast beams of various cross-sectional shapes of

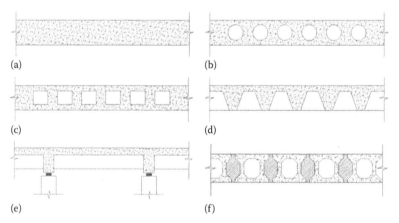

Figure 2.5 Slab bridge types: (a) solid slab; (b) circular void slab; (c) rectangular void slab; (d) corrugated slab; (e) precast beam slab; (f) *shear-key* slab.

constant or varying depth are joined together contiguously by the cast-in-place reinforced concrete and are transversely prestressed to make them act effectively as a void or solid slab with orthotropic properties.

4. *Shear-Key Slab* (Figure 2.5f). A shear-key slab is constructed of pre-stressed beams or reinforced concrete beams connected along their length by cast-in-place concrete to form joints but which are not pre-stressed transversely. The name *shear key* is used because the joints are not fully continuous for transverse moments. The distribution of loads between beams is by the torsional stiffness of individual beams and vertical shears between key joints.

The slab bridge can be simplified as a strip beam 1D model, 2D grillage model, or 3D plate/shell FEM model.

2.2.3 Beam–slab deck type

A beam–slab bridge consists of a number of longitudinal beams connected either compositely or noncompositely across the tops by a continuous slab. It is the most popular type for the small- to medium-span bridges.

Spaced beam–slab bridges, as shown in Figure 2.6, are usually made of beams spaced between 2 m (7 ft) and 4 m (13 ft) apart. The bridges can be designed with precast, prestressed concrete beams or steel beams acting compositely with the concrete deck. The deflection behavior is different from that of orthotropic plate. Beams along the longitudinal direction are taking most of the loads. Diaphragms are usually placed in the middle/end or other places to help distribute live loads laterally. When dead loads are the only concern, a 1D beam model can be used in analysis. When analyzing live loads, a 1D beam model (with two degree-of-freedom beam element) employed by live load distribution factors on influence lines can be used. However, with advanced computer technologies and widely available comprehensive analysis tools, a completed 2D grillage model with influence surface loading is preferred, which will be more accurate in live load analysis and also simplifies the procedures of live load distribution factor calculations.

2.2.4 Cellular deck type

Box (also called *cellular*) deck-type bridges consist of a box or boxes enclosed by slabs and webs. They contain one or a few large cells, attached or detached. Small- and medium-span concrete bridges are usually cast *in situ* or precast in segments. Long-span concrete or steel bridges are frequently constructed as segmented cantilevers. In addition to the less material used, light weighed, and high longitudinal bending stiffness, box girder

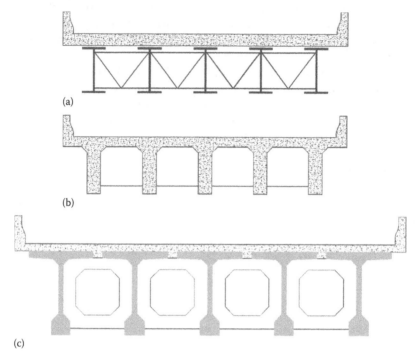

Figure 2.6 Beam–slab bridges: (a) steel composite; (b) cast-in-place concrete; (c) precast concrete.

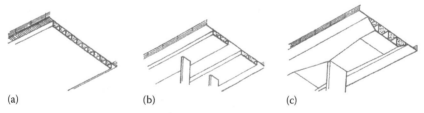

Figure 2.7 Box-girder bridges: (a) rectangular attached multicell bridge; (b) detached multicell box girder bridge; (c) trapezoidal attached multicell bridge.

bridges have the advantage of high torsional stiffness. Load distribution of a box girder bridge, with its strong torsional stiffness, is usually more uniform across the bridge width than that of an I-type beam–slab bridge with the same span length and width.

Figure 2.7 shows some box girder bridges as examples. Figure 2.7a shows a void slab with large attached cells (over 60% void ratio). Figure 2.7b is a detached multicell box girder bridge. Figure 2.7c is a void slab with inclined webs on the sides. If a single cell is used for this type of bridges, distortion should be considered in the analysis.

2.3 APPROXIMATE ANALYSIS METHODS

2.3.1 Plane frame analysis method

For the approximate analysis method (or so-called simplified method), a longitudinal girder, or a strip of unit width as in the case of a slab bridge, is isolated from the rest of the bridge and can be treated as a 1D beam or a plane frame structure in general.

For long-span bridges, the whole bridge may be considered as a 1D beam model. For straight multigirder bridges, this simplified method can also be adopted for determining the controlling force and longitudinal moments. A girder, plus its associated portion (effective width) of the slab, is subjected to dead and live loads where dead loads can be approximated by their tributary. However, live loads have to be maximized by loads' lateral position and girder influence lines, which are called live load envelopes. A study was made and summarized in the AASHTO load and resistance factor design (LRFD) code (2013) as live load distribution factors. Live loading results of one lane of design vehicles and/or lane load must be multiplied by live load distribution factors to consider the lateral distribution of live loads. Usually different specifications have different calculation methods for live load distribution factors. Some may have the same procedures. For example, AASHTO LRFD code (2013) and Ontario code (OHBDC 1991) are using the similar approach.

Bridge deck—It is structurally continuous in the orthogonal directions on the plane. The applied load on the deck will have 2D distributions of shear forces, moments, and torques. If 2D distribution is considered, it is definitely more complex than the one modeled as a 1D continuous beam.

In a refined analysis method, the transverse flexural stresses on the slab can be found from the computer results. However, if a simplified method is used, the transverse flexural stresses have to be checked separately. To check the transverse flexural stresses on the bridge slab, Westergaard equations are always referenced.

In AASHTO LRFD Specifications (2013), width of the equivalent strip, as shown in Figure 2.8, is taken as specified. Unlike fully and partially filled grids, where live load moments may be determined by an empirical formula, the strip of concrete deck slab shall be treated as continuous beams or simply supported beams between girders with dual wheels of design truck applied.

The following equivalent strip width for concrete deck (Equations 2.1a and 2.1b) is from AASHTO LRFD Specifications. It is a modified Westergaard equation in SI units for calculating transverse flexural stresses between girders.

$$+M : \to E = 660 + 0.55S \tag{2.1a}$$

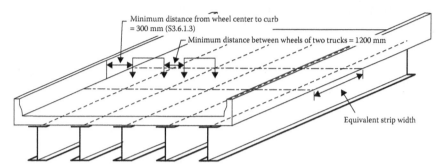

Figure 2.8 Transverse live load moments of a bridge deck.

$$-M :\rightarrow E = 1220 + 0.25S \tag{2.1b}$$

where:
    E is the equivalent strip width in mm
    S is the stringer spacing
    +M is the positive moment region
    −M is the negative moment region

The dead load effect can be obtained by treating dead load as stationary load acting on a continuous beam. Trucks, as the live load, have to be moved laterally to determine the maximum positive and negative moments. Multiple presence factor (1.2, 1.0, and 0.85 for one, two, and three trucks, respectively, as per AASHTO LRFD) shall be considered in the analysis. Figure 2.8 shows the lateral loading for determining transverse moments on a bridge deck.

The dominant failure modes of a bridge deck are either flexure shear or punching shear. The punching shear is not obtained easily by even the refined analysis method, such as FEM, unless a very fine meshed model is built around the critical location. Figure 2.9 adopted from laboratory tests by Hwang et al. (2010) shows the footprint of a truck wheel and its punching shear critical locations. As specified in the AASHTO LRFD Specifications (2013) for punching shear, the shear strength V_n, multiplied by the strength reduction factor φ, shall be larger than the ultimate shear produced by the wheels (Equation 2.2).

$$V_u \leq \varphi V_n \tag{2.2}$$

Without shear reinforcement, the shear strength of concrete V_n in Equation 2.3 is governed by AASHTO Equation 5.13.3.6.3-1 in metric form

(a)

(b)

Figure 2.9 Footprint of a truck wheel and its associated punching shear failure mode.
(a) Laboratory static test; (b) punching failure on the top of the deck. (Data
from Hwang, H. et al., *Engineering Structures*, 32, 2861–2872, 2010.)

$$V_n = \left(0.17 + \frac{0.33}{\beta_c}\right)\sqrt{f_c'}\, b_o d \leq 0.33\sqrt{f_c'}\, b_o d \qquad (2.3)$$

where:

β_c is the ratio of long side to short side of concentrated load or reaction area

b_o is the perimeter (= 2[(b + d) + (c + d)], as shown in Figure 2.9b) of the critical section defined in AASHTO LRFD Specifications (2013)

Live load distribution factors—When a single-beam model is used for analyses of a multiple girder bridge, unlike dead loads that usually distribute equally, live loads of one lane is not necessarily carried by one girder or one girder may carry more than one lane of live loads, depending on girder spacing and lateral distribution components such as diaphragms. Lateral distribution factors, which define the portion of live loads carried by an individual girder, simplify the analysis process to a beam analysis. Instead of modeling the bridge in both longitudinal and transverse directions, a single girder is isolated and subjected to loads comprising one line of wheels of the design vehicle multiplied by the distribution factor. Previously, AASHTO defined wheel load distribution factor as S/D, where S is the girder spacing and D, which uses units of length, is specified to a certain value according to the bridge type. In recently developed LRFD Specifications (2013), even though the form of S/D is still maintained for certain bridge types, the definition of distribution factor is modified drastically to include in the formula, besides girder spacing, deck thickness, span length, depth of beam, and number of beams. Another improvement is to distinguish the definition of exterior beams from interior beams, multilane from one design lane loaded, shears from moment, and correction factors for skew bridges.

Lateral live load distribution theories were developed before 1970s to provide engineers a practical way to count the uneven distribution of live loads in single-beam model analyses. The intent of applying live load distribution factors is to provide an envelope for all possible live load cases so the results may be conservative or, in some special occasions, even unconservative. As 3D spatial modeling, analyses and influence surface loading are widely available nowadays; a refined 3D analysis with influence surface loading is encouraged in modern bridge engineering practices.

Effective flange width (shear lag)—When a girder cross-section is under flexural stress, shear deformation on top plane will happen in flange, unlike the beam theory assumed. The thinner the flange, the bigger the shear deformation, and the farther away from the web, the bigger the longitudinal displacement of flange accumulated by shear deformation. This shear deformation will cause flexural stress changes along a flange. The local

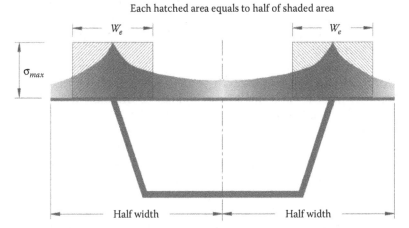

Figure 2.10 Shear lag effect on stress distribution and equivalent effective width.

increase/decrease of the deck longitudinal flexural stress near the intersection to the beam web is called *shear lag effect* (Figure 2.10). This effect can be taken into account during the stress calculation by assuming only a portion of flange working to resist bending moment, or so-called effective flange width. In AASHTO, the effective flange width is assumed constant along the bridge, although some may assume otherwise. The prior-to-2008 AASHTO LRFD Specifications for interior beam's effective flange width take the least of

- One-fourth of the effective span length (the effective span length may be taken as the actual span for simply supported spans and the distance between points of dead load contraflexure for continuous spans).
- 12 times the average slab thickness, plus the greater of web thicknesses, or one-half the width of the top flange of the girder.
- Average spacing of adjacent beams.

For exterior beam, the effective flange width may be taken as 1/2 the effective width of the adjacent interior beam, plus the least of

- One-eighth of the effective span length.
- 6 times the average slab thickness, plus the greater of one-half the web thickness, or 1/4 the width of the top flange of the girder.
- Width of the overhang.

The current AASHTO LRFD Specifications (2013) are using the full tributary areas of the girder, which is the third criterion shown previously.

Live load influence line—An influence line is defined as the variation of function, such as reaction, shear, bending moment, or stress, when a unit load is moving over the structure. Figure 2.11 shows examples of influence lines for a three-span continuous bridge.

Usually, influence line results from analyzing a beam model, with x as the distance and y as the ordinate. The influence lines of moment, shear, or reaction are recorded at a small interval. Having an influence line defined, a standard live loading process can determine the positions of a live load specified by a specification and thus the extreme live load results. Different specifications have different live load definitions and combinations. For example, AASHTO LRFD Specifications (2013) define the following loading combinations:

- Design tandem with design lane load
- One design truck with variable axle spacing with design lane load
- 90% of two design trucks with axles from two trucks spaced minimum 15,000 mm (two 145-kN axles spaced 4300 mm) with 90% of the design lane load

The illustrations of the live loading application and combinations are shown in Figure 2.12 with loads positioned longitudinally for extreme effect.

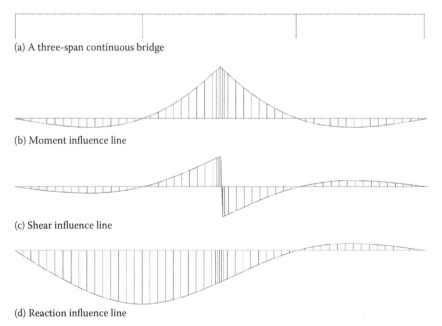

(a) A three-span continuous bridge

(b) Moment influence line

(c) Shear influence line

(d) Reaction influence line

Figure 2.11 (a–d) Examples of influence lines for a three-span continuous bridge.

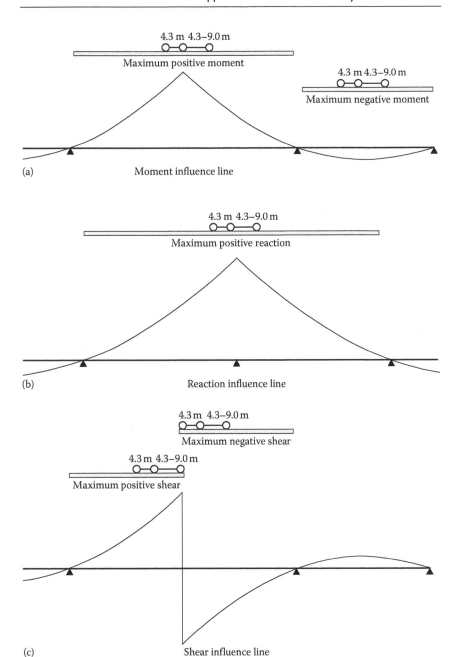

Figure 2.12 (a–c) Evaluation of extreme live loads.

2.4 REFINED ANALYSIS METHODS

2.4.1 Grillage analogy method

Grillage (or grid) analysis has been used by the bridge engineers for quite some time and is in the same category as the FEM (see Section 2.4.5). Grillage method can be regarded as a special case of FEM (Jategaonkar et al. 1985) with deck slab structure idealized as a 2D model consisting of beam elements. In this type of model, the deck is cut, theoretically, into pieces in both directions with each piece considered as a beam element (Hambly 1991). Choice of these imaginary cuts is based on experience and should be made with caution.

The general approach for the grillage analysis is to model the longitudinal girders as beam elements, straight or curved. If the intermediate diaphragms are present within the spans, the transverse beam elements are placed at the same locations as the diaphragms. If intermediate diaphragms are far apart or not present, deck slab is modeled as transverse beam elements with deck's moment of inertia based on a certain effective width. In this grillage model, each node has three degrees of freedom, one vertical translational and two planar rotational degrees of freedom.

2D grillage analysis is simulated by 2D grillage of beams with different section properties. The basic principle is the same as defined in Section 2.4.5. The difference from a generic type of finite element is that only vertical flexure and torsion of a beam are considered in a grillage element, as how most decks behave. Therefore, each node in a grillage model has a vertical translational displacement and two rotational displacements along axes in deck plane, and a grillage element has only vertical bending moment, vertical shear, and torque defined. When the grillage model is used, a suitable grillage mesh should be defined to get meaningful results. Figure 2.13 shows examples for four different types of bridges. Figure 2.13a shows stiffness to be about equal along the longitudinal and transverse directions and the beam elements coincide with the real longitudinal and transverse beams. Figure 2.13b shows longitudinal beams that are more predominant and coincident with the beam elements. The placing of the transverse beam elements is recommended that a proper aspect ratio be maintained between transverse and longitudinal elements, at diaphragm locations if diaphragms are present and at equal spacing to simulate the plate transverse distribution if no diaphragms are present. Figure 2.13c is a bridge with closely spaced beams. For practical purposes, each longitudinal beam element can represent more than one beam. The rule of thumb is to place longitudinal beam elements no farther apart than about one-tenth of the span (Hambly 1991). Figure 2.13d has wider beams with two longitudinal members per beam. Usually for this type of structure, the longitudinal members are much stiffer than the transverse members, which may be representing just the thin slab on top.

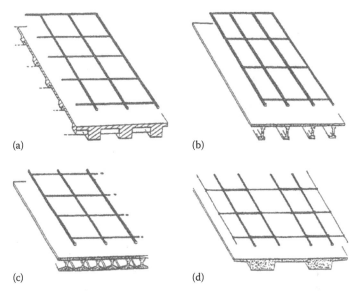

Figure 2.13 (a–d) 2D grillage meshes.

2.4.2 Orthotropic plate method

The orthotropic deck bridge is a special kind of deck, which can be solved by the orthotropic plate theory. The general differential equation given for the orthotropic plate can be found in any book discussing plate bending theory and is listed in Equation 2.4.

$$D_x \frac{\delta^4 w}{\delta x^4} + (D_{xy} + D_{yx}) \frac{\delta^4 w}{\delta x^2 \delta y^2} + D_y \frac{\delta^4 w}{\delta y^4} = p(x, y) \qquad (2.4)$$

where:

w is the deflection of the plate at any point (x, y)

D_x, D_y, D_{xy}, and D_{yx} are stiffness of the longitudinal flexure, the transverse flexure, longitudinal torsion, and transverse torsion, respectively

$p(x, y)$ is the loading intensity of any point

A simplified analysis is made by assuming:

- For decks with closed ribs: $D_y \cong 0$
- For decks with open ribs: $D_y \cong 0$, $D_{xy} \cong D_{yx} \cong 0$

Based on Hambly (1991), the moment and flexure relationships are shown in Equation 2.5a and principle stresses are shown in Equation 2.5b.

$$M_x = -D_x \left(\frac{\partial^2 w}{\partial x^2} + \upsilon_y \frac{\partial^2 w}{\partial y^2} \right)$$

$$M_y = -D_y \left(\frac{\partial^2 w}{\partial y^2} + \upsilon_x \frac{\partial^2 w}{\partial x^2} \right)$$

$$M_z = -2D_{xy} \left(\frac{\partial^2 w}{\partial x \partial y} \right)$$

where:

$$D_x = \frac{E_x t^3}{\left[12(1 - \upsilon_x \upsilon_y) \right]} \tag{2.5a}$$

$$D_y = \frac{E_y t^3}{\left[12(1 - \upsilon_x \upsilon_y) \right]}$$

$$D_{xy} = \frac{1}{2}(1 - v_x v_y)\sqrt{D_x D_y}$$

As engineers usually deal with stresses, principal stresses and Mohr's circle of stresses are expressed in Equation 2.5b and illustrated in Figure 2.14.

$$\sigma_1 = \frac{\sigma_x + \sigma_y}{2} - \sqrt{\left[\left(\frac{\sigma_x - \sigma_y}{2} \right)^2 + \tau_{xy}^2 \right]} \tag{2.5b}$$

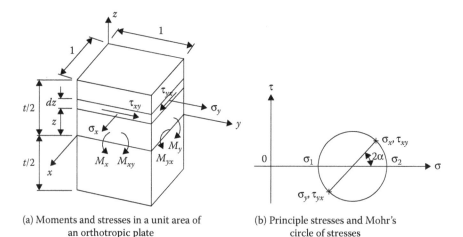

(a) Moments and stresses in a unit area of an orthotropic plate

(b) Principle stresses and Mohr's circle of stresses

Figure 2.14 (a, b) Moment and stress relationships of an orthotropic plate.

$$\sigma_2 = \frac{\sigma_x + \sigma_y}{2} - \sqrt{\left[\left(\frac{\sigma_x - \sigma_y}{2}^2 + \tau_{xy}^2\right)\right]}$$

$$\tan 2\alpha = \frac{2\tau_{xy}}{\sigma_x - \sigma_y}$$

When applying orthotropic plate model in deck analyses, deck is meshed into regular plate elements. However, unlike an isotropic plate element, a local coordinate system is required so as to define two directions that have different bending stiffness.

2.4.3 Articulated plate method

When the transverse distribution of loads is only through shear forces with no transverse prestressing forces, it is defined as articulated plate or *shear-key* slab with idealized articulated plate model, as shown in Figure 2.15. For this type of bridge that has small transverse bending stiffness, the transverse flexural and torsional stiffnesses, D_y and D_{yx}, respectively, in Equation 2.4 would approach zero; the longitudinal bending and torsional stiffnesses, D_x and D_{xy}, respectively, are defined for different types of bridges as (Jategaonkar et al. 1985; Bakht and Jaeger, 1985):

1. Slab bridge with solid block

$$D_x = \frac{Et^3}{12} \tag{2.6}$$

$$D_{xy} = \frac{Gt^3}{3} \quad \text{if } S > t$$

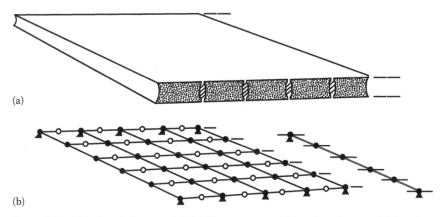

(a)

(b)

Figure 2.15 Articulated plate model. (a) Plates connected by shear keys and (b) articulated plate numerical model.

$$D_{xy} = GS\frac{t}{3} \quad \text{if } S < t$$

2. Slab bridge with rectangular void block

$$D_x = Et_1\frac{H^2}{2} \tag{2.7}$$

$$D_{xy} = G\frac{J}{S} \quad \text{and} \quad J = \frac{4A^2}{\oint \dfrac{ds}{t_1}}$$

3. Slab bridge with circular void block

$$D_x = E\left(\frac{t^3}{12} - \frac{\pi t_v^4}{64S}\right) \tag{2.8}$$

$$D_{xy} = G\frac{t^3}{3}\left[1 - 0.84\left(\frac{t_v}{t}\right)^4\right]$$

4. Box girder bridge

$$D_x = \frac{EI_g}{S} \tag{2.9}$$

$$D_{xy} = G\frac{J}{S} \quad \text{and} \quad J = \frac{4A^2}{\oint \dfrac{ds}{n_s t_1}}$$

where:
   S is the girder spacing for multigirder bridges or unit width for slab bridges
   t is the thickness of the solid slab
   t_v is the diameter of the circular hole of a void slab
   t_1 and H are the thickness of the wall and median height of the rectangular void slab, respectively
   I_g and A are moment of inertia and cross-sectional area of the box girder, respectively

Articular plate is a special case of orthotropic plate and can be solved using the same method as defined for the orthotropic plate theory. If a bridge with *shear key* is modeled as a beam, shear keys are considered when calculating live load lateral distribution factors. For example, AASHTO LRFD Specifications (2013) indicate that the factors for bridges with shear keys are different from the factors obtained for bridges with monolithic deck or with transverse post-tensioning.

2.4.4 Finite strip method

A simplified finite element with bridge deck modeled end to end is called *finite strip* (Figure 2.16). The displacement functions for in-plane and out-of-plane deformation of the strip are of the form

$$w, u, v = \sum f(y)\sin\left(\frac{n\pi x}{L}\right) \tag{2.10}$$

where:
    x is the direction along the structure
    y is the direction across the strip

The harmonic analysis is then performed. Further development on the finite strip analysis extends to the curved circular structures with harmonic function (Fourier series) used for variations along circular arcs. As finite strip method involves fewer nodes and a smaller matrix to solve, it is sometimes more economical than other methods such as finite element. There are several variations of finite strip method, semianalytical, spline, and boundary element. The conventional finite strip method, because of its formulation, may be very slow to converge with concentrated load and needs many series of terms to achieve acceptable accuracy. With today's available computer speed and memory, finite strip method is a plausible way to handle bridge problems.

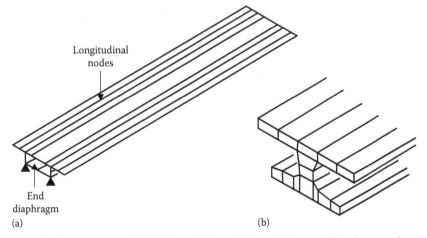

Figure 2.16 Finite strip model. (a) Strip division of a box girder and (b) a closeup of strip division of an I-girder.

2.4.5 Finite element method

The mathematical theory and formulation of the FEM are well documented in many textbooks, and they are not in the scope of this chapter. Regarding element types, a structure can be modeled using 1D, 2D, or 3D elements or even the combinations of these three elements (Jategaonkar et al. 1985).

Line elements—Line elements for modeling the bridge members include two main types. The first is the bar type with only axial tension or compression with one degree of freedom at each node; usually it is used for modeling a truss member, a bearing, or an individual member of the cross-frame. The second type is the beam element, as shown in Figure 2.17, which has six degrees of freedom. It is used usually to model the beam or column that has axial stiffness as well as bending stiffness. For more simplified line elements, certain degrees of freedom can be excluded for cases where only others are of concern or predominant. For example, a 3D frame element can be retrograded to a 2D beam element in cases that only two degrees of freedom, vertical translational and rotational displacements, are considered.

A grillage model, in Section 2.4.1, is another example of this simplification. For a grillage model, another degree of freedom, torsion, is added back into the model. Because of the translational bending of the slab and diaphragm action, the main beams will be under torque. For highly skewed, curved bridges, or with long overhang, this torsional action may be significant. Therefore, displacements of vertical translational, vertical flexural rotation, and axial torsional rotation are dominant in deck behavior, and when only these displacements are considered, a 3D frame element is retrograded to a grillage element. In this aspect, a grillage model and a plane frame model are the same, except that the exclusion of displacements is different.

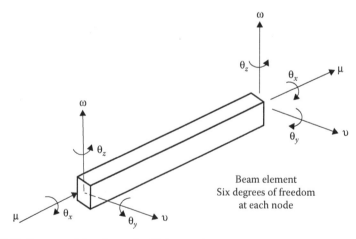

Figure 2.17 Degrees of freedom of a 3D frame element.

Area elements—Area elements in the finite element analysis also include two types, elements with in-plane effects only and elements with both in-plane and out-of-plane effects. The in-plane element, which is often referred as *membrane element*, may be either plane stress or plane strain element. Each node of a membrane element has two degrees of freedom (u, v) in the element plane. It has been used less than the second plate element type, which is used to simulate not only in-plane (membrane) action with degrees of freedom (u, v) but also plate bending (flexural) action with an additional three degrees of freedom (w, θ_x, θ_y) at each node. This type of combined plate element is often referred to as *plane shell element*, to differentiate a pure bending plate element. Plane shell element is so called as it can be used to assembe a true shell structure. As shown in Figure 2.18, these area elements may be triangular or rectangular in shape. The shell elements can be used to model many parts of bridge structures. Figure 2.19 shows the nodes and elements of a deck modeled by shell elements, and Figure 2.20 illustrates an actual structure with its idealized model.

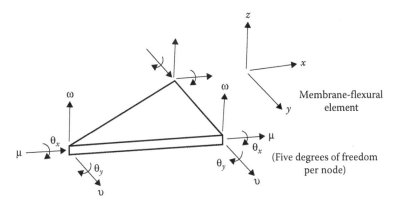

Figure 2.18 Degrees of freedom of a plane shell element.

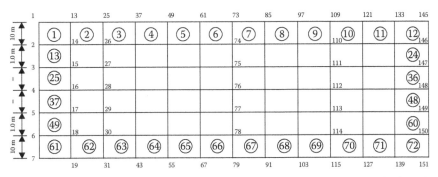

Figure 2.19 Finite element example of nodes and elements (numbers in circles) of a slab model.

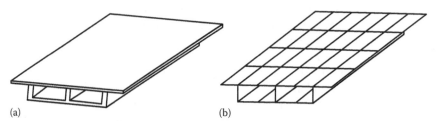

(a)   (b)

Figure 2.20 Idealized model using area elements: (a) actual structure; (b) structure idealized by 3D plane shell elements.

When applying plane shell elements in bridge analyses, it should be noted that each node has only five, rather than six, degrees of freedom. The rotational displacements along axis perpendicular to plate plane output from a finite element analysis are faked by a technique avoiding ill-conditioned stiffness matrix. The sixth rotational displacements are meaningful only at nodes connecting kinked plates, and these displacements are caused only because of geometric transformation from bending rotations in other planes.

Volume elements—A volume element is sometimes called *solid element* with three, four, eight, or more nodal points. Figure 2.21 shows an eight-node volume element as an example. In bridge superstructures, the model usually can be built up from line or area elements or combinations of these two types of elements. Volume elements are used rarely except for the substructures with massive concrete piers or abutments. Even for the substructures, the line (beam) elements are used more frequently than the solid elements because of easier usage and interpretation. If massive concrete is used, it may be modeled by *rigid link* elements to simulate the rigid body motion between two points.

For a typical 3D model of a slab–beam bridge, the slab is modeled as plates (area elements) with thicknesses equal to the slab thicknesses. If the beam is widely spaced, more nodes should be assigned between beams to simulate the higher shear-lag effects between beams and slab deck. A good representation

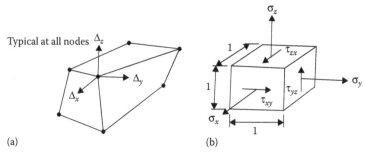

(a)   (b)

Figure 2.21 Finite element volume element. (a) Node displacements and (b) element stresses.

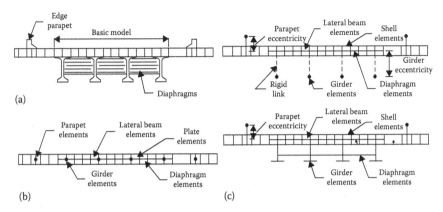

Figure 2.22 Beam on slab modeling. (a) Bridge cross section; (b) two-dimensional CGM finite element model cross section; and (c) three-dimensional EGM finite element model cross section.

for the beam itself is to use plane shell elements for both the web and the flanges. To get meaningful results along the beam web, at least three plate elements should be used for modeling the web. For the beam flanges, two plane shell elements on each side of the web are sufficient for good modeling. When the lateral distribution of longitudinal stress is of concern, more elements may be needed on flanges. Another requirement for the finite element mesh is to maintain a certain aspect ratio (close to unity) between the elements. Figure 2.22 shows several different modeling techniques for a beam–slab bridge, which can be in 1D, 2D (Figure 2.22b), or 3D (Figure 2.22c) model. Details will be discussed further in Chapters 4 through 8.

The advantage of using finite element is that the analysis can be carried out for a transition area, such as welds between the steel girder and the transverse stiffener. A local area can be finely modeled separately from whole model in brief so that efforts can then concentrate on the problem area in detail. This technique is called *subdivision method* (or *substructuring method*) and is used more in other industries, such as aerospace or ship structure. For bridge structures, it is useful for failure analysis but not recommended in rapid design work. Principles of finite element analysis and strategies to apply it in different situations are discussed in Chapter 3 in detail.

2.4.6 Live load influence surface

If a refined analysis method, as any method defined in Section 2.5, is used, influence surfaces are then generated, with x and y as the surface coordinates and z as the ordinate. To apply the live loads, influence surfaces of all sorts (moment, shear, torsion, deflection, reaction) are formed, such as the moment influence surface shown in Figure 2.23a. The conventional

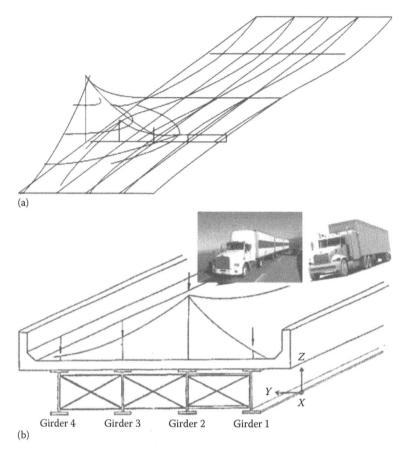

(a)

(b)

Girder 4   Girder 3   Girder 2   Girder 1

Figure 2.23 Live load on an influence surface: (a) moment influence surface; (b) placement of vehicles on an influence surface.

technique of using *influence surface* directly projects the ordinates of the axles' footprints, as shown in Figure 2.23b. A technique called *composite influence line* can be used in influence surface loading of 2D grillage model discussed in Chapter 7 (Fu 1994, 2013). Composite influence line is used to suppress the associated influence lines of adjacent girders to the primary girder. Before suppression, distribution factors are multiplied by their respective girder influence ordinates. Here, *distribution factor* is defined as the fraction of the wheel load, not from the S/D method defined by AASHTO. The advantages of using *composite influence lines* are in the saving of computer memory for influence surface and in easy access for future use (Fu 1994, 2013). Other than the *composite influence lines* method, a more sophisticated *3D FEA influence surface* method will be discussed in Chapter 3 in detail.

2.5 DIFFERENT TYPES OF BRIDGES WITH THEIR SELECTED MATHEMATICAL MODELING

If bridge structures need to be mathematically modeled, any piece-wise approximation needs to be established first. An approximate solution is reached by subdividing the structure into regions of interest. Substructure and superstructure can be decoupled into two different analyses if they are not constructed integrally.

Methods of analysis of highway bridges range from the simplified beam model with live load distribution factors defined by a design specification to the complicated 3D finite element model with influence surface loading. The simplified beam model, with the newly developed distribution factor (AASHTO 2013), is supposed to close the gap between these two extremes, but it is still on the conservative side. Unless a more accurate method is needed for rating or posting, the AASHTO method is still the most popular method used in design of bridges.

As for the refined analysis, several methods have been mentioned in Section 2.4. Among the methods, grillage analogy method is the most popular and 3D generic finite element is the most detailed. Comparing these two methods, there are two important differences mentioned by Jategaonkar et al. (1985), which are briefly discussed here:

1. *Conservative/nonconservative results.* The 3D generic finite element analysis is an approximation to the exact solution. It can be shown that as the number of finite elements in the model increases, provided that a conforming type of elements is used, the convergence to the exact solution is from below and the solution obtained from it is lower bound to deflection and stresses, which is not conservative. A grillage analysis, on the other hand, gives a theoretical solution based on the assumption of the grillage model and is not so critical of the mesh size.
2. *Accuracy.* The 3D generic finite element analysis can refine the mesh to obtain the local stress near the critical location, such as holes or sharp turns, or heavily loaded location, such as the position of the concentrated load. Grillage analysis can give accurate analysis results in terms of overall moments and shears (and thus the overall stresses) but not *local stresses*. In such cases, local stresses can be obtained by handbooks, such as design aids for concentration factor, or closed form solution, such as Westergaard method for deck bending in Section 2.1.

With this in mind, several mathematical models are suggested for different types of bridges, and they are listed in the following sections.

When modeling and analyzing a bridge, it should be noted that 3D modeling with influence surface live loading may produce more accurate results for a middle- or short-span bridge than a long-span bridge such as cable-stayed or suspension bridge. When nonlinear effects are of concern, material non-linear effect may play an important role for a middle- or short-span bridge, whereas geometric nonlinear effect may be essential for a long-span bridge. As modern modeling technique and analysis tools are widely available, 3D modeling with influence surface loading is always encouraged. For a long-span bridge, geometric nonlinear analysis should be considered in most cases.

2.5.1 Beam bridge and rigid frame bridge

To simplify the analysis, a beam-type bridge can be simply modeled by 2D beam elements, and a rigid frame-type bridge can be modeled by 2D frame elements. With this simple model and quick turnout, the model can be used to analyze construction staging (Figure 2.24), thermal loading due to differential temperature (Figure 2.25), prestressing loading as equivalent applying forces (Figure 2.26), and loading due to support movement (moment redistribution in Figure 2.27b for nonsettlement case versus Figure 2.27c for differential settlement case). Bridge with different soil conditions to simulate the support movement (Figure 2.28a) can be modeled as a three-spring foundation (Figure 2.28b). A frame structure with soil springs and their effects are shown in Figure 2.29, where the three-spring constants can be represented by

$$\text{Vertical spring:} \quad K_z = \frac{25GA^{0.5}}{(1-\nu)}$$

$$\text{Horizontal spring:} \quad K_x = 2G(1+\nu)A^{0.5} \tag{2.11}$$

$$\text{Rock spring:} \quad K_{zr} = \frac{2.5GZ}{(1-\nu)}$$

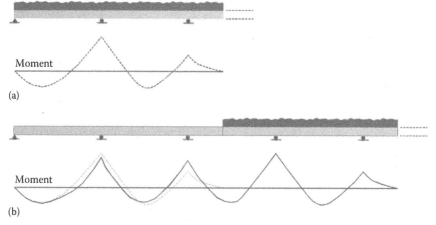

(a)

(b)

Figure 2.24 (a, b) Bridge construction staging.

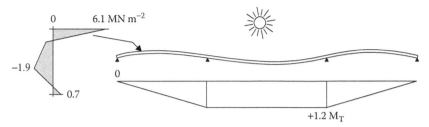

Figure 2.25 Bridge thermal loading due to differential temperature.

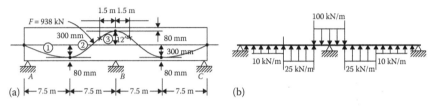

Figure 2.26 (a, b) Bridge prestressing loading as equivalent applying forces.

where:

E = Young's modulus of soil
G = Shear modulus of soil = $E/[2(1 + v)]$
v = Poisson's ratio of soil
A = Foundation area
Z = Foundation section modulus (Richart et al. 1970)

Structures with soil spring will be applied to earthquake analysis covered in Chapter 17.

2.5.2 Slab bridge

In the AASHTO LRFD Specifications (2013), a beam model with equivalent strip width can be built for the slab bridge. With one-lane loaded, the equivalent width of longitudinal strips is (AASHTO Equations 4.6.2.3-1 and -2)

$$E = 250 + 0.42\sqrt{L_1 W_1} \tag{2.12}$$

and with multilane loaded,

$$E = 2100 + 0.12\sqrt{L_1 W_1} \le \frac{W}{N} \tag{2.13}$$

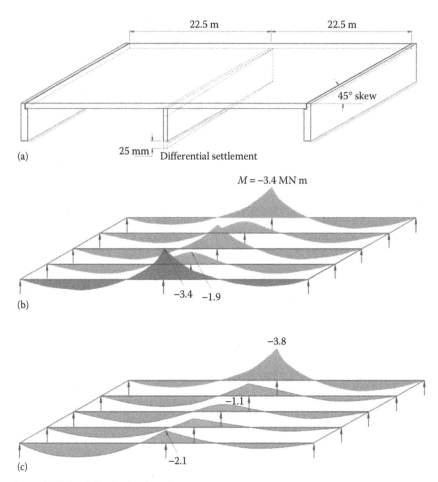

(a) 25 mm Differential settlement

$M = -3.4$ MN m

-3.4 -1.9

(b)

-3.8

-1.1

-2.1

(c)

Figure 2.27 (a–c) Bridge loading due to support movement.

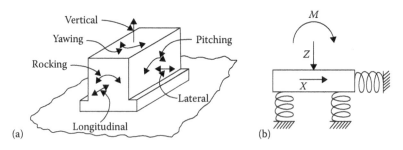

Figure 2.28 Bridge soil foundation with (a) support movements and (b) modeled as a three-spring foundation.

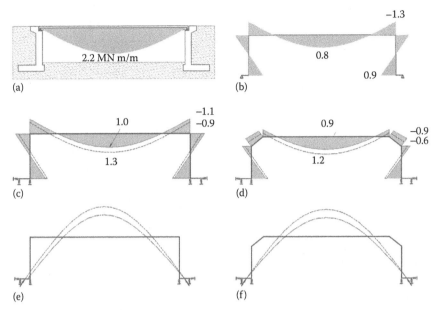

2.2 MN m/m

(a)

−1.3

0.8

0.9

(b)

−1.1
1.0
−0.9

1.3

(c)

0.9

−0.9
−0.6

1.2

(d)

(e)

(f)

Figure 2.29 (a–f) 2D frame structure modeled with soil springs.

where:
> E is the equivalent width (mm)
>
> L_1 is the modified span length taken equal to the lesser of the actual span or 18,000 mm
>
> W_1 is the modified edge-to-edge width of bridge taken equal to the lesser of the actual width or 18,000 mm
>
> W is the actual edge-to-edge width of bridge
>
> L is the physical length of bridge
>
> N is the number of design lanes

If a bridge is skewed, the longitudinal force effects may be reduced (AASHTO 2013). The simplified beam model yields reasonably good results for the bridge design work.

If a refined model, such as grillage analogy method, is used, a bridge may be divided into strips of equal widths, with idealized longitudinal beams lined up as shown in Figure 2.13a. For each longitudinal beam, the assignment of rigidities can be found using Equations 2.6 through 2.9.

2.5.3 Beam–slab bridge

Beam–slab bridge is the most popular bridge group and is well defined in the AASHTO LRFD Specifications (2013). Section 2.2 describes the approximation of the beam model by using the effective width, live load distribution factors, and influence lines. There are several examples in Chapter 4 for RC beam bridge, Chapter 5 for PC beam bridge, and Chapter 7 for steel I-girder bridge using this method.

There are several conditions to be met for a beam–slab bridge, and they are defined in the AASHTO LRFD Specifications (2013) as

- Width of the bridge is constant.
- Number of beams is not less than four.
- Beams are parallel and have approximately the same stiffness.
- Roadway part of the overhang does not exceed 1 m (3 ft).
- Curvature in plane is less than the limit specified in the AASHTO LRFD Specifications (2013).
- Cross section is consistent with one of the cross-sections shown in the AASHTO LRFD Specifications (2013).

If earlier conditions are violated, the refined methods, such as grillage analogy or FEM, are recommended. If the grillage analogy is used, the same procedures defined for beam model can be used for each longitudinal beam, and the transverse stiffnesses, D_y and D_{yx}, for the solid slab, as defined in Section 2.4.2, can be used for the transverse beam element. If more detailed information is required, finite element is a practical method. When applying finite element, however, it has to be cautious that mesh size, coordinates, loading directions, and boundary conditions affect on getting good results.

2.5.4 Cellular/box girder bridge

The beam model of this type can be built just like the beam–slab bridge with the effective widths defined for each web. For segmental concrete box and single-cell cast-in-place box beams, effective width is defined more elaborately. As defined in the AASHTO LRFD Specifications (2013), a beam model can be used as an approximate method with appropriate effective width and distribution factor, as mentioned in Section 2.2.

One of the differences between the detached box bridge and other girder-type bridges is distortion of the box due to eccentric loading (Figure 2.30a); the effect can be substantial for flexible section, such as steel. The EBEF (equivalent beam on elastic foundation) approach (Figure 2.30b) provides good approximation of the moments and stresses due to distortion and warping around the box section (Heins and Hall 1981). If distortion and warping are predominant, use of a more refined model, such as 3D FEM model, is suggested.

For a cellular deck, where the cells are either attached or detached, the principal modes of deformation are due to longitudinal bending, transverse bending, torsion, and distortion. If grillage analogy method is adopted,

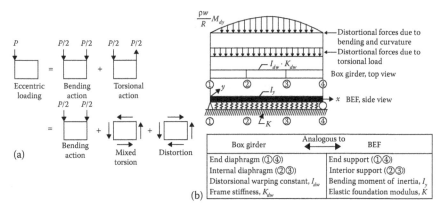

Figure 2.30 Distortion of the box girder and its analysis approach: (a) box due to eccentric loading; (b) equivalent beam on elastic foundation analysis.

to simulate the action, proper mesh has to be established with the grillage points subjected to continuity and equilibrium. If there is no diaphragm or cross-bracing, the transverse elements are formed by the slab, and they should be spaced with at least four elements between the dead load points of contraflexure. If internal or external diaphragms exist, the mesh joints should coincide with the locations of the diaphragms. The function of the internal diaphragms is maintenance of the shape of the box and reduction of distortion. The function of the external diaphragms is reduction of the differential displacement between boxes. Figure 2.31 shows different types of multicell deck and their mesh definitions with longitudinal lines along their respective ribs.

If finite element is adopted for the analysis, the same principle is applied as stated for the grillage analogy method. To obtain meaningful results, at least two elements should be used for the vertical or inclined web and at least two (maybe more, if the flange is wide) elements should be used for the top and bottom flanges. Figure 2.32 shows an example of using finite element modeling for a box girder bridge. More detailed coverage for steel box girder bridge is in Chapter 8.

2.5.5 Curved bridge

Horizontally curved bridges are commonly used. It has often been used in complex, multilevel interchanges, where the geometrics of a bridge structure are dictated by the roadway alignment.

There are two approximate methods that have been used to analyze curved girder bridges. The first method, called V-Load method, is used for curved I-girder bridges. The second method, called M/R method, is used for curved box girder bridges (FHWA/University of Maryland 1990).

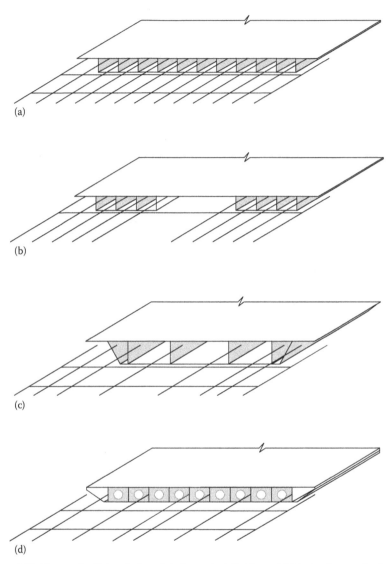

Figure 2.31 Types of cellular deck and their mesh definitions. (a) Mesh of a attached multicell box girder; (b) mesh of an detached multicell box girder; (c) mesh of a trapezoidal attached multicell box girder; and (d) mesh of multicell voided slab.

The theory of the V-Load method for curved I-girder bridges (Figure 2.33) is based on the statics of a curved flange carrying an axial stress or force. This then results in a radial distribution force on the flange, and this radial force is converted to a shear force across the diaphragms. Thus, it is called *V-Load method*. The M/R method (Figure 2.34) establishes three equilibrium equations first. These two methods are approximate and were

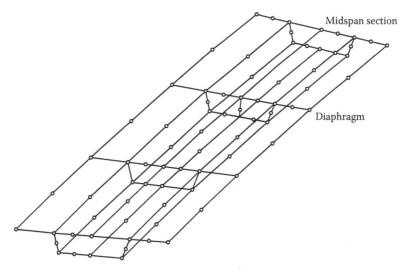

Figure 2.32 Finite element model of a box girder bridge.

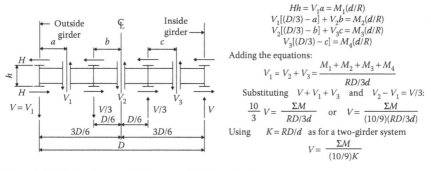

$$Hh = V_1a = M_1(d/R)$$
$$V_1[(D/3) - a] + V_2b = M_2(d/R)$$
$$V_2[(D/3) - b] + V_3c = M_3(d/R)$$
$$V_3[(D/3) - c] = M_4(d/R)$$

Adding the equations:

$$V_1 = V_2 + V_3 = \frac{M_1 + M_2 + M_3 + M_4}{RD/3d}$$

Substituting $V + V_1 + V_3$ and $V_2 - V_1 = V/3$:

$$\frac{10}{3}V = \frac{\Sigma M}{RD/3d} \quad \text{or} \quad V = \frac{\Sigma M}{(10/9)(RD/3d)}$$

Using $K = RD/d$ as for a two-girder system

$$V = \frac{\Sigma M}{(10/9)K}$$

Figure 2.33 V-Load method for curved I-girder bridges.

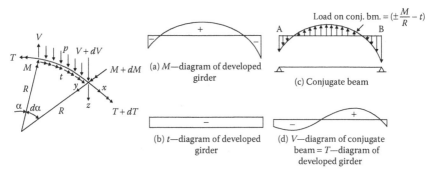

(a) M—diagram of developed girder

Load on conj. bm. $= (\pm\frac{M}{R} - t)$

(c) Conjugate beam

(b) t—diagram of developed girder

(d) V—diagram of conjugate beam = T—diagram of developed girder

Figure 2.34 M/R method for curved box girder bridges.

used to analyze and design curved bridges in the past. They may be used in the preliminary design but are not recommended for the final design, especially for heavily skewed support(s) and/or sharply curved span(s).

Currently, the most popular modeling method in applying finite element analysis is either 2D grillage analogy method or generic 3D modeling. When using 2D grillage analogy method, as discussed in Section 2.4.5, only vertical translational and planar rotational displacements are considered in an element, but geometry of an element could be straight or curved. When using a generic 3D modeling method, all displacements are considered and different types of elements can be used in a model. The same principle is defined for the beam–slab and box girder bridges. One exception is that shell elements have to be used for the web to follow the curved profile of the girders. Curved beam elements are also recommended (Hsu et al. 1990; Fu and Hsu 1995) for the grillage analogy method to eliminate the incompatibility and unbalanced forces at the joints. Curved concrete bridges are covered in Chapter 6, whereas curved steel bridges are discussed in Chapter 7 for I-girder and Chapter 8 for box girder, respectively.

2.5.6 Truss bridge

Truss members are joined by gusset plates at the panel joints, and their connections can be made by riveting, bolting, or welding. Usually, trusses are designed assuming that the members carry direct axial stresses only, which are termed *primary stresses*. However, bending stresses, referred to as *secondary stresses*, are also produced by truss distortion and joint rigidity. The axial forces in a pin-jointed truss can be found directly by the planar truss bridge analysis program (Fu and Schelling 1989) and may be used for analysis, rating, or design purposes. With truss joints rigidly connected, frame analysis with 3D modeling should be used. A refined frame analysis must include (1) composite action with the deck; (2) continuity among the truss components, where it exists; (3) force effects due to the weight of components, change in geometry due to deformation, and axial offset at the panel points; and (4) in-plane and out-of-plane buckling. Figure 2.35 shows the detailed 2D truss bridge model with an influence line for live load consideration.

In the United States, the practice is that, with proper care in sectioning and details, it is probably safe to assume that it is not necessary to compute secondary stresses. In Europe, the code specifies that, when considering the limit state of fatigue, or the limit state of serviceability, it is allowable to use either (1) assuming fixed joints in the analysis or (2) assuming pinned joints, which modify the analysis by the inclusion of flexural stresses due to axial deformation, self-weight of the members, and the stiffness of joints. More detailed discussions with 2D and 3D examples are covered in Chapter 10 for steel truss bridges.

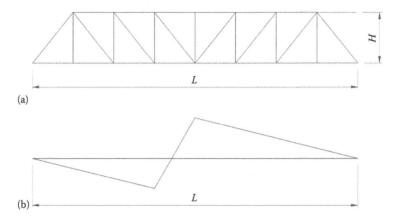

Figure 2.35 (a, b) Truss bridge 2D model with an influence line for the live load consideration.

2.5.7 Arch bridge

An arch bridge is defined with members shaped and supported in such a manner that intermediate vertical loads are transmitted to the support primarily through axial compressive force, the reverse of cables of the suspension bridge. If correctly designed, the self-weight of the arch structure induces mainly compressive forces. This is achieved by making the arch shape correspond as closely as possible to the line of thrust due to the dead loads. If it is a truss arch, the thrust is equally shared by the top and bottom chords of the truss arch members. The three fundamental equations of static equilibrium for 2D arch model are in the horizontal, vertical, and rotation directions.

For the three-hinged (two at supports and one at the crown) arch, the structure is statically determinate. For all other arch types, fixed arch or two-hinged arch, the unknowns exceed the equations of statics and are suited for computer analysis. The frame-type program can be used for assuming piece-wise linear beam elements with three degrees of freedom, corresponding to H, V, and M. Some computer programs have the curved beam elements and can give more accurate results. This type of analysis, without considering the axial deformation, is called a first-order arch analysis. In early development, to save computation time, the influence lines for moments, shears, axial force, and reactions can be generated by using a reciprocal relationship. The region of lane loading and location of truck loading should be placed properly to give the maximum live load effect. Figure 2.36 shows typical influence lines of a three-hinged arch.

The arches can be classified by their types as (1) open spandrel, (2) solid spandrel, (3) tied arches with bow-string, and (4) arch-like frames. It is convenient to perform the analysis in terms of unit width of ring and by dividing the ring into equal segments. For deck arches with columns and

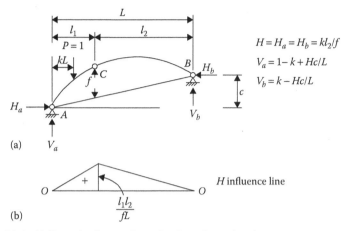

Figure 2.36 (a, b) Typical influence line of a three-hinged arch.

through arches with hangers, the segments should coincide with the panel joints. For solid spandrel arches, 20 segments between the spring lines should be enough, and the distribution of live and dead loads are best dealt with as discrete point loads. The fill pressure considers the soil at rest, and total loads on arch segment are shown in Figure 2.37. More detailed discussions on analysis and construction will be covered in Chapter 9.

2.5.8 Cable-stayed bridge

A cable-stayed bridge is a highly statically indeterminate structure. Cable-stayed bridge may be analyzed as a planar or space frame with consideration of its linear and nonlinear behavior.

Linear system—For a linear system, the deflections of the structural system under applied loads may be determined by applying the classical theory (or so-called first-order theory). By assuming Hook's law, linear superposition is applied to the internal forces, the displacements, and the stresses. However, for cable-stayed bridges, the linear assumption is on the nonconservative side for long-span bridges and can be used only for preliminary designs.

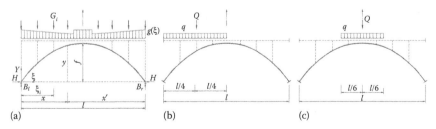

Figure 2.37 (a) Arch model and (b, c) critical loads on arch segment.

A simple solution for determining the force on the stiffening girder of linear equation by the classical theory uses the beam-on-elastic-support analogy (Troitsky 1988). If the shortening of both cables and tower is considered, the spring constants for the elastic support can be determined by Equation 2.14 as

$$K = \frac{1}{\left(H_t / A_t E_t\right) + \left(L_e / A_c E_c \sin^2 \alpha\right)} \tag{2.14}$$

where:

A_t, E_t, and H_t are the area, Young's modulus, and height of the tower, respectively

A_c, E_c, L_c, and α are the area, Young's modulus, length, and inclined angle of the cable, respectively (Figure 2.38b)

In early analysis of this system, the continuous stiffening girder on elastic supports is considered as the basic system (Figure 2.38a), and the cable forces are taken as being redundant.

For the preliminary analysis, a moment diagram may be constructed for the girder. The cable forces are obtained through the shear forces and then applied to the tower. The stresses at any section of the bridge system may be evaluated by computer. Calculation would determine the approximate cable stresses under dead load on the girder plus live load.

Nonlinear system—Nonlinearity of cable-stayed bridges generally can be categorized as the cable, stiffening girders, and towers. The nonlinearity of the cable is caused by the variation in sag with tensile force. To overcome this nonlinear effect, Ernst uses the equivalent modulus of elasticity E_i to replace the modulus of elasticity of straight cable, and it will be discussed more in Chapter 11, which is designated for cable-stayed bridges.

The nonlinearity of the stiffened girders and towers is subjected to the interaction of compressive axial force and bending moments. The girder

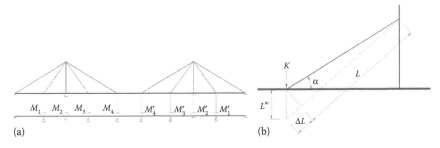

Figure 2.38 Basic cable-stayed system: (a) assumption of continuous stiffening girder on elastic supports; (b) moveable cable.

and tower in this case have to be analyzed as a beam column. The stiffness of the girder has less effect on the vertical deflection of the girder system. However, the towers are the most critical components of the system because the second-order moments may cause formation of a mechanism (plastic hinges) in a tower. Another aspect of nonlinearity is due to large displacements of the structure. Because it affects the stresses, the principle of superposition does not apply, and the problem has to be treated by the large displacement theory (or so-called second-order theory). The iteration process keeps modifying the geometry and maintaining the equilibrium of the system.

2.5.9 Suspension bridge

Structural analysis of suspension bridge is usually made for the combination of dead load, live load with impact, traction and bracing, temperature changes, settlement of supports, and wind (both static and dynamic effects). Figure 2.39 shows suspension bridge models with different arrangements.

In the early stages of development of the theory for the suspension bridge, elastic theory was used for the analysis. The suspension bridges were analyzed by the classical theory of structures, the so-called elastic (also known as first-order) theory of indeterminate analysis that ignores deformation of the structure. The elastic theory can be simply expressed as

$$M = M' - hy \qquad (2.15)$$

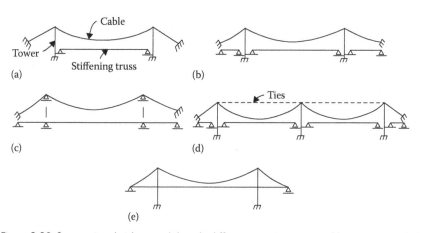

Figure 2.39 Suspension bridge model with different arrangements: (a) one suspended span with pinned stiffened truss; (b) three suspended spans with pin-ended stiffened trusses; (c) three suspended spans with continuous stiffened trusses; (d) multisuspended spans with pin-ended stiffened trusses; (e) self-anchored suspension bridge.

where:

M is bending moment in stiffening girder

M' is bending moment in the unsuspended girder for the live loads $(h = 0)$

h is horizontal tension in cable

y is cable sag

This theory is only used in a preliminary design for estimating cable quantities. As a consequence of large displacements of long suspension spans, the elastic theory results in underestimated moments, shears, and deflections. A deflection theory is then developed and referred to as second-order theory with the expression of

$$M = M' - hy - (H + h)v \qquad (2.16)$$

where:

v is cable deflection under loads

As displacements affect structural geometry, Equation 2.16 is not linear, and linear superposition technically is not applicable. There would be difficulty in using the influence line concept. For this type of analysis, programs that can handle large deflection and material nonlinearity should be used. Large deflection analysis is necessary for structures, such as suspension bridge, that undergo a large translation and rotation, and where their load-carrying path is altered as the load is increased. The nonlinear procedure for the suspension bridge is tedious and time consuming. With simplification to a quasi-linear theory, an average value of H (H_{max} and H_{min}) may be used as a basis of linearized influence line as in the case of first-order theory. There may be two sets of influence line generated, one by H_{max} and another by H_{min}, to establish the most critical live load effect. More detailed procedures to handle the nonlinearity in computation with modern technology, especially on live load, will be deferred to Chapter 12, which is designated for suspension bridges.

Chapter 3

Numerical methods in bridge structure analysis

3.1 INTRODUCTION

Numerical methods, such as the finite element method (FEM), are fundamental to bridge structure analysis. When analyzing or designing a bridge, different modern computer programs may be used. A deep understanding of the principles of the underlying methods to analyzing or designing a bridge is essential to properly conduct bridge analysis and design and is particularly important in building an appropriate computer model representing a bridge for different types of analyses. In this chapter, the principles of FEM, the automatic time incremental creep analysis method, and the influence line/surface live loading method are introduced to provide the basis for computational applications in bridge analysis and design.

FEM was first introduced in 1960s and is widely adopted in bridge engineering as the primary structural analysis approach. As modern computer science has advanced since the end of the twentieth century, FEM's application to bridge structure analyses, including its pre- and postprocessing techniques, has also greatly developed. FEM plays a critical role in modern bridges' analyses and designs. Although many generic FEM packages and more bridge-specific analysis systems are available and engineers or researchers do not need to develop a FEM package by themselves, a general understanding of FEM's principles, procedures, and its limitations will help to master its applications, including model preparation, result procession, and error identification.

Creep and shrinkage behaviors are part of the nature of concrete material. Most of these behaviors occur during the early stages, and there is less development as concrete ages. Therefore, their total effects are limited. However, the amount of both displacements and internal force redistribution due to creep and shrinkage has to be analyzed in certain concrete bridges, especially those built in multiple stages, or prestressed concrete bridges (Bažant et al. 2011). Dischinger and effective Young's modulus methods, as shown in Chapter 4, are commonly used in concrete creep analyses.

However, these methods are based on particular mathematical models of creep development. The implementation of these methods involves different element stiffness computations. Automatic incremental creep analysis method, developed by the authors and presented here, relies only on the linear assumption of creep effects and separates the time domain nonlinearity away from FEM itself. As long as the creep factor, a coefficient scalar to describe the proportion of creep strain to elastic strain, is not coupled with loads, this method is suitable for any creep development model, and its implementation can be separated from any FEM system.

The third topic discussed in this chapter is the influence line/surface live loading method. Searching for the extreme live load positions where internal forces or displacements at a particular point of interest are maximal or minimal is a unique analysis problem to bridge analysis and design. For some simple vehicle patterns defined by certain specifications, the extreme positions can simply be identified from influence lines. For some complex vehicle patterns in which only minimum vehicle spacing is defined, simple enumeration may not work. Dynamic planning is commonly used as a generic influence line live loading analysis method. Based on the longitudinal influence line live loading method, influence surface live loading can be further developed, with certain assumptions on traffic movements.

3.2 FINITE ELEMENT METHOD

3.2.1 Basics

FEM is an approximate approach to solve a global equilibrium problem with a continuum domain by a discrete system that contains a finite number of well-defined components or elements. With the fast computing power and large memory capacity of a modern digital computer, the discrete system can be used to solve a very large and complicated continuum problem. Due to the complexity of real engineering structural problems, often the continuous close-form solution is absent or impossible. With more advanced modern computer hardware and software technologies, the application of FEM becomes the obvious choice in structural analyses.

The principle of FEM is based on the minimization of total potential energy, which states that the sum of the internal strain energy and external works must be stationary when equilibrium is reached. In elastic problems, the total potential energy is not only stationary but also minimal. The stationary of total potential energy is equivalent to its variation over admissible displacements being zero and can be expressed as (Zienkiewicz et al. 1977)

$$\frac{\partial \Pi}{\partial a} = \left\{ \begin{array}{c} \dfrac{\partial \Pi}{\partial a_1} \\ \dfrac{\partial \Pi}{\partial a_2} \\ \vdots \end{array} \right\} = 0 \qquad (3.1)$$

where $a, a_1, a_2, \ldots$ denote displacements and Π is the total potential energy—the sum of the total internal strain energy and total potential energy of external loads is

$$\Pi = U + W \qquad (3.2)$$

In Equation 3.2, U and W are the total strain energy and the total potential energy of external loads, respectively. For a given domain, they can be subdivided into many regular or well-formed elements. The total strain energy U is the sum of strain energies of individual elements. Given any admissible displacements at the nodes of an element, if an appropriate displacement pattern can be assumed based on these nodal displacements, displacements at any point within the element can be expressed as a function of nodal displacements. Strain can then be derived as a function of nodal displacements. Considering the relationship between stress and strain, stress can be expressed as a function of nodal displacements as well. The total potential energy due to external loads is a simple function of nodal displacements. Therefore, the total potential energy Π is a function of nodal displacements. Applying variations over nodal displacements as in Equation 3.1 or the well-known Rayleigh–Ritz process piecewise over all elements, a relationship between unknown nodal displacements and known external loads can be established as

$$Ka = f \qquad (3.3)$$

where:
   K is the so-called global stiffness matrix
   f is external nodal loads
   a is nodal displacements

The procedures of applying FEM for structural analysis are standardized and can be summarized as follows:

1. Subdivide the continuum or structure into small elements. This step is also called *system discretization*. Element types and mesh density have to be determined in this step.
2. Determine an appropriate displacement pattern of an element. This is critical to the solution as it derives how displacements at any point

within an element are interpolated from nodal displacements. Together with mesh density, the displacement pattern affects the convergence of the FEM solution. Displacement pattern is defined by different types of elements. Therefore, once the types of elements used to discrete the system are decided, displacement patterns are automatically determined.

3. Compute the stiffness matrix of every element and assemble the global stiffness elements.
4. Prepare the global stiffness matrix according to known boundary conditions. As any arbitrary rigid movements can satisfy Equation 3.3, the global stiffness matrix K becomes singular. To solve Equation 3.3, K has to be condensed to contain only unknown nodal displacements.
5. Solve Equation 3.3.
6. Compute strains and stresses of each element. Once nodal displacements are solved, displacements at any point within an element can be interpolated by assumed displacement patterns. Furthermore, strains and stresses at any point of element can be obtained.

Theories and literatures on FEM are widely available. In this chapter, the key procedures like a generic FEM and some other special topics regarding its numerical application in bridge structural analyses will be discussed.

3.2.2 Geometric and elastic equations

When external loads are acting on an elastic body, displacements and deformations* will be induced. The displacement at any point a is described by its projection on the Cartesian axes, u,v,w, respectively, as shown in Figure 3.1. These three displacement components are functions of coordinates x,y,z, respectively.

$$a = \begin{bmatrix} u & v & w \end{bmatrix}^T \qquad (3.4)$$

The deformation at any point in the elastic body is described by three direct strains and three shear strains, which are the first derivations of displacements.[†]

$$\varepsilon = \begin{bmatrix} \varepsilon_x & \varepsilon_y & \varepsilon_z & \gamma_{xy} & \gamma_{yz} & \gamma_{zx} \end{bmatrix}^T$$
$$= \begin{bmatrix} \dfrac{\partial u}{\partial x} & \dfrac{\partial v}{\partial y} & \dfrac{\partial w}{\partial z} & \dfrac{\partial u}{\partial y} + \dfrac{\partial v}{\partial x} & \dfrac{\partial v}{\partial z} + \dfrac{\partial w}{\partial y} & \dfrac{\partial w}{\partial x} + \dfrac{\partial u}{\partial z} \end{bmatrix}^T \qquad (3.5)$$

---

[*] *Displacement* refers to translational or rotational movement along a direction and is used to measure the absolute geometric change at a point in structure. *Deformation* refers to shape change in a direction and is used to measure the strain at a point in the elastic body.
[†] When geometric nonlinearity is considered, the second order derivatives will be included as in Equation 3.35.

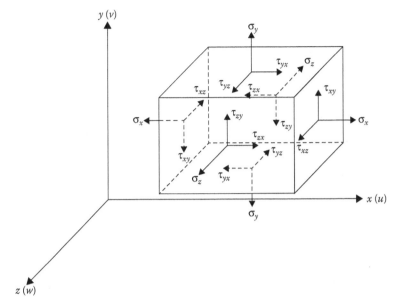

Figure 3.1 Stresses and denotations on an infinitesimal cube of any point in an elastic body.

Equation 3.5 is the geometric equation that defines the relationship between displacements and strains. When the displacements of an elastic body are known, its strains can be derived from the geometry equation. However, the displacements cannot solely be defined by known strains. Any global rigid displacement can produce the same strains according to Equation 3.5.

The displacements described by Equation 3.4 are generic for a point on an elastic body. When a particular type of element is discussed, components of displacements can be simplified or modified. For example, a two-dimensional (2D) stress or strain element will not have the w component. A spatial beam element will have rotational displacements along three Cartesian axes, and Equation 3.4 will become:

$$a = \begin{bmatrix} u & v & w & \theta_x & \theta_y & \theta_z \end{bmatrix}^T \tag{3.6}$$

For isotropic elastic materials, according to Hooke's Law, the relationship between stresses and strains is defined as

$$\varepsilon_x = \frac{\sigma_x}{E} - \mu \frac{\sigma_y}{E} - \mu \frac{\sigma_z}{E}; \varepsilon_y = \frac{\sigma_y}{E} - \mu \frac{\sigma_z}{E} - \mu \frac{\sigma_x}{E}; \varepsilon_z = \frac{\sigma_z}{E} - \mu \frac{\sigma_x}{E} - \mu \frac{\sigma_y}{E} \tag{3.7}$$

and

$$\gamma_{xy} = \frac{\tau_{xy}}{G}; \gamma_{yz} = \frac{\tau_{yz}}{G}; \gamma_{zx} = \frac{\tau_{zx}}{G} \tag{3.8}$$

where:

    E is the Young's modulus
    G is the shear modulus
    μ is the Poisson ratio

For isotropic materials, shear modulus can be derived from Young's modulus:

$$G = \frac{E}{2(1+\mu)} \tag{3.9}$$

Equations 3.7 and 3.8 are elastic equations. By solving stresses in these equations, another form of elastic equations can be written as Equation 3.10 or in matrix form as Equation 3.11.

$$\sigma_x = \frac{E(1-\mu)}{(1+\mu)(1-2u)}\left(\varepsilon_x + \frac{\mu}{1-\mu}\varepsilon_y + \frac{\mu}{1-\mu}\varepsilon_z\right),$$

$$\sigma_y = \frac{E(1-\mu)}{(1+\mu)(1-2u)}\left(\frac{\mu}{1-\mu}\varepsilon_x + \varepsilon_y + \frac{\mu}{1-\mu}\varepsilon_z\right),$$

$$\tag{3.10}$$

$$\sigma_x = \frac{E(1-\mu)}{(1+\mu)(1-2u)}\left(\frac{\mu}{1-\mu}\varepsilon_x + \frac{\mu}{1-\mu}\varepsilon_y + \varepsilon_z\right),$$

$$\tau_{xy} = \frac{E}{2(1+\mu)}\gamma_{xy}, \tau_{yz} = \frac{E}{2(1+\mu)}\gamma_{yz}, \tau_{zx} = \frac{E}{2(1+\mu)}\gamma_{zx}$$

$$\left[\sigma_x \quad \sigma_y \quad \sigma_z \quad \tau_{xy} \quad \tau_{yz} \quad \tau_{zx}\right]^T \tag{3.11}$$

$$= D\left[\varepsilon_x \quad \varepsilon_y \quad \varepsilon_z \quad \gamma_{xy} \quad \gamma_{yz} \quad \gamma_{zx}\right]^T \text{ or } \sigma = D\varepsilon$$

where D is the so-called elastic matrix as shown in Equation 3.12.

$$D = \begin{bmatrix} \lambda+2G & \lambda & \lambda & & & \\ \lambda & \lambda+2G & \lambda & & 0 & \\ \lambda & \lambda & \lambda+2G & & & \\ & & & G & 0 & 0 \\ & 0 & & 0 & G & 0 \\ & & & 0 & 0 & G \end{bmatrix} \tag{3.12}$$

λ is a constant related to Young's modulus and Poisson ratio as $\lambda = E\mu/[(1+\mu)(1-2\mu)]$.

In general, initial strains caused by shrinkage or temperature change and/or initial stresses due to existing condition may exist at any point. Only will the difference between actual and initial strains cause elastic stress changes, and the total stresses should be the sum of elastic stresses and initial stresses. The elastic Equation 3.11 can be rewritten in a generic form as

$$\sigma = D(\varepsilon - \varepsilon_0) + \sigma_0 \tag{3.13}$$

where:

$$\sigma = \begin{bmatrix} \sigma_x & \sigma_y & \sigma_z & \tau_{xy} & \tau_{yz} & \tau_{zx} \end{bmatrix}^T \tag{3.14}$$

are total stresses

$$\sigma_0 = \begin{bmatrix} \sigma_x^0 & \sigma_y^0 & \sigma_z^0 & \tau_{xy}^0 & \tau_{yz}^0 & \tau_{zx}^0 \end{bmatrix}^T \tag{3.15}$$

are initial stresses

$$\varepsilon = \begin{bmatrix} \varepsilon_x & \varepsilon_y & \varepsilon_z & \gamma_{xy} & \gamma_{yz} & \gamma_{zx} \end{bmatrix}^T \tag{3.16}$$

are total strains

$$\varepsilon_0 = \begin{bmatrix} \varepsilon_x^0 & \varepsilon_y^0 & \varepsilon_z^0 & \gamma_{xy}^0 & \gamma_{yz}^0 & \tau_{zx}^0 \end{bmatrix}^T \tag{3.17}$$

are initial strains

3.2.3 Displacement functions of an element

To apply Equation 3.5 to obtain the total strain energy of an element, displacements at any point within the element should be explicitly expressed by nodal displacements of the element. This expression is called *element displacement* or *shape functions*. Due to variations of geometry shape and mechanical behavior of an element, there is no general theoretical definition on how the displacement at a point is related to all nodal displacements of an element. Only certain types of element, for example, beam-bending element, have known theoretical displacement functions. As a generic approach of FEM, these relationships have to be assumed according to different types of elements. The definition of the displacement function for a certain type element plays a critical role in its behavior and convergence of a solution. It is easy to understand that the error in the calculation of strain energy due to an inaccurate or coarse displacement function can be minimized by reducing the size of the element. However, an accurate or fine displacement assumption can reduce the error of a large element so that even a coarse mesh still can get accurate and convergent results.

By using displacement functions, the displacement of any point in the element can be expressed as Equation 3.18.

$$u = \sum_{i=1}^{n} N_i u_i, \; v = \sum_{i=1}^{n} N_i v_i, \; w = \sum_{i=1}^{n} N_i w_i, \ldots \tag{3.18}$$

where:

n is the number of element nodes

u_i, v_i, w_i are the node displacements at node i

N_i is the displacement function of node i and describes how a known displacement at node i will influence or contribute to the displacement at any point within an element

From its definition, the displacement function must satisfy the following conditions:

1. $N_i = 1$ at node i and $N_i = 0$ at all other nodes
2. Ensures any of the unknown displacement is continuous at element boundaries, that is, displacements at any point on an element boundary interpolated by nodal displacements of any adjacent elements should be the same
3. Contains linear term so it is able to represent constant strain
4. $\sum_{i=1}^{n} N_i = 1$, so it can represent rigid displacement, that is, displacement at any point should be the same as that at any node when all nodes have the same displacements

In developing displacement functions for a type of element, the more complicated the shape of the element, the higher the polynomial order of the displacement function is required. An element with a higher order of displacement functions will lead to a higher accuracy. Therefore, a coarser mesh will produce relatively higher accurate results. Or, in other words, a finer mesh is needed when a simple element with a lower-order displacement function is used.

Taking a commonly used 2D rectangle element as an example, as shown in Figure 3.2; the displacement functions of a four-node element are

$$N_1 = \frac{1}{4}(1+\xi)(1+\eta), \; N_2 = \frac{1}{4}(1-\xi)(1+\eta)$$

$$\tag{3.19}$$

$$N_3 = \frac{1}{4}(1-\xi)(1-\eta), \; N_4 = \frac{1}{4}(1+\xi)(1-\eta)$$

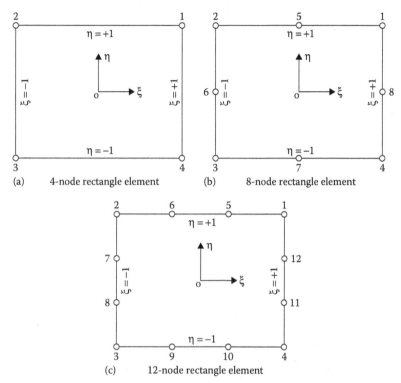

Figure 3.2 (a–c) Three different types of rectangle elements.

The displacement functions of an eight-node element are

$$N_1 = \frac{1}{4}(1+\xi)(1+\eta)(\xi+\eta-1)$$

$$N_2 = \frac{1}{4}(1-\xi)(1+\eta)(-\xi+\eta-1)$$

$$N_3 = \frac{1}{4}(1-\xi)(1-\eta)(-\xi-\eta-1)$$

$$N_4 = \frac{1}{4}(1+\xi)(1-\eta)(\xi-\eta-1)$$

$$N_5 = \frac{1}{2}(1-\xi^2)(1+\eta), \quad N_6 = \frac{1}{2}(1-\eta^2)(1-\xi)$$

$$N_7 = \frac{1}{2}(1-\xi^2)(1-\eta), \quad N_8 = \frac{1}{2}(1-\eta^2)(1+\xi)$$

(3.20)

Displacement functions at some nodes of a less-used 12-node element are

$$N_1 = \frac{1}{32}(1+\xi)(1+\eta)\left[9(\xi^2+\eta^2)-10\right]$$

$$N_2 = \frac{1}{32}(1-\xi)(1+\eta)\left[9(\xi^2+\eta^2)-10\right] \tag{3.21}$$

$$N_6 = \frac{9}{32}(1+\eta)(1-\xi^2)(1-3\xi), \quad N_7 = \frac{9}{32}(1-\xi)(1-\eta^2)(1+3\eta)$$

To investigate the characteristics of displacement function, three-dimensional (3D) views of some of the earlier functions are shown in Figures 3.3 through 3.5. The functions in Equations 3.19 through 3.21 are linear, square, and cubic, respectively. From the earlier definitions and 3D views shown in Figures 3.3 through 3.5, it can be seen that the conditions in 1, 3, and 4 are met. To check the continuous condition, the element edge 1–5–2 of the eight-node element can be used as an example. Displacement functions of all nodes other than 1, 5, and 2 are 0, which means the interpolation of any displacement along the edge merely depends on nodal displacements at nodes 1, 5, and 2. Therefore, any displacement at any point along the edge will obtain the same value by interpolation from either of the adjacent elements.

3.2.4 Strain energy and principles of minimum potential energy and virtual works

When applying external forces, strains and stresses will be present over the entire elastic body. The total strain energy accumulated by increasing external loads from zero to a given load will be used to measure the internal

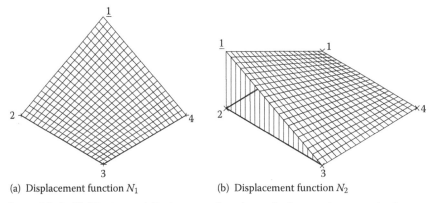

(a) Displacement function N_1 (b) Displacement function N_2

Figure 3.3 (a, b) 3D views of displacement functions of a four-node rectangle element (linear function).

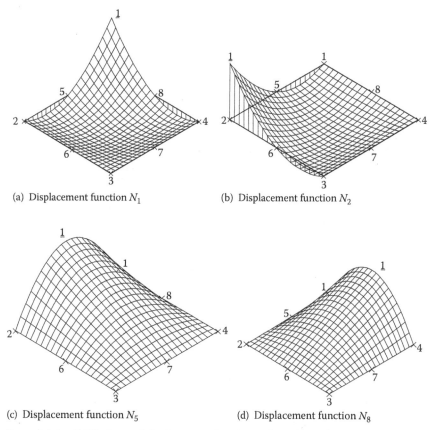

(a) Displacement function N_1

(b) Displacement function N_2

(c) Displacement function N_5

(d) Displacement function N_8

Figure 3.4 (a–d) 3D views of displacement functions of a eight-node rectangle element (square function).

works that the external loads are done. The multiplication of stress and strain at any point gives the strain energy density $\bar{U}$. Taking the spatial strain and stress problem, as shown in Figure 3.1, as the example to illustrate a generic approach, the accumulated strain energy density starting from the beginning to any equilibrium point is the shaded area as shown in Figure 3.6. It can be expressed as follows:

$$\bar{U} = \int_0^{\varepsilon_x} \sigma_x d\varepsilon_x + \int_0^{\varepsilon_y} \sigma_y d\varepsilon_y + \int_0^{\varepsilon_z} \sigma_z d\varepsilon_z + \int_0^{\gamma_{xy}} \tau_{xy} d\gamma_{xy}$$

$$+ \int_0^{\gamma_{yz}} \tau_{yz} d\gamma_{yz} + \int_0^{\gamma_{zx}} \tau_{zx} d\gamma_{zx} = \int_0^{\mu} \sigma^T d\varepsilon$$

(3.22)

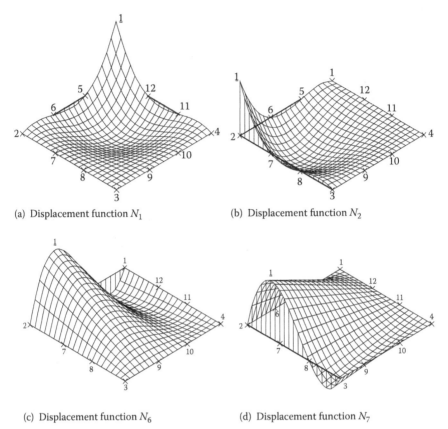

(a) Displacement function N_1

(b) Displacement function N_2

(c) Displacement function N_6

(d) Displacement function N_7

Figure 3.5 (a–d) 3D views of displacement functions of a 12-node rectangle element (cubic function).

where ε is the strain at any equilibrium point. When elastic is assumed, the curve in Figure 3.6 will become a straight line, and Equation 3.22 can be simplified as

$$\bar{U} = \frac{1}{2}\sigma^T\varepsilon \tag{3.23}$$

Substituting σ with Equation 3.11, the strain energy density can be obtained in Equation 3.24.

$$\bar{U} = \frac{1}{2}\varepsilon^T D\varepsilon \tag{3.24}$$

The total strain energy is the integration of strain energy density over the entire elastic body as shown in Equation 3.25.

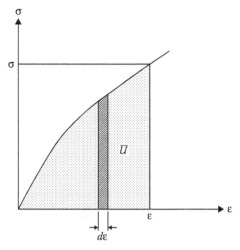

Figure 3.6 Strain energy density.

$$U = \frac{1}{2}\int \varepsilon^T D\varepsilon \, dv \qquad\qquad (3.25)$$

The works done by external loads to increase strains from 0 to ε are the products of nodal displacements and external loads. Therefore, the total potential energy of external loads is

$$W = -a^T f \qquad\qquad (3.26)$$

According to the minimum potential energy principle (Equations 3.1 and 3.2), the global equilibrium equations, as shown in Equation 3.27, can be obtained by substituting Equations 3.25 and 3.26 into Equations 3.1 and 3.2.

$$\frac{1}{2}\frac{\partial \left(\int \varepsilon^T D\varepsilon \, dv \right)}{\partial a} = f \qquad\qquad (3.27)$$

Substituting Equation 3.18 into Equation 3.5, the strains at any point of an element can be expressed by all its nodal displacements:

$$\varepsilon = \left[B_1 \, B_2 \ldots B_n \right] a = Ba \qquad\qquad (3.28)$$

where n is the number of element nodes and B_i is expressed as

$$B_i = \begin{bmatrix} \dfrac{\partial N_i}{\partial x} & 0 & 0 \\[2ex] 0 & \dfrac{\partial N_i}{\partial y} & 0 \\[2ex] 0 & 0 & \dfrac{\partial N_i}{\partial z} \\[2ex] \dfrac{\partial N_i}{\partial y} & \dfrac{\partial N_i}{\partial x} & 0 \\[2ex] 0 & \dfrac{\partial N_i}{\partial z} & \dfrac{\partial N_i}{\partial y} \\[2ex] \dfrac{\partial N_i}{\partial z} & 0 & \dfrac{\partial N_i}{\partial x} \end{bmatrix}$$

Equation 3.28 expresses the relationship between strains and displacements, for both an individual element and the entire domain. When the entire domain is considered, the node number n will be the total nodes meshed in the domain. Thus, Equation 3.27 becomes Equation 3.29 when substituting ε with Equation 3.28,

$$\int B^T DB dv\, a = f \tag{3.29}$$

or the global equilibrium Equation 3.3 where

$$K = \int B^T DB dv \tag{3.30}$$

K is the so-called global stiffness matrix. When a domain of an individual element is considered, the results of Equation 3.30 will be the stiffness matrix of an element.

The global equilibrium equation 3.29 or 3.3 can also be derived from the principle of virtual works. Given any equilibrium state of a system, small fictitious displacements—the virtual displacements—are assumed. The virtual displacement will cause internal virtual strains. The virtual work principle states that the virtual work done by actual external forces during the virtual displacements is equal to the internal strain energy increased at actual internal stresses due to the virtual strains:

$$\int \delta\varepsilon^T \sigma dv = \delta a^T f \tag{3.31}$$

where $\delta\varepsilon$ denotes internal virtual strains corresponding to external virtual displacements δa. Applying Equation 3.28 into 3.31, the equilibrium equation can be obtained as

$$\psi(a) = \int B^T \sigma dv - f = 0 \tag{3.32}$$

where ψ is the sum of general internal and external forces. Equation 3.32 can be stated as that at any equilibrium point internal forces due to internal stresses should balance external loads that cause internal strains. Furthermore, when physical equation 3.13 is substituted into Equation 3.32, a more generic equilibrium equation can be obtained as

$$\psi(a) = \int B^T DB dv\, a - \int B^T D\varepsilon_0 dv + \int B^T \sigma_0 dv - f = 0 \tag{3.33}$$

Equation 3.33 illustrates the balance between internal and external forces when initial strains and initial stresses exist.

Each element's stiffness matrix K_e can be obtained by integration over the entire element body. The physical meaning of any element at row i and column j of K_e, $K_e(i,j)$ is the force caused at ith degree of freedom because of a unit displacement happening at jth degree of freedom, as the existence or contribution of the element. The variables i and j are the order number of degree of freedom of an individual element. Because the total strain energy of a continuum is the sum of strain energies of subdivided elements, assembling all elements' stiffness matrices in an appropriate order can form the global stiffness matrix in Equation 3.30. Obviously, if all elements connected at a global node have the same local coordinate systems as the global coordinate system, stiffness elements corresponding to this global node in K can be obtained by summing the contributions $(K_e(i,j))$ from all connected elements. This process is the assembly of global stiffness matrix, which reveals the implementation of the approach by meshing a continuum into finite regular-shaped elements.

3.2.5 Displacement relationship processing when assembling global stiffness matrix

As discussed in the Section 3.2.4, an element stiffness matrix will be assembled into a global stiffness matrix. The assembly is done by matching element nodes with their global order. For example, an element has two nodes, i and j, and its element stiffness matrix is shown in Figure 3.7b. When assembling, each submatrix in Figure 3.7b will be added into its corresponding submatrix in the global matrix in Figure 3.7a. It should be noted that the element stiffness matrix must be transformed into the global coordinate system before adding it into the global matrix. The element stiffness matrix is established in its local coordinate system, which is often different from the global coordinate system. Because stiffness of a degree of freedom is a vector in space, the transformation of the stiffness matrix can be taken as a simple standard space transformation process.

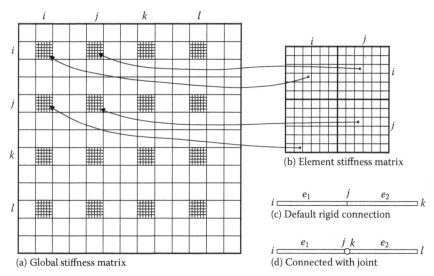

(b) Element stiffness matrix

(c) Default rigid connection

(d) Connected with joint

(a) Global stiffness matrix

Figure 3.7 (a–d) Assembling global stiffness matrix and processing displacement relationship.

All displacements at the connection of adjacent elements are continuous by default, or the connection is rigid from element to element as shown in Figure 3.7c. It is obvious that the global stiffness of node j will be the sum of submatrices of both elements $(e_1$ and $e_2)$. These results are due to one-to-one mapping of element stiffness and global stiffness during assembling matrices. However, the relationship between element stiffness and global stiffness does not have to be one to one. When this happens, a matrix-processing technique, *the displacement relationship*, will be used. Taking the simulation of commonly used joints as example, the principle of displacement relationship is discussed briefly next in this section.

As shown in Figure 3.7d, two beam elements are connected with a joint. Four nodes, i, j, k, and l, in the global matrix will be needed to have enough degrees of freedom to represent the extra rotation at the joint. If each node is assumed to have six (6) degrees of freedom, node j and k will be sharing five (5) of them and each node has one rotation independent of one another. The relationships of displacements between nodes j and k will be that the five (5) shared displacements of node k are mapped to those of node j, and their rotation is separated. When assembling e_1, it is a usual summing process. When assembling e_2, matrix elements corresponding to shared displacements at node k will be added to node j instead, rather than to node k as is normally done. Only the rotation matrix elements will be added to its own position, node k in global. This type of relationship is often called the master–slave relationship.

The displacement relationship and its processing are an important part of FEM. In addition to beam joints mentioned earlier, this process can be used to simulate many other complicated mechanics situations, such as spring or rigid body connections.

3.2.6 Nonlinearities

In the prior derivations of the global equilibrium equation, both the geometry relationship (Equation 3.5) and the material relationship (Equation 3.13) are in linear forms. When displacements are small and strains are within the linear range with stresses, as for most engineering problems, linear solutions (Equation 3.13) are accurate and adequate. However, large displacements and/or nonlinear constitutive material problems widely exist in engineering practices. The geometric nonlinearity of long-span cable bridges, discussed in Chapter 11, and the plastic behavior of middle- and short-span bridges, discussed in Chapters 14, 15, and 17, are two typical examples of these problems in bridge structural analyses. The approach to the respective geometric nonlinear and material nonlinear problems is an important part of FEM.

In general, when material nonlinearity is considered, the stresses and strains relationship (Equation 3.13) would be

$$\sigma = \sigma(\varepsilon) \tag{3.34}$$

When geometric nonlinearity is considered, the strains will contain the second order of displacement derivatives as

$$\varepsilon = \begin{Bmatrix} \varepsilon_x \\ \varepsilon_y \\ \varepsilon_z \\ \gamma_{xy} \\ \gamma_{yz} \\ \gamma_{zx} \end{Bmatrix} = \begin{Bmatrix} \dfrac{\partial u}{\partial x} + \dfrac{1}{2}\left[\left(\dfrac{\partial u}{\partial x}\right)^2 + \left(\dfrac{\partial v}{\partial x}\right)^2 + \left(\dfrac{\partial w}{\partial x}\right)^2\right] \\[2ex] \dfrac{\partial v}{\partial y} + \dfrac{1}{2}\left[\left(\dfrac{\partial u}{\partial y}\right)^2 + \left(\dfrac{\partial v}{\partial y}\right)^2 + \left(\dfrac{\partial w}{\partial y}\right)^2\right] \\[2ex] \dfrac{\partial w}{\partial z} + \dfrac{1}{2}\left[\left(\dfrac{\partial u}{\partial z}\right)^2 + \left(\dfrac{\partial v}{\partial z}\right)^2 + \left(\dfrac{\partial w}{\partial z}\right)^2\right] \\[2ex] \dfrac{\partial u}{\partial y} + \dfrac{\partial v}{\partial x} + \dfrac{\partial u}{\partial x}\dfrac{\partial u}{\partial y} + \dfrac{\partial v}{\partial x}\dfrac{\partial v}{\partial y} + \dfrac{\partial w}{\partial x}\dfrac{\partial w}{\partial y} \\[2ex] \dfrac{\partial v}{\partial z} + \dfrac{\partial w}{\partial y} + \dfrac{\partial u}{\partial z}\dfrac{\partial u}{\partial y} + \dfrac{\partial v}{\partial z}\dfrac{\partial v}{\partial y} + \dfrac{\partial w}{\partial z}\dfrac{\partial w}{\partial y} \\[2ex] \dfrac{\partial w}{\partial x} + \dfrac{\partial u}{\partial z} + \dfrac{\partial u}{\partial x}\dfrac{\partial u}{\partial z} + \dfrac{\partial v}{\partial x}\dfrac{\partial v}{\partial z} + \dfrac{\partial w}{\partial x}\dfrac{\partial w}{\partial z} \end{Bmatrix} \tag{3.35}$$

Thus, the strains and displacements relationship (Equation 3.28) becomes

$$\varepsilon = Ba = \left(B_0 + B_L[a] \right) a \tag{3.36}$$

where B_0 is the same matrix as when geometric nonlinearity is not considered and $B_L(a)$ is due to the second order of displacement derivatives and relates to current displacements.

When nonlinearities are considered, the solution of Equation 3.32 has to be approached by incremental method, in which changes of $\psi(a)$ respective to a small increment of a are to be noted.

$$d\psi = \left(\int \frac{dB^T}{da} \sigma dv + \int B^T \frac{d\sigma}{da} dv \right) da = K_T da \tag{3.37}$$

In Equation 3.37, K_T is the tangential stiffness, respective to small increment of displacements. Taking the geometric nonlinearity as an example, the tangential stiffness can be derived as

$$K_T = K_0 + K_\sigma + K_L \tag{3.38}$$

where:

$K_0 = \int B_0^T D B_0$ represents the usual stiffness when displacements are small

K_σ is the first term in Equation 3.37, which reflects the stiffness due to the existence of stresses, that is, the initial stress or geometric matrix:

$$K_\sigma = \int \frac{dB^T}{da} \sigma dv = \int \frac{dB_L^T}{da} \sigma dv \tag{3.39}$$

K_L is the stiffness due to large displacements:

$$K_L = \int \left(B_0^T D B_L + B_L^T D B_L + B_L^T D B_0 \right) dv \tag{3.40}$$

When material nonlinearity is considered as well, the elastic matrix D should be evaluated at strains due to current displacements.

The solutions of nonlinear problems can be reached by iterations on Equations 3.33 and 3.37. Given initial estimated displacements a_0, which are obtained as linear solution, their corresponding internal strains can be computed. Furthermore, the internal stresses can be obtained by either linear or nonlinear stress and strain relationship. As shown in Equation 3.33, the initial unbalanced general forces $\psi(a_0)$ can be determined. The unbalanced general forces reveal that the internal forces cannot balance the external forces due to the effects of nonlinearities. The displacements have to be adjusted by Equation 3.37. Tangential stiffness K_T will first be formed at current displacements (a_0). Taking $\psi(a_0)$ as $d\psi$ in Equation 3.37, the displacement adjustment can be solved. Once an adjustment is obtained, new displacements a_1 are established. The iteration process will keep looping till the unbalanced general forces $\psi(a_n)$ become significantly small. To ensure the convergence of this iteration process, external loads are often loaded incrementally, with each step containing only a fraction of the total loads.

3.2.7 Frame element

Frame components, which work as both beams in bending/shearing and also as truss members in axial tension/compression, are very common in structural engineering, and the development of a frame element is fundamental in FEM. This section will briefly introduce its displacement functions, elastic stiffness matrix, and initial stress matrix.

The total strain energy of a frame element is the sum of the axial tension/compression strain energy and the bending strain energy. Therefore, when developing the elastic stiffness matrix, the axial tension/compression and the bending behaviors can be separated. The beam-bending theory assumes that the cross section at any point along the beam axis will remain a plane after bent. Based on this assumption, bending strain energy along a cross section can be expressed as the product of bending moment and rotation angle of a cross section or the second-order derivative of vertical deflection. For a two-node frame element as shown in Figure 3.8, according to the requirements in Section 3.2.3, the displacement functions can only be linear. It is not enough to describe the bending deflection, as the second-order derivative does not exist. Two additional rotational displacements (ϕ_1 and ϕ_2) have to be added. Although they belong to the same nodes (nodes 1 and 2, respectively), a two-node beam element has four independent nodal displacements and is truly working as a four-node line element.

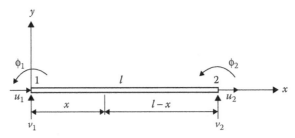

Figure 3.8 Two-node frame element.

The nodal displacements of a frame element as shown in Figure 3.8 are

$$a_e = \begin{bmatrix} u_1 & v_1 & \phi_1 & u_2 & v_2 & \phi_2 \end{bmatrix}^T \tag{3.41}$$

The strains of a frame element contain the axial tension/compression strain and bending strain as

$$\varepsilon_e = \begin{Bmatrix} \varepsilon_a \\ \varepsilon_b \end{Bmatrix} = \begin{Bmatrix} \dfrac{du}{dx} \\[2mm] -y\dfrac{d^2v}{dx^2} \end{Bmatrix} \tag{3.42}$$

where:
  y is the vertical distance of a fiber layer to the neutral axis of a cross section
  u and v are axial and vertical displacements, respectively

Their interpolation functions are

$$\begin{Bmatrix} u \\ v \end{Bmatrix} = \begin{bmatrix} N_1 & 0 & 0 & N_4 & 0 & 0 \\ 0 & N_2 & N_3 & 0 & N_5 & N_6 \end{bmatrix} a_e \tag{3.43}$$

where:

$$N_1 = L_1, \quad N_2 = L_1^2(3-2L_1), \quad N_3 = L_1^2 L_2 l, \quad N_4 = L_2,$$

$$N_5 = L_2^2(3-2L_2), \quad N_6 = -L_1 L_2^2 l, \quad L_1 = 1 - \frac{x}{l}, \quad L_2 = \frac{x}{l} \tag{3.44}$$

Knowing $\phi = dv/dx$, it can be easily verified that the earlier functions satisfy the conditions of a displacement function in Section 3.2.3.
  The matrix B in Equation 3.28 is

$$B = \begin{bmatrix} -\dfrac{1}{l} & 0 & 0 & \dfrac{1}{l} \\[2mm] 0 & -y\left(\dfrac{12x}{l^3}-\dfrac{6}{l^2}\right) & -y\left(\dfrac{6x}{l^2}-\dfrac{4}{l}\right) & 0 \end{bmatrix}$$

$$\begin{bmatrix} 0 & 0 \\[2mm] y\left(\dfrac{12x}{l^3}-\dfrac{6}{l^2}\right) & -y\left(\dfrac{6x}{l^2}-\dfrac{2}{l}\right) \end{bmatrix} \tag{3.45}$$

When integrating over the entire element as in Equation 3.30, the beam-bending assumption and a prismatic cross section can be taken into consideration.

The elastic matrix D is one single constant as E. The elastic stiffness element can be derived as

$$
K_0 =
\begin{bmatrix}
\dfrac{EA}{l} & 0 & 0 & \dfrac{-EA}{l} & 0 & 0 \\[2mm]
0 & \dfrac{12EI}{l^3} & \dfrac{6EI}{l^2} & 0 & \dfrac{-12EI}{l^3} & \dfrac{6EI}{l^2} \\[2mm]
0 & \dfrac{6EI}{l^2} & \dfrac{4EI}{l} & 0 & \dfrac{-6EI}{l^2} & \dfrac{2EI}{l^2} \\[2mm]
\dfrac{-EA}{l} & 0 & 0 & \dfrac{EA}{l} & 0 & 0 \\[2mm]
0 & \dfrac{-12EI}{l^3} & \dfrac{-6EI}{l^2} & 0 & \dfrac{12EI}{l^3} & \dfrac{-6EI}{l^2} \\[2mm]
0 & \dfrac{6EI}{l^2} & \dfrac{2EI}{l^2} & 0 & \dfrac{-6EI}{l^2} & \dfrac{4EI}{l}
\end{bmatrix}
\tag{3.46}
$$

where:
  A denotes the cross-sectional area of an element
  $I = \int y^2 dA$ denotes the moment inertia to the neutral axis of the cross section

When geometric nonlinearity is considered, the axial strain will be coupled with bending deflection. Equation 3.42 will become

$$
\varepsilon_e = \begin{Bmatrix} \varepsilon_a \\ \varepsilon_b \end{Bmatrix} = \begin{Bmatrix} \dfrac{du}{dx} + \dfrac{1}{2}\left(\dfrac{dv}{dx}\right)^2 \\[3mm] -y\dfrac{d^2v}{dx^2} \end{Bmatrix}
\tag{3.47}
$$

Following similar procedures, the initial stress matrix of a frame element can be derived by Equation 3.39:

$$
K_\sigma = \dfrac{F_x}{30l}
\begin{bmatrix}
0 & 0 & 0 & 0 & 0 & 0 \\
0 & 36 & 3l & 0 & -36 & 3l \\
0 & 3l & 4l^2 & 0 & -3l & -l^2 \\
0 & 0 & 0 & 0 & 0 & 0 \\
0 & -36 & -3l & 0 & 36 & -3l \\
0 & 3l & -l^2 & 0 & -3l & 4l^2
\end{bmatrix}
\tag{3.48}
$$

3.2.8 Elastic stability

As shown in Equation 3.39, stiffness may be enhanced or reduced by K_σ—the initial stress stiffness due to existing stresses when large displacements are considered. When total stiffness is reduced by the initial stress stiffness, as in columns or plates under compression, there will be a critical point in which stiffness in one or many degrees of freedom reaches 0 (i.e., $K_0 + K_\sigma$ becomes singular). This phenomenon is the so-called elastic stability problem, in which a critical point clearly defines the entry to an unstable state. In addition to the elastic problem, stability problems can further be classified as plastic stability and excessive displacements according to the reason of singularity of the total stiffness (tangential stiffness $K_0 + K_\sigma + K_L$). For example, if the stability problem is due to the elastic matrix D, it is plastic stability problem; if it is due to large displacements, it is the excessive displacements problem. It is obvious that both are nonlinear problems and are the same in a mathematical view. When nonlinear stability is of a concern, both plastic and large displacements should be considered together. When excessive displacements happen, some components may have entered plastic range, and when some components enter plastic range, displacements may become large. The approach to nonlinear stability solutions is the same as normal nonlinear problems as illustrated in the previous section. In this section, only the elastic stability is discussed, as it gives the upper limits of critical loads and is more essential to structural analyses. For instance, during preliminary designs of bridges in which compression and bending are dominating (i.e., arch bridges and cable-stayed bridges), elastic stability is usually analyzed first. The upper limit will guide the adjustment to structure dimensions and component sizes. Further discussion and application on stability is discussed in Chapter 14.

The solution to an elastic stability problem can be categorized into an eigenvalue problem. When only initial stress is considered, the following equilibrium equation can be derived from either the global equilibrium equation 3.33 or the tangential equilibrium equation 3.37:

$$\left(K_0 + K_\sigma \right) a = f \tag{3.49}$$

K_σ is proportional to the current axial tension/compression stress as shown in Equation 3.48. The search for critical loads in elastic stability can be simplified by amplifying K_σ until the total stiffness matrix in Equation 3.49 becomes singular, which is equivalent to the following general eigenvalue problem:

$$\left| K_0 + \lambda K_\sigma \right| = 0 \tag{3.50}$$

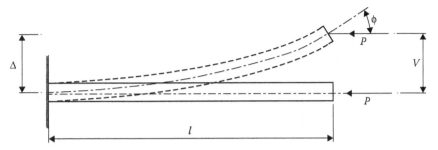

Figure 3.9 Example of elastic stability problem—a beam under compression.

Taking the frame element discussed in the previous section and a typical $P-\Delta$ problem as shown in Figure 3.9 as an example, the FEM approach to the elastic stability problem can be compared with theoretical solutions. The cantilever beam in Figure 3.9 is fixed at one end, so that the unknown displacements are only υ and ϕ. The elastic stiffness matrix in Equation 3.46 and initial stress stiffness matrix in Equation 3.48 can be condensed to Equations 3.51 and 3.52, respectively:

$$K_0 = \begin{bmatrix} \dfrac{12EI}{l^3} & \dfrac{6EI}{l^2} \\ \dfrac{6EI}{l^2} & \dfrac{4EI}{l} \end{bmatrix} \tag{3.51}$$

$$K_\sigma = \dfrac{-P}{30l} \begin{bmatrix} 36 & 3l \\ 3l & 4l^2 \end{bmatrix} \tag{3.52}$$

The total stiffness matrix is

$$K = \dfrac{EI}{l^3} \begin{bmatrix} 12 - 36\omega & 6l - 3\omega l \\ 6l - 3\omega l & 4l^2 - 4\omega l^2 \end{bmatrix} \tag{3.53}$$

where:

$$\omega = \dfrac{Pl^2}{30EI} \tag{3.54}$$

The roots, $w_1 = 0.08287$ and $w_2 = 1.073$ of the following equation, will make the total stiffness matrix in 3.53 singular:

$$4(12 - 36\omega)(1 - \omega) - (6 - 3\omega)^2 = 0 \tag{3.55}$$

Substituting the two roots into Equation 3.54, two critical loads can be obtained as

$$P_{cr}^1 = \frac{0.252\pi^2 EI}{l^2} \text{ and } P_{cr}^2 = \frac{3.262\pi^2 EI}{l^2} \tag{3.56}$$

Comparing the first critical loads with the theoretical solution ($P_{cr}^T = 0.250\pi^2 EI/l^2$; Zhu 1998), the FEM approach can produce very accurate solutions. It should be noted that the previous solution is based on one element (two degrees of freedom). If the number of elements in the beam meshes increases, the accuracy improves accordingly.

3.2.9 Applications in bridge analysis

When applying FEM to bridge analysis, there are some common questions and issues that engineers have to clarify. These issues include (1) what types of element should be used in a bridge model; (2) when a 2D model is sufficient and when a 3D model is necessary; and (3) how to correctly interpret FEM results from bridge engineering perspectives, especially when a bridge is modeled as plate or shell elements.

In Sections 3.2.1 through 3.2.8, only generic principles and procedures of FEM are briefly illustrated, aiming at helping engineers to understand the theories behind an FEM package. And, as an example, only 2D frame element is discussed in detail. In general, truss, frame, and shell elements can cover most bridge analyses.

Truss element, like a member in a truss bridge, is a line element with only two nodes. It has only axial strain/stress, and the most important feature is that its strain/stress is constant over the entire element. Truss element is also called link element. Bridge bearings, hangers, prestress tendons, cables, and so on, can be modeled as truss elements.

Frame element, like a member in a frame structure, is a line element with only two nodes. It behaves as a beam but could be under axial tension/compression or a combination of beam and truss elements. Most FEM packages combine behaviors of beam, truss, and torsional element into one as a frame element—the most commonly used element type in bridge analysis. In line models, girders, stringers, diaphragms, pylons, columns, piers, and so on are usually modeled as frame elements.

Shell element combines in-plane stress/strain behavior together with bending of a plate, either as a thin plate or as a thick plate. When a bridge component is modeled into the plate level, such as a box girder or steel I-girder, shell element could be used. Some components that behave in-plane, such as webs, can be simplified as shells to streamline the modeling.

Nowadays, whether or not to model in 3D is no longer a question because modern graphical pre- and post-processing tools are widely available.

For detailed analysis, most bridges should be modeled in 3D, not only for better accuracy but also for simplification of component simulations. Even a long-span bridge, such as a suspension bridge discussed in Chapter 12, is preferable to be modeled in 3D rather than 2D because the stiffness of components such as pylons and truss members of stiffening girder can be easily computed and thus be simulated accurately in 3D. For certain analysis purposes, such as extreme live loads analysis of floor beams in truss bridges, 3D model becomes inevitable.

When dimensions in longitudinal and transverse axes are comparable, such as middle- and short-span girder bridges, an intermediate model, or the so-called grid model, is widely used. The element in a grid model is retrograded from a 3D frame element by ignoring two translational displacements on the grid plane and one rotational displacement along the axis perpendicular to the grid plane. Thus, each node of an element has only vertical displacements, bending rotation and torsional displacements. Element internal forces contain bending and torsional moments plus shear, accordingly. A grid model can easily analyze distributions in the longitudinal direction of a girder and in the transverse direction among girders while maintaining the same number of degrees of freedom. Therefore, the grid model is very common in girder bridge analyses. Furthermore, the grid model can be expanded to simulate a wide box girder, in which webs are not connected directly by separate diaphragms, but by flanges (Hambly 1991). However, a true 3D model with shell elements is encouraged when lateral distributions among webs of a box girder are of interest. Many behaviors of a wide thin-walled box girder, such as warping when torsion is restrained, distortion when insufficient diaphragm is used, and shear lagging due to longitudinal shear deformations of flanges, cannot be represented in a grid model.

Most component design theories and code checking are based on internal forces over a cross section of a component. For example, when designing rebar quantities of a frame member, bending moment, shear, and axial forces should be known. When a bridge component is modeled as truss or frame elements, internal forces output from FEM analyses can be used directly for engineering design and code checks. When a component is meshed into shell elements, such as a web in box girder as shown in Figure 3.10a, results from FEM have to be translated into the perspective of a bridge component, or the original FEM results are not meaningful and cannot be used in design or code checks. This is because the stress results from FEM analysis are in each element's local coordinate system, which may vary from one element to another. Stresses have to be transformed to a unique axis that is meaningful to engineering, like the longitudinal axis of a component. When in curve segments, elements have to be unfolded along curves and stress results can then be plotted on flat regions. As shown in Figure 3.10b, for example, the horizontal stresses

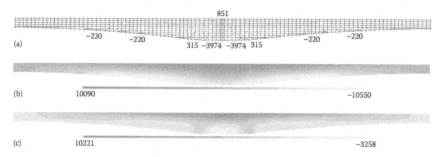

(a)

(b)

(c)

Figure 3.10 Interpretation of stresses as engineering perspectives. (a) Shell elements of a web in a box girder and vertical shear stresses. (b) Stresses along horizontal direction after unfolded. (c) Major principal stresses.

along web curves are transformed from two axial stresses and one shear stress of all involved shell elements. What is shown in Figure 3.10b can be defined as axial stress perpendicular to a cross section, which is one of the dominating stresses and is what a bridge engineer looks for. Further, the major/minor principal stresses,[*] which are transformed from stress components at any point, are needed more often and more meaningful than their original stress components in each element's local coordinate system. Figure 3.10c shows the major principal stress of the same web as in Figure 3.10a and b.

When a bridge is modeled as shell elements, or further as 3D block elements, engineers often want to compare the stress distribution obtained from shell elements to that from a simple model as frame elements so that the differences from the beam theory can be better understood. Special functions in postprocessing in this regard are particularly important to bridge analysis, or 3D detailed modeling will be greatly limited in bridge analysis and design. For example, Figure 3.11 shows a special function in a postprocessing package that can first transform stress components to axial stress perpendicular to any predefined cross section and then integrate this stress over the cross section to obtain equivalent sum forces over the section. The equivalent forces, which are shown at the bottom of Figure 3.11, can then be used to compute axial stress distribution by beam-bending theory. The stress comparison, as shown in both top and bottom flanges, can help engineers to understand effects such as warping, distorting, and shear lags.

---

[*] The two or three result stresses at any point on plane or in spatial that are transformed from its three or six stress components as shown in Figure 3.1

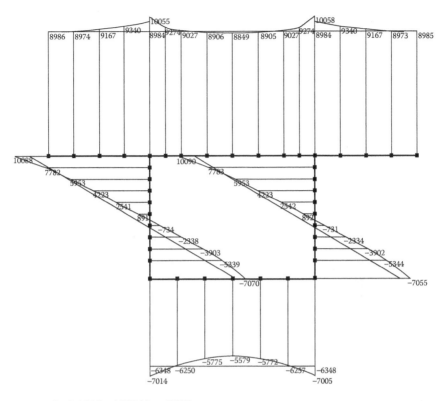

$Fx = 11.8939$ $Fy = 0.498916$ $Fz = -3.72129$
$Mx = -672.528$ $My = -702.866$ $Mz = 311822$

Figure 3.11 Stress integration over a cross section comparing with beam theory. Curves—axial stresses distribution from a shell element model. Straight lines—axial stresses distribution recomputed from beam-bending theory by using equivalent internal forces obtained from stress integration.

3.3 AUTOMATIC TIME INCREMENTAL CREEP ANALYSIS METHOD

After elastic strains instantly occurred with loads on a concrete structure, creep strains will later be developed. The development of creep strains depends on the age of concrete when loads are applied and the time of observing. However, the creep strains are always proportional to the initial elastic strains that cause them. Creep strains affect a structure in two ways: (1) extra displacements would be developed after construction and (2) extra displacements would cause load redistributions. For concrete or composite bridge structures built in multiple stages, creep analyses are important as loading and concrete aging history can be complicated. Together with creep strains, concrete material will also develop shrinkage strains, which have a similar behavior as creep strains in terms of time history. However, shrinkage strains are elastic strain independent or are not related to loads. Therefore, shrinkage analysis is simpler than creep analysis. In general, these two types of time domain issues are considered concurrently with bridge structural analysis.

Generically, the Young's modulus of concrete varies as aging and the creep strain developed not only depends on concrete age and observation time, but also couples with concrete stress. When analyzing creep effects in perfect accuracy, integration over the entire observation time span is inevitable. Therefore, the analysis method is complicated and its procedures are closely related to a particular creep and shrinkage model. As a result, the evolution of the FEM system is tied to the mathematic model of creep and shrinkage, and similarly, the adoption of a new creep and shrinkage model is limited by an existing FEM system. When considering most common concrete bridge situations, such as low service stress (<40% of concrete strength) and no-unloading in terms of predominate structural weight, the creep strain is proportional to the elastic strain that happened at a given age, and a constant Young's modulus of 28 days can be taken. Thus, nonlinearity of creep can be limited in the time domain only, and the relationship to loads can still be linear. Further, the time history can be divided by many small time steps and the stress within each step can be treated as constant. The Automatic Time Incremental Creep Analysis Method introduced in this section is a simplified method based on the above assumptions. As illustrated by an example in this section, the results by the simplified method are very close to other complicated integration method, and the error is engineering acceptable.

As revealed in Equation 3.58 that the creep strains are proportional to elastic strains and the development of such a creep strain factor in time domain is separated from external loads and the structure itself, it can be concluded that the superposition of loads is still valid when creep is considered. Based on the principle of superposition, the automatic time incremental creep analysis method first computes the creep effects at all time steps in the future due to external loads and creep redistribution loads at the current time. A simple accumulation of analysis results can then produce the final creep effects at any observation time (Wang 2000).

3.3.1 Incremental equilibrium equation in creep and shrinkage analysis

When creep and shrinkage are considered in a constant stress scenario,

$$\varepsilon = \varepsilon_e + \varepsilon_c + \varepsilon_s = \varepsilon_e + \varepsilon_e \varphi(t, \tau) + \varepsilon_s(t, \tau) \tag{3.57}$$

where ε_e, ε_c, and ε_s denote the regular elastic strain, creep strain, and shrinkage strain, respectively. Both creep and shrinkage strains depend on the age of concrete and the observation time where the age for creep is the duration after applying loads and the age for shrinkage is the duration after concrete is allowed to dry. The creep strain also is proportional to the elastic strain as

$$\varepsilon_c = \varepsilon_e \varphi(t,\tau) \tag{3.58}$$

$\varphi(t,\tau)$ is the creep factor, which may be expressed by many different mathematical models. The time origins of t and τ are the same as when the concrete starts to cure. No matter what model is used to describe creep development, the creep factor $\varphi(t,\tau)$ can be explained as at observation time t, the total creep due to an elastic strain at τ divided by the elastic strain. $\varepsilon_s(t,\tau)$ is the total shrinkage at time t, which is independent to the elastic strain ε_e.

Given an external load acting on time τ, at time t the system is balanced and the equilibrium equation is written as Equation 3.32. Considering a small time increment dt, the variation of elastic strain can be obtained from Equation 3.57 as

$$d\varepsilon_e = d\varepsilon - \varepsilon_e d\varphi - d\varepsilon_s \tag{3.59}$$

The internal stresses will have a change of $d\sigma$, and the incremental equilibrium equation can be obtained from Equation 3.32 as

$$d\psi(a) = \int B^T d\sigma dv = 0 \tag{3.60}$$

Substituting Equations 3.59 and 3.11 into Equation 3.60, the incremental equilibrium equation of creep and shrinkage can be derived as

$$K da = \int B^T \sigma d\varphi dv + \int B^T D d\varepsilon_s dv \tag{3.61}$$

where K is the global stiffness matrix as shown in Equation 3.30.

The physical meaning of Equation 3.61 is simple and clear: Incremental creep and shrinkage will cause equivalent loads and will be balanced by incremental displacements. By solving Equation 3.61, the incremental displacements at the next time step due to creep and shrinkage can be obtained. The total and elastic incremental strains can be computed from Equations 3.28 and 3.59, respectively. The incremental stresses can further be obtained. By accumulating all incremental values for each incremental time, the total internal stresses and displacements at any time due to creep and shrinkage can be solved.

It should be noted that the most available creep and shrinkage models are based on experiments on axial compression components. However, creep and shrinkage factors can be treated the same in all directions, including shear strains. Therefore, when computing the equivalent loads as Equation 3.61, the incremental creep and shrinkage factors can be isolated from matrix operations.

3.3.2 Calculation of equivalent loads due to incremental creep and shrinkage

The development of concrete shrinkage at a given observation time depends only on the concrete age when it is allowed to dry and is independent to stresses. Thus, the equivalent loads due to shrinkage (the second term on the right side of Equation 3.61) are straightforward. The computation of creep equivalent load, however, is complicated because it depends on both stresses and the concrete age when stresses are loaded. Figure 3.12 shows generic stress changes of one component at different time steps. Each stress change could be caused by external loads or creep/shrinkage redistribution. As time and the concrete age are considered when each stress change applies, this diagram represents a typical loading history. Assuming stress change at each time step is $\Delta\sigma_i$, the time ordinate at each time step is t_i, and the concrete age is τ_0 when the first stress change $\Delta\sigma_0$ is loaded, the total stress at any time step t_i is

$$\sigma_i = \sum_{j=0}^{i} \Delta\sigma_j \tag{3.62}$$

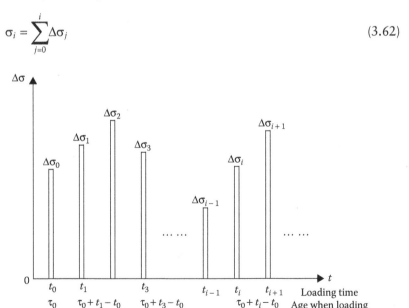

Figure 3.12 Stress changes and loading history of a concrete component.

and the creep equivalent stress at the next time step is

$$\sigma_i d\varphi = \sum_{j=0}^{i} \Delta\sigma_j \left[\varphi\left(t_{i+1}, \tau_0 + t_j - t_0\right) - \varphi\left(t_i, \tau_0 + t_j - t_0\right) \right] \qquad (3.63)$$

Considering $\int B^T \Delta\sigma_i dv = K_e \Delta a_i$, the creep equivalent nodal loads of an element at time step t_i can be written as

$$F_i^e = K_e \sum_{j=0}^{i} \Delta a_j \left[\varphi\left(t_{i+1}, \tau_0 + t_j - t_0\right) - \varphi\left(t_i, \tau_0 + t_j - t_0\right) \right] \qquad (3.64)$$

where Δa_j is the incremental displacements at time t_j corresponding to the stress change of $\Delta\sigma_j$. From Equation 3.64, it is obvious that the calculation of creep equivalent load is separated from the element stiffness matrix. Given the history of displacement changes due to any loading types, including redistribution loads of creep and shrinkage themselves, creep equivalent nodal loads at the next time step can be simply obtained by Equation 3.64, and the displacement changes at the next time step can be solved from Equation 3.61. Iterating this process through the entire observation history (from the first loading time to a future time) with a small time step, displacements and internal forces due to creep and shrinkage at any time can be analyzed. When applying this method to bridge analysis, causes of stress changes at any time, as shown in Figure 3.12, include different types of external loads such as construction loads, structural weights, stage changes, prestressing, and redistribution of creep and shrinkage themselves.

3.3.3 Automatic-determining time step

Considering the behavior of concrete creep and shrinkage, these effects may need to be analyzed at five years or even 50 years after the structure is built (Bazant et al. 2011). The small time step used in the previous iteration should be determined based on the performance and accuracy. As all creep theories assert that the creep development will decrease gradually and cease eventually, the time step can be increased from a smaller one at an earlier age to a large time span at a more matured age. This adjustment can be done automatically by detecting a small displacement change. With today's advancement of modern computers, when a bridge is modeled as a spatial frame, performance degraded due to short time steps, such as a week or even shorter time, would not be a consideration.

3.3.4 A simple example of creep analysis

A three-span continuous bridge that is built span by span will be used as an example here to illustrate the concrete creep behavior and the application of time incremental analysis method. The example has three equal spans with a span length of 30 m (Fan 1998). The first construction stage is the casting of the first 36-m girder segment with the support of falseworks (first span plus 6-m cantilever). The falseworks are removed after the concrete is cured for one week. The second stage is to cast the next 30-m girder segment. After the concrete is cured for the same number of days (one week), the last 24-m segment is cast as the last stage. The bridge is completed after the last segment is cured for one week. The structural weight is 100 kN per meter. In this model 3D frame elements are used.

Figure 3.13 shows moment distributions after the bridge is built when creep effects are not considered. The moments at the first and second interior bearings are –4,928 and –7,005 kN-m respectively. For comparison, both would be –9,000 kN-m if the three-span bridge is built all at once. Figure 3.14 shows the final moment distributions eight years after the bridge is built. Due to creep effects, moments at the interior bearings become –8,283 and –8,926 kN-m, respectively, revealing the tendency of concrete creep that the internal forces distributions would eventually be

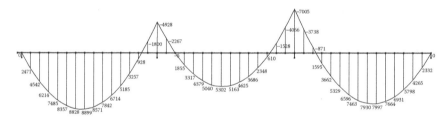

Figure 3.13 Moment distribution of a three-span continuous bridge built span by span, without consideration of concrete creep considered (kN-m).

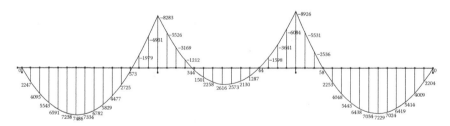

Figure 3.14 Moment distribution of a three-span continuous bridge eight years after built span by span, with consideration of concrete creep considered (kN-m).

close to what it should be when the bridge was built in one time. More detailed 2D and 3D illustrated examples, including creep and shrinkage, are shown in Chapter 5 for PC bridges.

3.4 INFLUENCE LINE/SURFACE LIVE LOADING METHOD

Live load analysis is a unique problem to bridge analysis and design. As some bridge design specifications define that many vehicles with minimum spacing are allowed to present in a lane, the simple search of maximum or minimum positions by moving axles along an influence line will not work well in general. A generic and effective influence line loading method that is suitable for any type of live load definition is important in bridge analysis and design. The traffic lane layouts in many bridges, such as interchanges or curved bridges, can be complicated, and, therefore, spatial live loads analysis becomes inevitable. Based on influence line loading, influence surface loading with multiple traffic areas is another important topic in live loading analysis, especially nowadays with advanced computer technologies, spatial analyses become essential to bridge designs.

In this section, the application of dynamic planning method in influence line loading and the principle of multiple traffic areas in influence surface loading will be introduced.

3.4.1 Dynamic planning method and its application in searching extreme live loads

Live loads usually contain a single concentrated load, uniformed (or called lane) loads, and vehicle loads. Searching for extreme positions of vehicle loads is complicated in live load analyses. Locating the positions or areas where a single concentrated load or uniformed loads reach the extreme is simple. In this section, vehicle loads are used as examples to illustrate the principle of dynamic planning method in search of extreme positions.

Different bridge specifications define different vehicle loads, and these definitions may be changed per traffic demands. Figure 3.15 shows a single vehicle model and two typical vehicle processions. As shown in Figure 3.15a, a vehicle can be described as a number of axles with constant axle weights and spacing. Because both axle weights and spacing are fixed, given only the location of its front axle on the influence line, its influence value can be obtained. Therefore, it can be simplified as a concentrated load as shown in Figure 3.15b. Figure 3.15c shows a typical vehicle procession that contains identical vehicles as illustrated in Figure 3.15a with a minimum leading

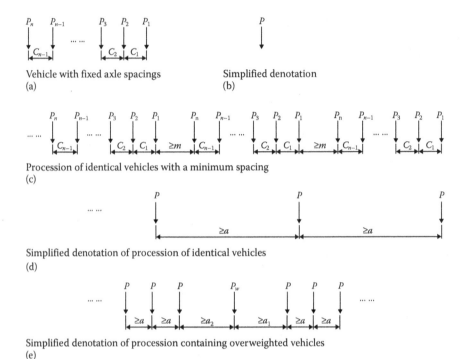

Vehicle with fixed axle spacings
(a)

Simplified denotation
(b)

Procession of identical vehicles with a minimum spacing
(c)

Simplified denotation of procession of identical vehicles
(d)

Simplified denotation of procession containing overweighted vehicles
(e)

Figure 3.15 (a–e) Typical vehicle loads and vehicle processions.

and trailing spacing between other vehicles. When determining the extreme positions of such a procession, these spacing are variables in addition to the location of the first vehicle. As each vehicle can be treated as constant, this type of procession can be simplified as shown in Figure 3.15d. Further, a procession may contain one and only one overweight vehicle with different leading and trailing spacing to other regular vehicles. Similarly, it can be simplified as shown in Figure 3.15e.

Figure 3.16a shows an example of the influence line. The goal of searching extreme live loads is to find the number and positions of vehicles on the influence line that makes the influence value maximal or minimal. Considering the minimum can be reached by the same procedures as the maximum after reversing influence value signs, the following procedures are illustrated for reaching maximum values only.

An extreme value function $e(x)$ is introduced in the dynamic planning method (Shi et al. 1987). The value of $e(x)$ is the extreme influence value of a particular vehicle or vehicle procession in the loading range from 0 to x. As a longer range will not produce less influence value than a shorter range, $e(x)$ is a monotonically increasing function as shown in Figure 3.16b. Taking a

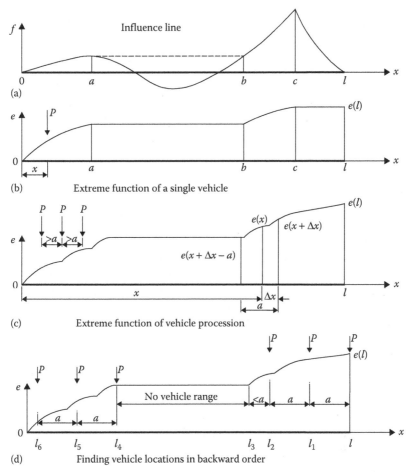

Figure 3.16 (a) Influence line, (b,c) extreme influence value, and (d) finding vehicle locations.

single vehicle as an example, Figure 3.16b shows its extreme value function corresponding to the influence line shown in Figure 3.16a. When moving the vehicle from 0 to a, the influence value keeps increasing until reaching a peak at a, whereas the curve segment of $e(x)$ from 0 to a keeps increasing accordingly. When moving the vehicle farther from a, the influence value stops increasing as a range with less or even negative values is reached. After passing point b, where the influence line has a value greater than that at point a, the curve resumes increasing until it leaves point c, from which the influence line turns lower again.

Given an extreme function of a single vehicle within a range $[0,l]$, extreme position can be easily located by numerating $e(x)$ in a backward order (i.e., going back from l to 0, the first position where $e(x)$ starts decreasing is the extreme position). In the example shown in Figure 3.16b, location c is the first point from which $e(x)$ starts decreasing. Therefore, the extreme position for a single vehicle on the influence line as shown in Figure 3.16a is c.

When determining the extreme function of a vehicle procession, iteration is needed as a minimum spacing between vehicles is introduced. Assuming the extreme value at current position x, $e(x)$ is known, and the iteration process to evaluate the extreme value at $x + \Delta x$ is

$$e(x + \Delta x) = \begin{cases} e(x), & \text{if } e(x + \Delta x - a) + v(x + \Delta x) \le e(x) \\ e(x + \Delta x - a) + v(x + \Delta x), & \text{if otherwise} \end{cases} \qquad (3.65)$$

where $v(x + \Delta x)$ stands for the influence value of a single vehicle at position $x + \Delta x$. Equation 3.65 would be clearer by attempting to place a vehicle at $x + \Delta x$. As there is a mandatory minimum vehicle spacing a, the preference for whether or not a vehicle is placed at $x + \Delta x$ (to produce more influence value) depends on the total effect of this vehicle and the maximum loading value on range $[0, x + \Delta x - a]$, that is, $e(x + \Delta x - a)$. If the total effect is increasing from the current position, use it as the extreme value at the next position. Otherwise, keep the extreme value the same for the next position.

Once the extreme function of a vehicle procession is determined within a range $[0,l]$, the number of vehicles and their positions that cause the maximum influence value can be located in a similar manner as searching for a single vehicle. Taking the extreme function as shown in Figure 3.16d as an example, the first decreasing point is l and the search has to keep going further back as more vehicles may be present. The second decreasing point is l_1 after a spacing of a away from the first vehicle. After the third vehicle is placed at l_2, there would not be an allowed point at l_3, even though it keeps decreasing as it is less than a minimum distance from l_2. No vehicle should be placed in the range $[l_4, l_3]$ as the extreme value does not decrease in this area. All six vehicles can be located in this way as shown in Figure 3.16d.

The process to determine the extreme positions for a procession that may contain an overweight vehicle, as shown in Figure 3.15e, can be established based on the earlier procedures for a procession that contains only normal vehicles. As illustrated in Figure 3.15e, the total effect of this kind of procession is the sum of the influence values of overweight vehicles, following normal vehicles and leading normal vehicles. The following vehicles can be located by searching for $e(x - a_2)$ according to the definition of the extreme

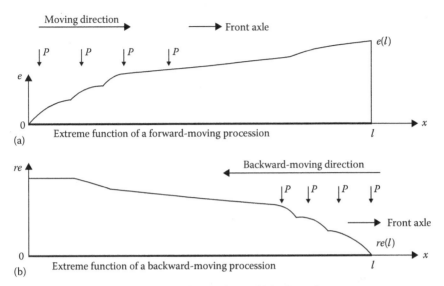

Figure 3.17 Extreme functions of (a) forward- and (b) backward-moving processions.

function. To determine extreme values due to leading vehicles, a similar extreme function $re(x)$ is introduced. As shown in Figure 3.17b, $re(x)$ defines the extreme value within the range $[x,l]$ for a procession moving backward from l to 0.

Having the extreme function for reverse-moving procession established, searching the location and maximum value of a procession that contains an overweight vehicle is equivalent to finding the maximum value of the following equation:

$$L(x) = e(x - a_2) + o(x) + re(x + a_1) \tag{3.66}$$

where $o(x)$ is the influence value of the overweight vehicle at position x. Simply moving the overweight vehicle from 0 to l will give the maximum value by Equation 3.66. The influence values of following and leading normal vehicles can simply be obtained from forward and backward extreme functions, respectively. However, it should be noted that the finding of following vehicles' positions on $e(x)$ is from $x - a_2$ to 0, and the leading vehicles' positions on $e(x)$ is from $x + a_1$ to l.

When implementing this method, the following issues should be taken into consideration: (1) the length of the original influence line has to be extended at both ends to ensure the last axle is moving out of range; (2) the extreme positions and values obtained are based on moving vehicles from 0 to l (this value should be compared with that of moving vehicles from l to 0 which can be simply obtained by reversing the influence line);

(3) when a procession contains an overweight vehicle, the maximum value obtained from Equation 3.66 should be compared with a procession that contains only normal vehicles, for possible mandated long leading and/or trailing spacing of overweight vehicle; (4) the minimum values and positions can be solved in the same way with reversing signs of influence values; and (5) an appropriate vehicle-moving step should be determined to maintain an accurate and a better-solution performance. In general, one-third to one-half of a meter (1/3–1/2 m) is suitable for most longitudinal live loading analyses, and one-fourth to one-third of a meter (1/4–1/3 m) is accurate enough for transverse live loading discussed in Section 3.4.2.

3.4.2 Transverse live loading

When influence surface loading is needed or in some transverse distribution analyses, transverse live loading analyses will be required. Most bridge specifications have the transverse placement of a vehicle load defined, which can be summarized as a series of fixed-axle vehicles moving along a given range. The same concept and principles of the extreme function introduced in the Section 3.4.1 can also be applied in transverse loading. As a multilane discount may be applied when multiple lanes present per a particular specification, each number of lanes should have a separate extreme function as shown in Figure 3.18. When determining the extreme value with an attempt of adding a new lane, it should be compared with what it was without adding a new lane, as the multilane discount may be higher if added. Another special issue in transverse live loading is the restrictions on vehicle moving, for example, a minimum distance to curb is usually defined in most specifications.

3.4.3 Influence surface loading

As spatial analyses became essential in bridge analysis and design, traditional lateral load distribution theories and simplified calculation methods are gradually substituted by spatial structural analyses and influence surface loading. Particularly for bridges with irregular shapes such as interchanges, spatial analysis and influence surface loading are inevitable.

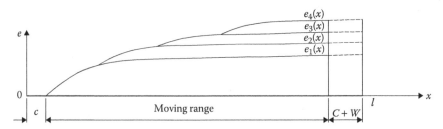

Figure 3.18 Extreme functions of transverse loading.

Fx

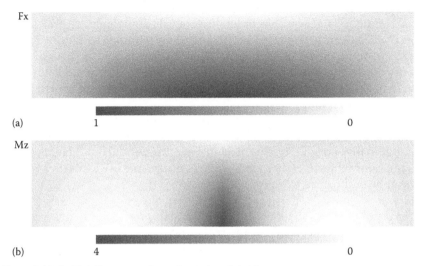

(a)   1   0

Mz

(b)   4   0

Figure 3.19 (a, b) Influence surface of a tied-arch bridge.

As an example shown in Figure 3.19, influence surface is a function of planar coordinates. Based on the influence line loading method introduced in Section 3.4.2, the influence surface loading method can be developed with certain assumptions.

The deck of a bridge with an irregular shape may be divided into different traffic areas. Figure 3.20a, for example, shows the plane view of a generic bridge deck. On a plane, traffic regions may be overlapped as seen in interchanges. A region on a plane can be defined by its centerline, left width, and right width, and both widths are constant along the entire region. Although the centerline of a region may be curved in reality as regions Ω_2 and Ω_3 shown in Figure 3.20a, it is assumed that

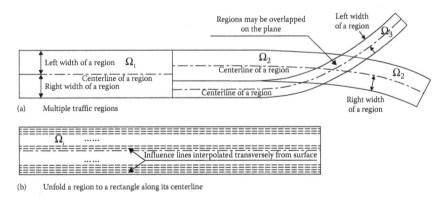

(a)   Multiple traffic regions

(b)   Unfold a region to a rectangle along its centerline

Figure 3.20 (a) Multiple traffic regions and (b) unfolding region to rectangle.

the approximation by unfolding the region along its curved centerline as a rectangle is acceptable in engineering. For example, region Ω_2 in Figure 3.20a can be represented by a rectangle similar to Figure 3.20b, which is obtained by unfolding its curved centerline to a straight line. Points on the straight centerline can be mapped to its curved centerline one to one. Both left and right widths of the unfolded region are the same as the curved region.

To be mathematically feasible and also considering the fact that traffic is maintained within a lane, a vehicle procession of a traffic lane is lined up longitudinally; no staggered vehicle in lateral is considered. Having these assumptions set forth earlier, the searching of extreme live loads on a region Ω_i can be outlined as follows:

1. Unfold the region along its centerline to a rectangular region
2. Divide the total width of the region into steps and establish longitudinal influence lines at each transverse step by interpolating from its original influence surface
3. Search maximum and minimum live load values and their corresponding positions for each longitudinal influence lines in step (2)
4. Two transverse influence lines regarding maximum and minimum values are formed
5. Search extreme live loads laterally by applying transverse live loading on these two influence lines in step (4)

Once the extreme live loads positions and influence values on all regions are found, the total extreme values and their positions are the sum of these overall regions. More precisely, the lane discount in surface loading should be considered regionally, rather than globally. For example, when loading on a region Ω_i, the discount is determined according to lane combinations only in this region. The concept of influence surface is applied to many different types of bridge discussed in Chapters 5 through 12.

Part II

Bridge behavior
and modeling

Chapter 4

Reinforced concrete bridges

4.1 INTRODUCTION

Reinforced concrete (RC) was first introduced into bridge engineering in the late nineteenth century, and it has become a major material for bridges ever since then for its versatility, flexibility, and durability. RC bridges were widely used during the reconstruction of Europe after World War II. In general, a bridge that mainly uses RC for its major structural components can be categorized as an RC bridge. For example, RC arch bridges, RC beam–slab bridges, and RC rigid frame bridges are all considered as RC bridges. Because of cracking, only partial of a concrete section is intact and functional, the RC sectional strength to resist moment, shear, and tensile is much lower than that of a prestressed concrete (PC). The cracking in the tensile area, which is allowed in RC and actually does exist in services state, poses potential corrosion risk on reinforce steels and thus deterioration of a cross section as a whole. The spanning capacity of an RC bridge is limited to short to middle spans, and its application also depends on the site environment.

Due to RC's special material behavior and the existence of cracking, several distinctive issues arise in both the structural analysis and the component design of an RC bridge. For example, how to count for the variation of sectional modulus from location to location when conducting structural analyses, as effective area of a cross section is related to moment it resisted, and when behaviors of concrete and steel have to be considered in separation are common questions an engineer may ask when modeling or designing an RC bridge. To be more practical, cracking and steel reinforcement to cross sections can be simply ignored in most generic structural analyses for obtaining component design forces. Sectional modulus variation due to cracking loss and steel reinforcement is minor with regard to global load distributions. Having obtained design forces, special principles and codes should be strictly followed when coming to component design phase. When the ultimate capacity of an RC bridge is of interest, which is more often the case for short- to medium-span RC bridges than medium- to long-span non-RC bridges, a full material nonlinear analysis is required. In such an analysis, material behaviors of concrete and steel are considered in great detail. For example, a specific constitutive relationship for steel RC as a whole may be used, special concrete elements with consideration of the existence of reinforcing steels can be developed, or concrete and steel are separately modeled in the structural level.

Fiber-reinforced concrete (FRC) is a kind of concrete that contains fibrous material for reinforcement to increase the structural integrity. FRC contains short discrete fibers that are uniformly distributed and randomly oriented. Fibers include steel, glass, synthetic, and natural fibers, which give different structural properties. Several ultrahigh-performance concrete (UHPC) bridges using FRC have been built in the United States (Fu and Graybeal 2011). The addition of fiber to concrete was aimed primarily at enhancing the tensile strength and postcracking behavior of concrete. FRC behaves as regular concrete but with higher strength, especially tensile strength. For highway bridge structures, FRC can be applied to overlays in bridge decks, seismic- and explosion-resisting structures, and recently UHPC bridges.

On the other hand, fiber-reinforced polymer (FRP) is a composite material made of a polymer matrix reinforced with usually glass, carbon, basalt, or aramid. FRP bars and grids have been commercially produced for reinforcing concrete structures for over 30 years. FRP bars have been developed for prestressed and non-prestressed (conventional) concrete reinforcement. FRP has been used for strengthening structural members of RC bridges that are structurally deficient or functionally obsolete due to changes in use or consideration of increased loadings (Kachlakev 1998). Many researchers have found that FRP composites applied to such members provide reliable and cost-effective rehabilitation. FRP composites are orthotropic materials with two constituents, that is, reinforcing and matrix phases. The reinforcing phase material is fiber, usually carbon or glass, which is typically stiffer and stronger, whereas the matrix phase material is generally continuous, less stiff, and weaker. The behavior of FRP-strengthened concrete structural members can be analyzed using finite element method (FEM).

As detailed RC cracking analysis, most early finite element models of RC were based on a predefined crack pattern. The recently developed smeared cracking approach overcomes these limitations of unpredicted predefined cracks and has been widely adopted for predicting the nonlinear behavior of concrete. It uses isoparametric formulations to represent the cracked concrete as an orthotropic material. More details of this subject are discussed in Section 4.4.2—Nonlinear Modeling.

In this chapter, RC bridge behavior at the material level, especially the coworking of concrete and steel; characteristics of skewed slabs as a common application of RC bridges; and different modeling methods are discussed in detail. Also, different analysis examples of beam–slab bridges by using different modeling methods and analysis packages are included. In Section 4.7, a study on a skewed, transversely post-tensioned slab bridge, including nonlinear analysis, field survey and monitoring, and comparison, is presented.

4.2 CONCRETE AND STEEL MATERIAL PROPERTIES

RC is made of concrete and steel, two materials with different physical and mechanical behavior. Concrete exhibits nonlinear behavior even under low-level loading due to nonlinear material behavior, environmental effects, cracking, biaxial stiffening and strain softening, and time-dependent effects such as creep and shrinkage (Darwin 1993). Reinforcing steel acts linearly in the working stress range until yielding, and it interacts with concrete in a complex way. Sophisticated finite element analysis (FEA) techniques can be used to accurately represent the behavior of RC structures. Cracking, softening in compression, yielding of steel, and bond slip are taken into account in modifying the analysis procedure.

Because of the difference in the short- and long-term behavior of constituent materials of RC, the popular method of representing RC consists of developing separate models for concrete and steel and combining those models either at the element level, through the addition of constitutive matrices, or at the structure level, through the use of different elements for each material. The presence of steel modifies the behavior of concrete in a way that evolved into the technique of tension stiffening, in which constitutive models for cracked concrete are modified to account for the ability of concrete within the composite to carry tensile stress after cracking, in contrast to a simple concrete element in which the stress-carrying capacity drops rapidly following the formation of crack.

The stress–strain relationship of concrete elements in compression is nonlinear up to the ultimate strain and beyond. Several models for the stress–strain relationship of concrete have been proposed in the past. At low levels of stress, transverse reinforcement (stirrups) is hardly stressed; the concrete behaves much like *unconfined* concrete. At stresses close to the uniaxial strength of concrete, internal fracturing causes the concrete to dilate and bear out against the transverse reinforcement, then causing a confining action in the concrete. This *confined* concrete with suitable arrangement of transverse reinforcement increases the strength and ductility of the concrete. The enhancement of strength and ductility by confining the concrete is an important aspect that needs to be considered in the design of structural concrete members, especially for extreme events such as seismic activity, blast effects, or vehicle crashes.

The stress–strain relationships corresponding to unconfined concrete, confined concrete, and longitudinal steel reinforcement are discussed in the Sections 4.2.1 through 4.2.3.

4.2.1 Unconfined and confined concrete

Numerous stress–strain relationships for unconfined and confined concrete were developed. The two most popular ones based on their usage are listed here. Kent and Park (1971) proposed a stress–strain equation for both unconfined and confined concrete, in which Hognestad's (1951) equation was generalized to describe the postpeak stress–strain behavior in a more complete manner. In this model, the ascending branch is represented by modifying the Hognestad second-degree parabola by replacing $0.85f'_c$ with f'_c and strain at peak stress for unconfined concrete ε_{co} with 0.002. Kent and Park modified their model again in 1982 as shown in Figure 4.1.

$$f_c = f'_c \left[\frac{2\varepsilon_c}{\varepsilon_{co}} - \left(\frac{\varepsilon_c}{\varepsilon_{co}} \right)^2 \right] \tag{4.1}$$

Mander et al. (1988a) first tested circular, rectangular, and square full-scale columns at seismic strain rates to investigate the influence of different transverse reinforcement arrangements on the confinement effectiveness and overall performance. Mander et al. (1988b) went on to model their experimental results. It was observed that if the peak strain and stress coordinates (ε_{cc}, f'_{cc}) could be found, then the performance over the entire stress–strain range was consistent, regardless of the arrangement of the

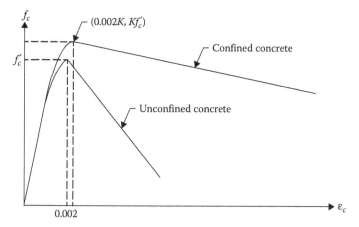

Figure 4.1 Stress–strain behavior of compressed concrete confined by rectangular steel hoops. (Data from Kent, D.C. and Park, R., *J Struct Div.*, 97(ST7), 1969–1990, 1971.)

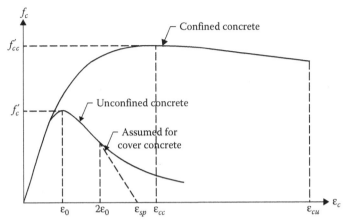

Figure 4.2 Stress–strain relation for monotonic loading of confined and unconfined concrete. (Data from Mander, J.B., Priestley, M.J.N., and Park, R., *J Struct Eng.*, 114(8), 1804–1826, 1988b.)

confinement reinforcement used. The equations are listed here and shown in Figure 4.2.

$$\frac{f_c}{f'_{cc}} = \frac{n(\varepsilon_c/\varepsilon_{cc})}{(n-1)+(\varepsilon_c/\varepsilon_{cc})^n} \tag{4.2a}$$

in which

$$n = \frac{E_c}{E_c - E_{sec}} \tag{4.2b}$$

$$E_c = 5000\sqrt{f'_c} \text{ (both in a unit of MPa)} \tag{4.2c}$$

$$E_{sec} = \frac{f'_{cc}}{\varepsilon_{cc}} \tag{4.2d}$$

where ε_{cc} is the strain at the maximum compressive strength of confined concrete f'_{cc}

$$\varepsilon_{cc} = \varepsilon_{co}\left[1+5\left(\frac{f'_{cc}}{f'_c}-1\right)\right] \tag{4.2e}$$

and f'_{cc}, the compressive strength of confined concrete, is given as

$$f'_{cc} = f'_c\left(-1.254+2.254\sqrt{1+\frac{7.94f'_t}{f'_c}}-2\frac{f'_t}{f'_c}\right) \tag{4.2f}$$

in which f_t' is given by

$$f_t' = \frac{1}{2} k_e \rho_s f_{yh} \tag{4.2g}$$

where:

ρ_s is the ratio of the volume of transverse confining steel to the volume of confined concrete core
f_{yh} is the yield strength of transverse reinforcement
k_e is the confinement coefficient

For circular hoops

$$k_e = \frac{\left[1 - (s'/2d_s)\right]^2}{1 - \rho_{cc}} \tag{4.2h}$$

For circular spirals

$$k_e = \frac{1 - (s'/2d_s)}{1 - \rho_{cc}} \tag{4.2i}$$

where:

ρ_{cc} is the ratio of the area of longitudinal reinforcement to the area of core of the section
s' is the clear spacing between spirals of hoop bars
d_s is the diameter of spiral

Due to its generality, the Mander et al. (1988b) model (Figure 4.2) has enjoyed widespread use in design and research despite a few shortcomings.

4.2.2 Reinforcing steel

The stress–strain relation of reinforcing steel exhibits an initial linear elastic portion, a yield plateau, a strain-hardening range in which the stress again increases with strain, and finally a range in which the stress drops off until fracture occurs. The length of the yield plateau and strain-hardening regions decreases as the strength of the steel increases. For monotonic loading, reinforced steel is represented as either an elastic–perfectly plastic material or an elastic strain-hardening material. It can also be represented using a trilinear stress–strain curve or a complete stress–strain curve. Most often elastic–perfectly plastic representation is selected (Darwin 1993).

In the analysis of moments and axial loads, two different models of the stress–strain performance of the reinforcing steel may be adopted. For nominal

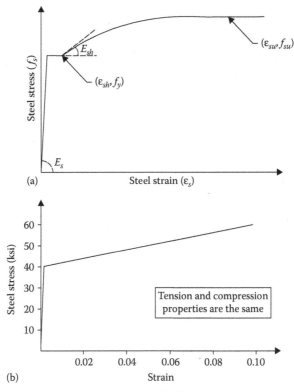

Figure 4.3 Stress–strain curve for steel. (a) True stress–strain curve for steel. (b) Idealized steel stress–strain relationships.

design capacities, an elastoplastic model is customarily adopted to provide a dependable estimate for design. For the *exact* analysis of existing RC members, a realistic stress–strain model should be applied using expected values of the control parameters. Such a model (Figure 4.3a), conveniently posed in the form of a single equation, is given as:

$$f_s = \frac{E_s \varepsilon_s}{\left\{ 1 + \left| E_s \varepsilon_s / f_y \right|^{20} \right\}^{0.05}} + (f_{su} - f_y) \left[1 - \frac{\left| \varepsilon_{su} - \varepsilon_s \right|^P}{\left\{ \left| \varepsilon_{su} - \varepsilon_{sh} \right|^{20P} + \left| \varepsilon_{su} - \varepsilon_s \right|^{20P} \right\}^{0.05}} \right] \qquad (4.3a)$$

where constant P can be represented as

$$P = \frac{E_{sh}(\varepsilon_{su} - \varepsilon_{sh})}{(f_{su} - f_y)} \qquad (4.3b)$$

For the longitudinal steel, a bilinear stress–strain relationship was estimated and employed (Figure 4.3b).

4.2.3 FRC and FRP

As mentioned in Section 4.1, the cracking behavior of FRC can be studied using the smeared crack approach. To determine the material properties of steel–FRC (SFRC), the inverse analysis techniques can be used to establish the stress–strain response of SFRC. This technique obtains the flexural response from bending tests to back calculate the stress–strain relationship. Both M–ϕ (moment–curvature) and P–δ (force–displacement) responses can be obtained from the test. The measured M–ϕ or P–δ responses reflect the influence of the steel fiber parameters and the concrete matrix.

4.2.3.1 Inverse analysis method

For this method, a three-step procedure is used to calculate the P–δ response of SFRC beams (Elsaigh et al. 2011a):

1. Assume a σ–ε relationship for the SFRC.
2. Calculate the M–ϕ response for a section.
3. Calculate the P–δ response for an element.

At the end of either step (2) or (3), the results from the analysis are compared to experimental results and adjustments are made to the σ–ε response until the analytical and experimental results agree within acceptable limits.

Based on the study by Elsaigh et al. (2011a and b), the tensile σ–ε response and results obtained from the nonlinear FEA of the beam were used in the analysis involving an SFRC slab manufactured using a similar material of the

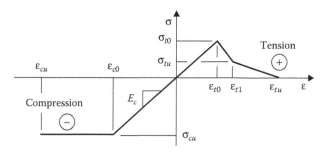

Figure 4.4 Stress–strain response of SFRC.

beam. Figure 4.4 shows the shape of the proposed σ–ε relationship used in this analysis. The mathematical form of the σ–ε relationship is expressed as follows:

$$\sigma(\varepsilon) = \begin{cases} \sigma_{cu} & \text{for } (\varepsilon_{cu} \leq \varepsilon \leq \varepsilon_{c0}) \\ E\varepsilon & \text{for } (\varepsilon_{c0} \leq \varepsilon \leq \varepsilon_{t0}) \\ \sigma_{to} + \psi \quad (\varepsilon - \varepsilon_{t0}) & \text{for } (\varepsilon_{t0} \leq \varepsilon \leq \varepsilon_{t1}) \\ \sigma_{tu} + \lambda \quad (\varepsilon - \varepsilon_{t1}) & \text{for } (\varepsilon_{t1} \leq \varepsilon \leq \varepsilon_{tu}) \end{cases}$$

(4.4a)

where:

$$E = \frac{\sigma_{cu}}{\varepsilon_{c0}}$$

$$\psi = \frac{\sigma_{tu} - \sigma_{t0}}{\varepsilon_{t1} - \varepsilon_{t0}}$$

(4.4b)

$$\lambda = \frac{-\sigma_{tu}}{\varepsilon_{tu} - \varepsilon_{t1}}$$

In Figure 4.4, σ_{t0} and ε_{t0} represent the cracking strength and the corresponding elastic strain, respectively; σ_{tu} and ε_{t1} represent the residual stress and the residual strain, respectively, at a point where the slope of softening tensile curve changes; ε_{tu} is the ultimate tensile strain; E is Young's modulus for the SFRC; σ_{cu} and ε_{c0} are the compressive strength and the analogous elastic strain, respectively; and ε_{cu} is the ultimate compressive strain.

As mentioned in Section 4.2.1, in compression, the concrete stress–strain relationship can be divided into ascending and descending branches. The behavior of FRP-confined concrete for flexural members can be assumed as similar to that of stirrup-confined concrete. Hence, confinement has no effect on the slope of the ascending part of the stress–strain relationship, and it is the same as for unconfined concrete, but not in the descending part in Figure 4.2. The compressive flexural strengths for both unconfined and confined concrete are the same and equal to the cylinder compressive strength. Figure 4.5 shows the uniaxial stress–strain curve for carbon and glass FRP composites in the fiber direction. For a more generalized expression, many studies show that instead of maintaining constant after compressive strength σ_{cu}, the σ–ε relationship may descend and the rate of descending is dependent on $V_f l/d$, where V_f is the volume ratio of fiber to concrete, l is the fiber length, and d is the fiber diameter (Gao 1991).

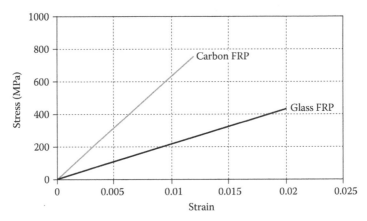

Figure 4.5 FRP uniaxial stress–strain curve for carbon and glass FRP composites in the fiber direction.

4.3 BEHAVIOR OF NONSKEWED/SKEWED CONCRETE BEAM–SLAB BRIDGES

Skew effect occurs in all types of bridge. It is discussed here because the effect is especially true and can be easily interpreted for concrete slab bridges. Nonskewed bridges, also known as straight, normal, or right bridges, are built with the longitudinal axis of the roadway normal to the abutment and therefore have a skew angle of 0°. As described in the American Association of State Highway and Transportation Officials (AASHTO) Load Resistance Factor Design (LRFD) Bridge Design Specifications (2013a), the skew angle of a bridge is defined as the angle between the longitudinal axis of the bridge and the normal to the abutment or, equivalently, as the angle between the abutment and the normal to the longitudinal axis of the bridge as shown in Figure 4.6. Skewed bridges are often built due to geometric restrictions, such as obstacles, complex intersections, rough terrain, or space limitations (Menassa et al. 2007).

As early as 1916, design recommendations were made to avoid building skewed bridges because of the many difficulties that arose when designing them, such as complex geometry and load distributions. However, because of increasingly complex site constraints, an increasing number of skewed bridges have been built. In addition to the complex geometry and load distributions caused by the skew, the skew angle can affect the performance of the substructure in conjunction with the superstructure, causing a coupling of transverse and longitudinal modes because of wind and seismic loads. Skew angles, in addition to the length-to-width ratio, also affect whether the bridge undergoes beam bending or plate action. As the skew increases or the length-to-width ratio of a bridge decreases, the bridge behaves more similarly to a plate than a beam.

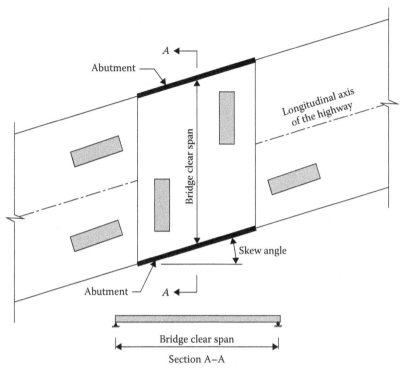

Figure 4.6 Description of a skew angle using a skewed bridge over a highway. (Data from Menassa, C. et al., *Journal of Bridge Engineering*, 12, 205–214, 2007.)

A nonskewed bridge deck behaves in flexure orthogonally in the longitudinal and transverse direction. The principal moments are also in the traffic direction and in the direction normal to the traffic. The slab of this type of bridges bends longitudinally leading to a sagging (or called positive) moment as it is shown in Figure 4.7. The load from the slab is transferred to reaction line directly through flexure. There will be a small amount of twisting moment because of the bidirectional curvature, and it will be negligible.

The force flow between the support lines in skew slabs is through the strip of area connecting the obtuse-angled corners, and the slab primarily bends along the line joining the obtuse-angled corners. The width of this primary bending strip is a function of skew angle and the ratio between the skew span and the width of the deck (aspect ratio). The areas on either side of the strip do not transfer the load to the supports directly but transfer the load only to the strip as cantilever as shown in Figure 4.8c. Hence the skew slab is subjected to twisting moments. This twisting moment is not small and hence cannot be neglected (Rajagopalan 2006). Because of this, the principal moment direction also varies, and it is the function of skew angle

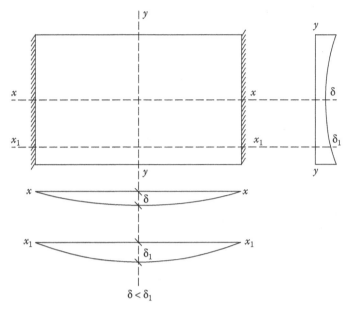

Figure 4.7 Deflection profile of a nonskewed deck.

and width-to-span ratio. The load is first transferred from the strip to the support over a defined length along the support line from the obtuse-angled corners. Later the force gets redistributed for full length. The force flow is shown in Figure 4.8a and b where the thin lines in Figure 4.8a indicate deformation shape. The distribution of reaction forces along the length of the supports is shown on both the support sides.

For skewed bridges, the deflection of the slab is not uniform or symmetrical as in the case of nonskewed deck. There will be warping that leads to higher deflection near obtuse-angled corner areas and less deflection near acute-angled corner areas. For small skew angles, both free edges will have downward deflection but differing in magnitude. For large skew angles, the maximum deflection is near the obtuse-angled corners. Near the acute-angled corner, there could be even negative deflection resulting in S-shaped deflection curve with associated twist. Increase in skew angle decreases bending moments but increases twisting moments (Rajagopalan 2006).

The characteristic differences between the behavioral aspects of a skewed deck and a nonskewed deck are as follows (Rajagopalan 2006):

- High reaction at obtuse corners.
- Possible uplift at acute corners, especially in the case of slab with very high skew angles.

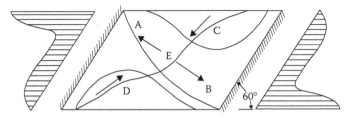

(a) Load transferred from zone C and D to E and then to the supports

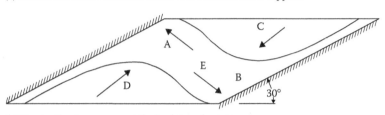

(b) Greater the skew, narrower the load-transfer strip

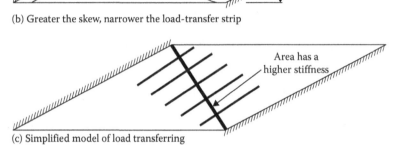

(c) Simplified model of load transferring

Figure 4.8 (a–c) Force flows in a skewed deck.

- Negative moment along the support line and high shear and high torsion near obtuse corners. Sagging moments orthogonal to abutment in the central region.
- At free edges, maximum moment nearer to obtuse corners rather than at the center.
- The points of maximum deflection toward obtuse-angled corners (the bigger the skew angle, the more shift of this point toward the obtuse corner).
- Maximum longitudinal moment and also the deflection reduce with the increase of skew angle for a given aspect ratio of the skew angle.
- As skew increases, more reaction is thrown toward obtuse-angled corners and less on the acute-angled corner. Hence the distribution of reaction forces is nonuniform over the support line.
- For a skew angle up to 15° and skew span-to-right width ratio up to 2, the effect of skew on principal moment values and its direction is very small.
- For a skew angle more than 15°, the behavior of the slab changes considerably.

Illustrated examples are provided in Sections 4.5 through 4.7.

4.4 PRINCIPLE AND MODELING OF CONCRETE BEAM–SLAB BRIDGES

The selection of the most appropriate modeling scheme depends on the nature of the information that is required. The first type of analysis could be performed with a linear elastic model, and the second type could be conducted with a more sophisticated RC model that is selected to represent the key aspects of nonlinear behavior for a particular structure or structural member.

4.4.1 Linear elastic modeling

The simplest form of an RC bridge is the RC slab bridge of solid sections or void sections. As described in Chapter 2, it can be simplified as a beam or a grid. The solid section can be idealized as an isotropic plate with the equivalent stiffness calculated from Equation 2.5. The voided section is idealized as an orthotropic plate, that is, a continuous medium with differing stiffness in directions parallel and perpendicular to the voids. The equivalent stiffness can be calculated from Equation 2.6 for rectangular void block or from Equation 2.7 for circular block (Sen et al. 1994).

Analysis of slab bridge decks using FEM involves the modeling of a continuous bridge slab as a finite number of discrete segments of slab or *elements* (Hambly, 1976). Generally all elements lie in one plane and are interconnected at a finite number of points known as nodes. The most common types of elements used are quadrilateral in shape, although triangular elements are sometimes also necessary (O'Brien and Keogh 1999). Some types of element, such as plate element, do not model in-plane distortion and consequently the nodes have only three degrees of freedom, namely, out-of-plane translation, and rotation about both in-plane axes (Timoshenko and Woinowsky-Krieger 1959). No particular problem arises from using elements that allow in-plane deformation in addition to out-of-plane bending, but the support arrangement chosen for the model must be such that the model is restrained from free body motion in either of the in-plane directions or rotation in that plane. Such analyses are necessary only if they are specifically required to model in-plane effects, such as axial prestress.

Finite element models, in which the elements are not at all located in one plane, can be used to model bridge decks, which exhibit significant 3D behaviors. The elements used for the modeling of slab bridge decks are flat shell elements, which can model out-of-plane bending in combination with in-plane distortion. The material properties of the elements are defined in relation to the material properties of the bridge slab. In case of bridges that are idealized as isotropic plates, only two elastic constants need to be defined for the finite elements, E and v. Geometrically orthotropic bridge decks are frequently modeled using materially orthotropic finite elements.

For materially orthotropic finite elements, five elastic constants, E_x, E_y, G_{xy}, v_x, and v_y, need to be specified.

In the third illustrated example with slab bridge decks, in-plane orthotropy was disregarded as the analysis tool used only permits bending orthotropy. However, in-plane (axial) and out-of-plane (bending) effects are uncoupled, and therefore this approximation does not affect comparisons for live load effects (bending) obtained from the model tests.

As described in Chapter 3, the finite element response to applied loading is based on an assumed displacement function. This function may be applicable only to the elements of certain shape; quite often the program will allow the user to define the elements that do not conform to this shape. Recommendations for FEA (O'Brien and Keugh 1999) are listed as follows:

1. Regular-shaped finite elements should be used wherever possible. These should trend toward squares in the case of quadrilateral elements and toward equilateral triangles in the case of triangles. In the case of quadrilateral elements the perpendicular lengths of the sides should not exceed 2:1 and no two sides should have an internal angle greater than 135°.
2. Mesh discontinuities should be avoided.
3. The spacing of elements in the longitudinal and transverse directions should be similar.
4. Elements should be located so that nodes coincide with the bearing locations.
5. Supports to the finite element model should be chosen to closely resemble those of the bridge slab.
6. Shear forces near points of support tend to be unrealistically large and should be treated with skepticism. However, results at more than a deck depth away from the support have been found in many cases to be reasonably accurate.

A beam-and-slab or cellular bridge deck may require a 3D FEA. It is possible to approximate the behavior of slabs and webs to thin flat shells, which can be arranged in 3D assemblage. At every intersection of shells lying in different planes, there is an interaction between the in-plane forces of one shell and the out-of-plane forces of the other, and vice versa. For this reason it is essential to use finite elements, which can distort under plane stress as well as plate bending. Because it is assumed that for flat shells, in-plane and out-of-plane forces do not interact within the plate, the elements are in effect the same as a plane stress element in parallel with a plate (or flat shell) bending element.

There is no logical limit to the cellular complexity, structural shape, or support system of a bridge that can be analyzed with a 3D flat shell model.

4.4.2 Nonlinear modeling

In nonlinear modeling of a RC structure, reinforcing steel can be modeled as the following:

1. Equivalent uniaxial material that is distributing throughout the finite element; often referred to as *smeared* steel (smeared model)
2. Discrete bars connected to the nodes in the finite element model (discrete model)
3. Uniaxial element that is embedded in a larger finite element (embedded model)

All three techniques involve the assumption of a perfect bond between steel and concrete, and in general its selection is based on the ease of application. The discrete and smeared representations were used more often. Surveyed by Darwin (1993), all models represent the steel and concrete as separate materials, whereas some consider the presence of steel in the development of the concrete material model, but all add the steel constitutive or stiffness matrix to the element or global matrix stiffness, respectively, as a separate uniaxial material. Although it is understood that bond slip will occur locally in the vicinity of flexural and shear cracks, members are designed so that the reinforcing steel is adequately anchored and thus the anchorage does not play a role in the strength of members in practice. Many models have been developed that totally ignored slip between the reinforcing steel and the concrete.

For models with smeared steel, the perfect bond relationship is the easiest to represent because it simply involves overlaying the constitutive matrix of the steel with the concrete element. For models with discrete steel, perfect bond also represents an easy solution, because the displacement of the nodal points is the same for both the steel and the concrete.

Bond slip can be modeled using both the discrete and distributed representation. Bond stress–slip relationships may be linear or nonlinear. Special link or bond zone elements are usually used in conjunction with discrete steel representations, whereas constitutive laws are used to model bond slip with distributed steel representations.

4.4.2.1 Cracking and retention of shear stiffness

The smeared cracking model procedure represents cracked concrete as an orthotropic material. After cracking occurred, the modulus of elasticity of the material is reduced to zero perpendicular to the principal tensile stress direction. This procedure has the effect of representing many finely spaced (or smeared) cracks perpendicular to the principal direction. The smeared crack concept fits the nature of the finite element displacement method, as the continuity of the displacement field remains intact.

The use of shear modulus, βG (with $0 < \beta \leq 1$), known as shear retention, improved most of numerical difficulties and improved the realism of cracking phenomena generated during the FEAs. A variable value of reduction factor has been selected to represent changes in shear stiffness (Darwin 1993).

As a general rule, moderately sophisticated elements such as four-node, two-dimensional isoparametric elements and eight-node, three-dimensional brick elements worked well. These elements usually provide the best results when used in conjunction with four- and eight-point Gauss integration, respectively. Higher-order elements provide locally more realistic deformation and strain fields. For macroscopic representation, element size and consideration of strain softening (fracture considerations) may not be important.

For better understanding the behavior of structures including general crack locations as well as concrete and steel stresses, it is advised to have a more refined mesh and a model that includes fracture considerations for concrete. Also, to capture the nonlinear behavior, the load step size must be kept small.

4.4.3 FRC/FRP modeling

Researchers have studied the behavior and modeling of RC members strengthened with FRP composites using FEM. The finite element model uses a smeared cracking approach for the concrete and 3D-layered elements to model FRP composites.

For research or forensic study purposes, 3D RC elements and layered solid elements can be used to simulate the behavior of FRP-strengthened RC structural elements (e.g., beams) using nonlinear FEM packages, such as ANSYS (2005). For RC, the 3D solid element (SOLID65 in ANSYS) with eight nodes and three degrees of freedom at each node, translations in the nodal x, y, and z directions, can be used. This element is capable of plastic deformation, creep, crushing in concrete, and cracking in three orthogonal directions at each integration point. Solid elements simulate the nonlinear material behavior with a smeared crack approach. When cracking occurs at an integration point, material properties are adjusted to effectively model a *smeared band* of cracks, rather than discrete cracks.

When a principal stress at an integration point in a concrete element exceeds the tensile strength, stiffness is reduced to zero in that principal direction perpendicular to the cracked plane. Cracking can be simulated at each integration point in three directions. FRP composites are modeled with 3D-layered structural solid elements (SOLID46 in ANSYS) having the same number of nodes and degrees of freedom as the concrete elements. The solid element allows for different material layers with different orientations and orthotropic material properties in each layer. Steel reinforcement bar can be modeled with a 3D truss (or spar) element (LINK8 in ANSYS). The truss element has two nodes and three degrees of freedom at each discrete node, translations in the nodal x, y, and z directions.

4.5 2D AND 3D ILLUSTRATED EXAMPLES: THREE-SPAN CONTINUOUS SKEWED CONCRETE SLAB BRIDGES

This example of a three-span continuous skewed RC slab bridge was extracted from a reference book by S. H. Park (2000). Plane, elevation, and cross-sectional views of the bridge are shown in Figure 4.9, and its computer rendering by Merlin-DASH® (Fu 2012) is shown in Figure 4.10. For production modeling of load-rating purpose, conventional elastic sectional modeling technique, instead of nonlinear modeling as described in Section 4.3.2, is used.

Two different linear elastic models were built for comparison. First-line *strip* model was built by Merlin-DASH, customized bridge software for RC, PC, and steel girders. For a slab bridge, a unit width of 12″ (300 mm), shown as a straight strip line in Figure 4.10, is assumed for modeling and analysis purpose.

Based on AASHTO LRFD specifications (2013a), the equivalent width of longitudinal strips per lane for both shear and moment with more than one lane loaded, which is the case here for comparison, is

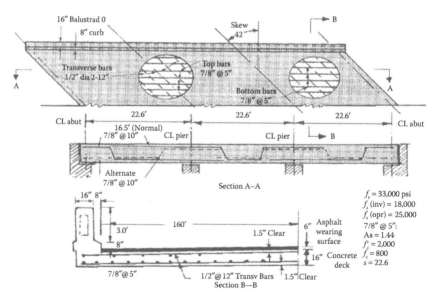

Figure 4.9 Plane, elevation, and cross-sectional views of the skewed RC slab bridge.

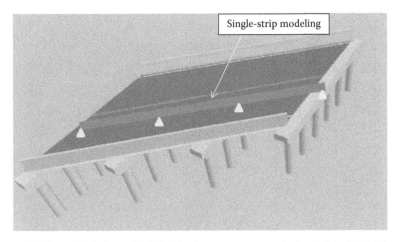

Figure 4.10 Example I skewed RC slab bridge computer rendering by Merlin-DASH.

$$E = 84.0 + 1.44\sqrt{L_i W_i} \leq \frac{12.0W}{N_L} \qquad (4.5a)$$

where:

  E is the equivalent width (in)
  L_i is the modified span length, less than 60′ (22.6′ or 6.9 m in this case)
  W_i is the modified edge-to-edge width, less than 60′ (36′ or 11 m in this case)
  W is the physical edge-to-edge width of the bridge
  N_L is the number of design lanes (two lanes in this case)

In this example one lane of loading is distributed within the equivalent width of 125.07″, 10.42′ (3.18 m). For skewed bridge, the longitudinal force effect may be reduced by the factor r with a skew angle of θ in degrees (42° in this case):

$$r = 1.05 - 0.25 \tan\theta \qquad (4.5b)$$

This correction factor calculated is 0.825. Because the model is a one-foot strip, the live load distribution factor within this one-foot strip can be considered as 0.0791 (=0.825 × 1′/10.42′) for this example.

The second model as shown in Figure 4.11 is a more sophisticated finite element model, with the whole bridge, including its skewness, modeled by CSiBridge®. The bridge is a three-span skewed concrete slab bridge. It consists of 16″ (406-mm) thick flat slab supported by abutments and bents. The slab is connected at its bottom to the abutments and bents. Abutments are supported by fixed foundation springs. Bents consist of bent caps and columns. The column bases are fixed and moments are released at the top.

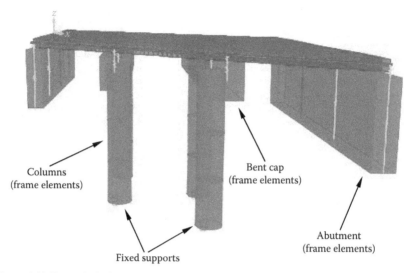

Figure 4.11 Example I skewed RC slab bridge modeling by CSiBridge.

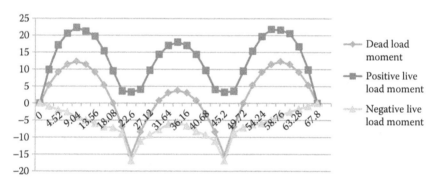

Figure 4.12 Merlin-DASH example I dead and live load moment results (kip-ft).

The concrete slab is modeled with shell elements, whereas abutments are modeled with frame elements. The bent caps and columns are also modeled with frame elements. Their comparison results are shown in Figure 4.12 for Merlin-DASH line strip model and in Figure 4.13 for CSiBridge finite element model, respectively. From Merlin-DASH program the maximum positive moment for dead load is found to be 12.3 kip-ft/ft (54.7 kN-m/m), whereas CSiBridge finite element model shows 286 kip-ft across the whole normal width of 26.8′ (8.17 m), which is about 10.7 kip-ft/ft (47.6 kN-m/m). As it is known that dead load distributions of a straight bridge should be very close from one model to another or from one program to another, the difference of this skewed example from a 2D to 3D model reveals that the 2D modeling adopting an equivalent width is conservative

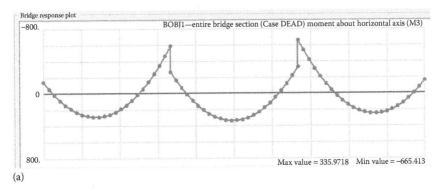

(a)

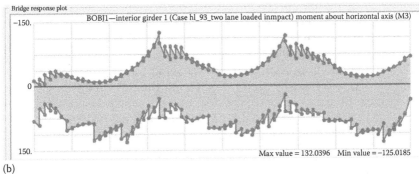

(b)

Figure 4.13 CSiBridge results. (a) Dead load moment diagram. (b) Live load moment envelop.

in the analysis of a skewed bridge. Another benefit of using 3D finite element model is that nonuniform moment distribution of the entire bridge due to skewness can be obtained, not by assuming the uniform distribution cross section-wise but by the line *strip* model. Other moment comparisons are shown in Table 4.1. It is concluded that for this three-span continuous concrete slab bridge, finite element model assuming two lanes loaded with HL-93 vehicle on each lane provides more accurate results, whereas

Table 4.1 Comparison of moments based on line strip and FEM methods

Dead/live moment	Line strip moment in kip-ft/ft (kN-m/m)	FEM moment in k-ft/ft (kN-m/m)	Location in ft (m)
Dead positive	12.3 (54.7)	286/26.8 = 10.7 (47.6)	9.24 (2.8)
Dead negative	−15.3 (−68.1)	−665/32 = −20.8[a] (−92.4)	22.6 (6.9)
Live positive	22.3 (99.2)	333/32 = 10.4 (46.3)	11.3 (3.4)
Live negative	−16.9 (−75.2)	−647/32 = −20.2[a] (−89.9)	22.6 (6.9)

[a] If the width along the skewed line 43′ is used, dead negative moments are 15.5 and 15.0 kip-ft/ft, respectively.

simulated line beam model gives more conservative results for most of the locations.

4.6 2D AND 3D ILLUSTRATED EXAMPLES: RC T-BEAM BRIDGE

This example of a single-span RC T-beam bridge was adopted from AASHTO LRFD manual (2013b). Cross-sectional view of the bridge is shown in Figure 4.14. The simple span concrete T-beam bridge consists of 6″ thick concrete deck and four monolithically casted concrete beams. This is a non-skewed bridge and ideal for line girder modeling. The beams are supported at the ends by the abutments connected at the bottom, and its computer rendering by Merlin-DASH (Fu 2012) is shown in Figure 4.15. For production modeling of load-rating purpose, conventional elastic sectional modeling technique, instead of nonlinear modeling as described in Section 4.3.2, is used.

Two different linear elastic models were built for comparison. First-line girder model was built by Merlin-DASH. For the T-beam bridge example, a T-beam with flange width of 6′-6 1/4″ (2.0 m), shown as a straight beam line (Figure 4.15), is considered for modeling and analysis purpose.

Based on AASHTO LRFD specifications (2013a), the distribution of live loads for moment in interior beams with more than one lane loaded, which is the case here for comparison, is

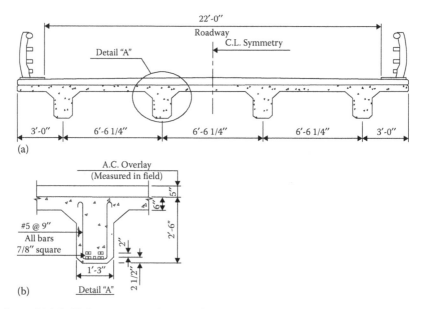

Figure 4.14 (a, b) Cross-sectional views of the T-beam bridge.

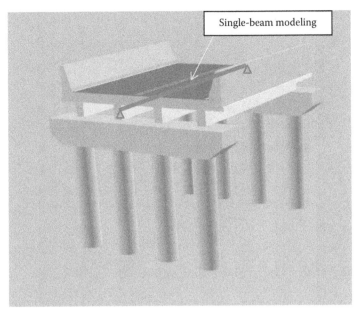

Figure 4.15 T-beam bridge computer rendering by Merlin-DASH.

$$0.075 + \left(\frac{S}{9.5}\right)^{0.6} \left(\frac{S}{L}\right)^{0.2} \left(\frac{K_g}{12.0Lt_s^3}\right)^{0.1} \qquad (4.6a)$$

where:

S is the spacing of the beam in feet (6.52′ or 2.0 m in this case)

L is the span of the beam in feet (26′ or 7.9 m in this case)

t_s is the depth of concrete slab in inches (6″ for the flange thickness or 152 mm in this case)

K_g is the longitudinal stiffness parameters in in⁴ (calculated as 98,280 in⁴ or 4.09×10^{10} mm⁴ in this case)

n is the modulus ratio between the beam and deck

I is the moment of inertia of the beam

e_g is the distance between the centers of gravity of the beam and deck

$$K_g = n(I + Ae_g^2) \qquad (4.6b)$$

The second model as shown in Figure 4.16 is a more sophisticated finite element model with the whole bridge modeled by CSiBridge. The concrete beam and deck are modeled with shell elements. The abutment is frame elements. The abutment is fixed at the bottom.

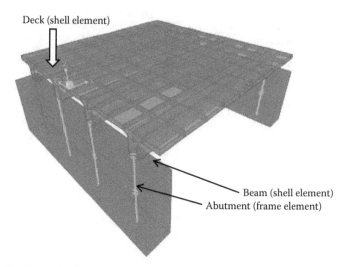

Figure 4.16 T-beam bridge computer model by CSiBridge.

The comparison results are shown in Figure 4.17 for Merlin-DASH line girder model and in Figure 4.18 for CSiBridge finite element model, respectively. From Merlin-DASH program the maximum positive moment for dead load is 112.6 kip-ft (152.7 kN-m), whereas CSiBridge finite element model shows 102.2 kip-ft (138.6 kN-m). In the live load analysis, the line girder program is assuming HL-93 vehicle(s) with AASHTO distribution factor, whereas finite element model has two lanes loaded with HL-93 vehicle on each lane. It can be seen that for a normal girder bridge, the single-beam model is comparable with the 3D finite element model (Table 4.2).

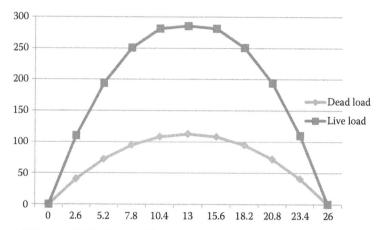

Figure 4.17 Merlin-DASH dead and live load moment results (kip-ft).

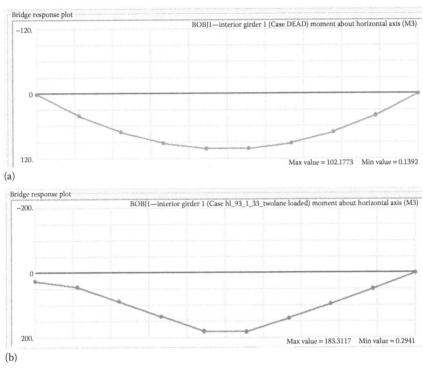

Figure 4.18 CSiBridge results. (a) Dead load moment diagram. (b) Live load moment envelop.

Table 4.2 Comparison of moments based on line girder and FEM methods

Dead/live moment	Line strip moment in kip-ft (kN-m)	FEM moment in kip-ft (kN-m)	Location in ft (m)
Dead positive	112.6 (152.7)	102.2 (138.6)	13 (4)
Live positive	285.5 (387.1)	183.3 (248.6)	13 (4)

4.7 3D ILLUSTRATED EXAMPLES: SKEWED SIMPLE-SPAN TRANSVERSELY POST-TENSIONED ADJACENT PRECAST-CONCRETE SLAB BRIDGES— KNOXVILLE BRIDGE, FREDERICK, MARYLAND

This illustrated example is a transversely post-tensioned bridge, which is nonlinear modeled to study the ultimate behavior after shear key joints' cracking. This is a two-lane simply supported single-span bridge with a 6.78-m span and a 31.4° skew angle. The superstructure consists of eight adjacent 1.22 m × 0.381 m × 7.12 m PC beams and a typical 127-mm minimum thick composite concrete deck, as shown in Figure 4.19a. The beams were

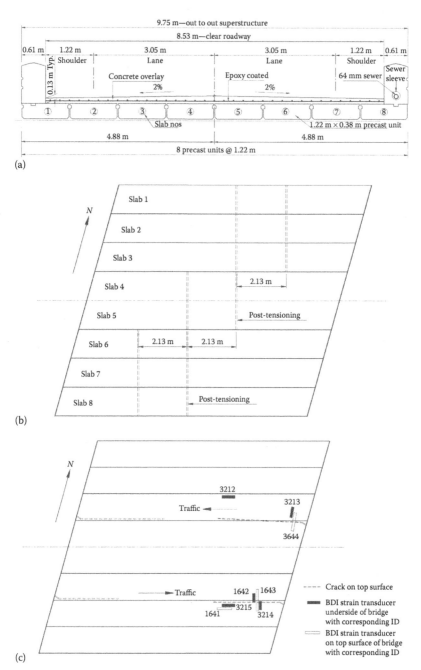

Figure 4.19 Existing post-tensioned Knoxville slab bridge, Maryland. (a) Cross section. (b) Plan view of precast beams and post-tensioning tie rods. (c) Plan view of the strain transducer locations.

transversely post-tensioned using four 25.4-mm diameter tie rods tensioned to 355.9 kN and placed normal to the beams as shown in Figure 4.19b. Load test was performed to measure the short-term live load strains on the bottom and top surfaces of the bridge as a test vehicle drove over the bridge. The strain data from the FEA model were compared to the strain data from the field test, and then the model was refined based on the varying material strengths until the results were sufficiently close to the field data.

Four main components composed the FEA model of the bridge: the precast prestressed solid concrete beams, the prestressing strands, the transverse post-tensioning, and the concrete overlay. The precast-concrete beams and the concrete overlay were modeled with solid brick elements, and the pretensioning strands in the precast-concrete beams and the post-tensioning tie rods were modeled with link elements (Fu et al. 2011). In the first stage analysis, concrete is assumed cracked between beams along the bonding so nonlinear analysis was adopted. For simulating the effect of shear friction after crack of the shear keys, contact elements (CONTA174 & TARGE170), in the finite element program ANSYS, were employed at the location of interface between beams. Contact friction is a material property that is used with the contact elements and is specified through the coefficient of friction, which was taken as 0.6 for the interface between slab beams. Both the solid brick and the link elements have three degrees of freedom (translations) at each node. Because this is for study and State of Maryland (U.S.) standard-generating purpose, very refined models were made to line up all skewed angles, rod orientations, and beam details. There were 46,080 solid brick elements and 3,520 link elements for a total of 49,600 elements for this skewed bridge.

The transverse strain from the FEA model, as shown in Figure 4.20, shows a close fit to the field data with strain transducers marked 3215 (underneath) and 1641 (top side) along the longitudinal direction in Figure 4.19c. The stress distribution at the concrete overlay–beam interface and the top surface was then analyzed to examine the cause of the cracks on the top surface of the concrete overlay. Generally, the greatest transverse tensile stresses, with a potential of concrete cracking, exist near the abutments and between the beams along the shear keys. With the model proved valid, a series of parametric study of different post-tensioning forces and configurations were conducted. The first stage is to study the level of post-tensioning forces (Fu et al. 2010) by nonlinear analysis. Figure 4.21 indicates that each beam behaves independently under wheel loads without a transverse post-tensioning force. Therefore, only beams with applied wheel loads show displacements, while the initial displacements of other beams experience zero. Eight FEM bridge models with different span lengths (6.10, 7.62, 9.14, 10.67, 12.19, 13.72, 15.24, and 16.67 m) were generated. As the transverse post-tensioning force of rods increases, displacement significantly decreases and is stabilized approximately at 338 kN of the post-tensioning force.

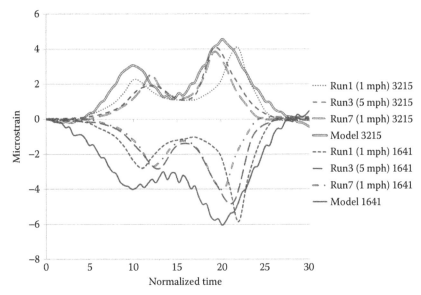

Figure 4.20 Strain transducer nos. 3215 and 1641 parallel to the slabs on the bottom surface of beam 7.

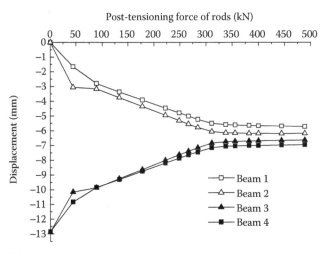

Figure 4.21 Displacements at midspan for different transverse post-tensioning forces.

Then, a second-stage parametric study on the orientation of the tendons was conducted (Fu et al. 2012) on skewed bridges. In Figure 4.22, transverse stress contour between concrete overlay and precast beam interface for a 7.6 m, 30° skewed bridge is shown. For comparison purpose, two types of tendon arrangement were studied. On Figure 4.22a, third points with skewed tendon parallel to the support are post-tensioned. On Figure 4.22b, four

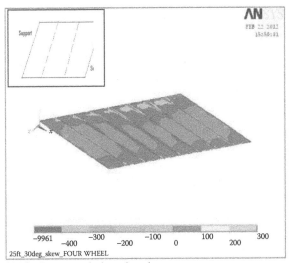

(a) Third points with skewed tendons

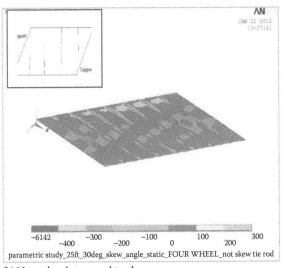

(b) Normal and staggered tendons

Figure 4.22 (a, b) Transverse stress at concrete overlay–beam interface of a 7.6 m, 30° skewed bridge (insertions show the tendon locations).

tendons oriented normal to traffic and staggered, similar to the test bridge, were used for this case. Insertions of both figures show the tendon locations of their respective arrangements. It can be seen that the left figure shows better results with less tension so the tendons, two close to the abutments and one at the center, parallel to the skewed supports were recommended. Similar parametric study was done on other span length arrangements (Fu et al. 2012).

Chapter 5

Prestressed/post-tensioned concrete bridges

5.1 PRESTRESSING BASICS

There are two prestressing methods available for prestressed concrete girders—pretensioning and post-tensioning. The main objective of prestressed concrete girders is to increase the load-carrying capacity for both strength and serviceability of concrete girders. Both prestressing methods and their modeling techniques will be discussed in the following sections.

For the pretensioning method, the process of producing prestressed concrete girders is similar to that of reinforced concrete. However, unlike reinforced concrete, special steel strands are used and pretensioned prior to placing the concrete. Prestressed concrete bridge girders are typically designed to resist high tensile stresses in the bottom flange of the girders at midspan. This is achieved by placing the pretensioning steel strands in the lower portion of the girders (McDonald 2005).

One consequence in attaining this desired strength at midspan is that tensile stresses at the ends of the member in the top flange exceed design code limits. Figure 5.1a provides a brief overview of the loading stages for prestressed concrete girders. Figure 5.1b demonstrates the linear stress distribution at the various stages of the prestressed concrete girder fabrication to the final installed condition. Stage 3, location 1 (at transfer length), which is the primary concern, illustrates the tensile stress that develops in the top flange of the girder.

For precast prestressed concrete girders, two techniques have been available for handling the tensile stresses that develop at the release of the prestressing force. These two techniques are based on the position and pattern of the prestressing strands. The two strand patterns consist of (1) all straight strands with debonding at the ends of the member or (2) straight strands with a certain number of the strands deflected upward at the ends of the girder. Figure 5.2 illustrates the strand profiles for these two detensioning techniques. Figure 5.3 illustrates the harping technique and displays the hold-down devices used prior to the placement of formwork.

Stage 1: Tensioning of prestressing strands in stressing bed before casting concrete.

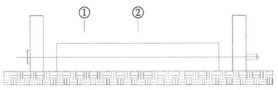

Stage 2: Placement of concrete in forms and around tensioned strands.

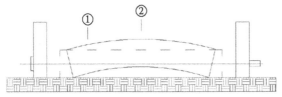

Stage 3: Release of strands causing shortening of member.
Stage 4: Member placed on piers and/or abutments and deck slab, if any, cast.
Stage 5: Full-service load after all prestress losses.
(a)

	1	2	3	4	5
Stage	Tensioning of prestressing strands	Concrete placement	Release of strands	Member installation	Full load
Location 1 (at transfer length)	———	———	◿	◿	◿
Location 2 (at midspan)	———	———	◿	◸	◿

(b)

Figure 5.1 (a) Prestress loading stages. (b) Stress distribution at various loading stages.

Different from pretensioning, post-tensioning is the application of a compressive force to the concrete at some point in time after casting. Post-tensioning tendons may be installed through voids formed by ducts cast into the concrete—in which case, they are *internal* tendons—or they may be installed outside the concrete itself—in which case, they are *external*

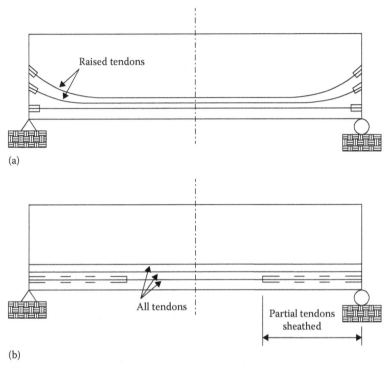

Figure 5.2 Prestressing strand profiles. (a) Harped strands. (b) Debonded strands. The dashed lines indicate debonding material around prestressing strand.

tendons. In the most common technique of internal post-tensioning, cables are threaded through ducts in the cured concrete and the stressed tendons are locked with mechanical anchors. These cables are stressed to design values by hydraulic jacks, and the ducts are thoroughly grouted up with cement grout after stressing has occurred. Figures 5.4 and 5.5 show two different types of post-tensioning. Figure 5.4 illustrates a post-tensioned beam before concrete pouring and post-tensioning to show its rebar cages and conduits. Figure 5.5 shows a perspective view of a typical precast balanced cantilever segment with various types of tendons (FLDOT 2002).

Also illustrated in FLDOT (2002), Figure 5.6 shows a typical layout of cantilever tendons that are anchored on the face of the precast segments, which do not allow later inspection of the anchor head following tendon grouting. An alternate approach is to anchor the cantilever tendons in blisters cast with the segments at the intersection of the top slab and web where anchorages of these tendons can be inspected at any time. The same arrangement can be made for bottom continuity tendons at midspan. Figure 5.7 shows a typical layout of span-by-span tendons for an interior span where all tendons deviate at a common deviation saddle.

Figure 5.3 Harped prestressing strands. (Data from FLDOT/Corven Engineering, Inc., *New Directions for Florida Post-Tensioned Bridges*, Volume 1: Post-Tensioning in Florida Bridges, Florida Department of Transportation, Tallahassee, FL, February 2002.)

Figure 5.4 Post-tensioned beam before concrete pouring and post-tensioning. (Data from FLDOT/Corven Engineering, Inc., *New Directions for Florida Post-Tensioned Bridges*, Volume 1: Post-Tensioning in Florida Bridges, Florida Department of Transportation, Tallahassee, FL, February 2002.)

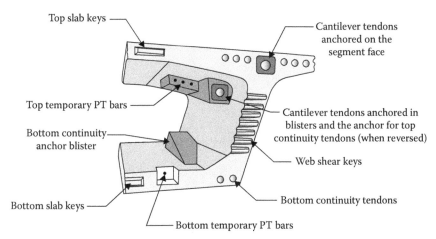

Figure 5.5 Typical balanced cantilever segment. (Data from FLDOT/Corven Engineering, Inc., *New Directions for Florida Post-Tensioned Bridges*, Volume 1: Post-Tensioning in Florida Bridges, Florida Department of Transportation, Tallahassee, FL, February 2002.)

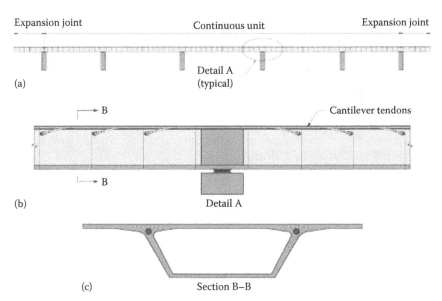

Figure 5.6 (a–c) Cantilever post-tensioning tendons anchored on the segment faces. (Data from FLDOT/Corven Engineering, Inc., *New Directions for Florida Post-Tensioned Bridges*, Volume 1: Post-Tensioning in Florida Bridges, Florida Department of Transportation, Tallahassee, FL, February 2002.)

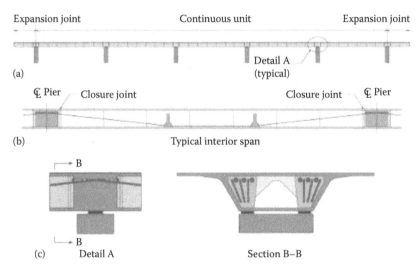

Figure 5.7 (a–c) Interior span post-tensioning for span-by-span construction. (Data from FLDOT/Corven Engineering, Inc., *New Directions for Florida Post-Tensioned Bridges*, Volume I: Post-Tensioning in Florida Bridges, Florida Department of Transportation, Tallahassee, FL, February 2002.)

5.2 PRINCIPLE AND MODELING OF PRESTRESSING

Any modeling method that satisfies the requirements of equilibrium and compatibility and utilizes stress–strain relationships for the proposed material can be used in the analysis. As it is commonly known, the prestressing force used in the stress computation does not remain constant with time. The collective loss of prestress is the summation of all individual losses, which may be examined individually or considered a lump sum loss. The four most critical conditions in the structural modeling of tendons are (Fu and Wang 2002) the following:

- *Immediate loss of stress in tendon*—Friction between the strand and its sheathing or duct causes two effects: (1) curvature friction and (2) wobble friction. The retraction of the tendon results in an additional stress loss over a short length of the tendon at the stressing end. Loss will also happen due to tendon slip before full grip of the anchorage. The combined loss is commonly referred to as the friction and seating loss.
- *Elastic shortening*—The elastic shortening of the concrete due to the increase in compressive stress causes a loss of prestressing force in tendons.
- *Long-term losses*—Several factors cause long-term losses: (1) relaxation of the prestressing steel, (2) shrinkage in concrete, and (3) creep in concrete. In grouted (bonded) post-tensioning systems, creep strain

in the concrete adjacent to the tendon causes a stress decrease in tendons. For unbonded tendons, the decrease in stress along the tendons due to creep in concrete is generally a function of the overall (average) precompression of the concrete member.

- *Change in stress due to bending of the member under applied loading*—For a rigorous evaluation of the affected member, change in stress must be taken into account, particularly when large deflections are anticipated.

In general, pretensioning/post-tensioning tendon modeling and its analysis can be categorized into two major groups (which are described in the following sections): (1) tendon modeled as applied loading and (2) tendon modeled as load-resisting elements (Fu and Wang 2002).

5.2.1 Tendon modeled as applied loading

When tendons are modeled as applied loading, there are three modeling techniques:

- *Simple load balancing*—The force of the tendon on the concrete is considered to balance (offset) a portion of the load on the member, hence the *load balancing* terminology. The shortcoming of this method is that the immediate and long-term stress losses in prestressing must be approximated and accounted for separately.
- *Tendon modeling through primary moments*—The primary moment M_p due to the prestressing force P at any location along a member is defined as the prestressing force P times its eccentricity e. The eccentricity of the force is the distance between the resultant of the tendon force and the centroid of the member. The primary moment may be used as an applied loading in lieu of the balanced loading for structural analysis. Bridge designers use this modeling technique more commonly than building designers. In practice, the primary moment diagram is discretized into a number of steps. Each discrete moment is equal to the change in the value of moment between two adjacent steps in the primary moment diagram.
- *Equivalent load through discretization of the tendon force*—The force distribution is represented by a series of straight segments. Hence, the force distribution would be represented by a series of sloping lines with steps at the discretization points. The force distribution can be further simplified by considering the force in each tendon segment to be equal to the force at the midpoint of the segments (Figure 5.8).

As demonstrated in PCI (2011), for continuous bridges, support reactions caused by restrained deformations due to post-tensioning result in

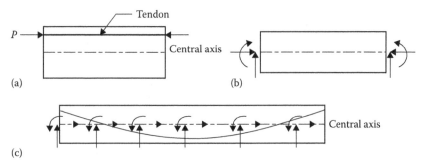

Figure 5.8 Equivalent load through discretization of the tendon force. (a) Tendon as external force of an element. (b) Equivalent tendon force of an element. (c) Equivalent tendon forces along the central axis of the beam.

additional moments called *secondary moments*. A common approach to evaluate secondary moments due to post-tensioning is to model the effect of the post-tensioning tendon as a series of equivalent uniformly distributed loads. Figure 5.9 shows the required equations for the calculation of the equivalent loads for a typical end span of a post-tensioned beam.

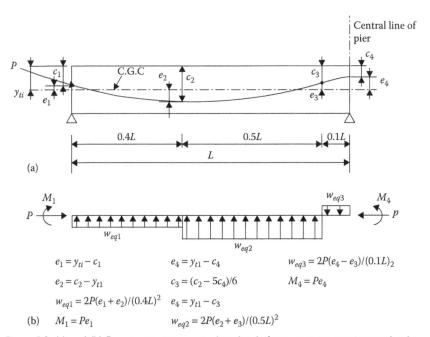

$$e_1 = y_{ti} - c_1$$

$$e_2 = c_2 - y_{t1}$$

$$w_{eq1} = 2P(e_1 + e_2)/(0.4L)^2$$

(b) $$M_1 = Pe_1$$

$$e_4 = y_{t1} - c_4$$

$$c_3 = (c_2 - 5c_4)/6$$

$$e_4 = y_{t1} - c_3$$

$$w_{eq2} = 2P(e_2 + e_3)/(0.5L)^2$$

$$w_{eq3} = 2P(e_4 - e_3)/(0.1L)_2$$

$$M_4 = Pe_4$$

Figure 5.9 (a) and (b) Post-tensioning equivalent loads for two-span continuous bridge. (Data from Precast/Prestressed Concrete Institute, *Precast Prestressed Concrete Bridge Design Manual*, 3rd Edition, 2011.)

5.2.2 Tendon modeled as load-resisting elements

The tendon is not considered to be removed from the concrete member. Rather, it is modeled as a distinct element linked to the concrete member (Figure 5.10). The change in the prestressing force is automatically accounted for in the equilibrium equations set up for the analysis of the segment.

For tendons modeled as resisting elements, four post-tensioning analysis types are shown in Figure 5.11: (1) beam type, (2) tendon type, (3) plane stress type, and (4) solid type. The former two are used in routine bridge analyses, whereas the latter two with more detailed modeling technique are used more in research or forensic analysis (LUSAS 2012). For post-tensioning, the tendons can be either external or internal where internal tendons can be either bonded or unbonded (Figure 5.12).

5.2.3 2D and 3D modeling

Based on the discussion in Section 2.4.5, two-dimensional (2D) or three-dimensional (3D) models can be generated based on the project's needs. For a 2D model, only one beam is considered and section properties of that beam are based on the locations of their respective neutral axes. Two 2D beam models representing two different stages of noncomposite and

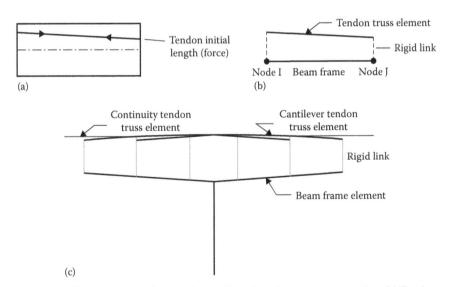

Figure 5.10 Tendon modeled as an element linked to the concrete member. (a) Tendon as element. (b) Tendon element geometry. (c) Finite element modeling of the segmentally erected bridge with post-tensioning tendons.

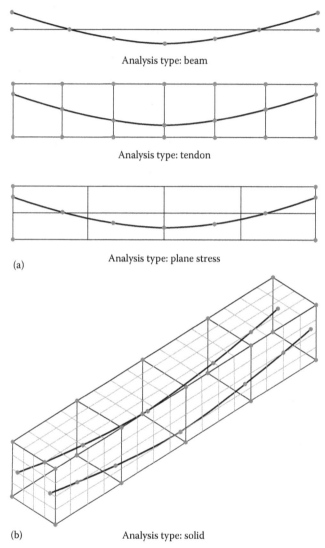

Analysis type: beam

Analysis type: tendon

Analysis type: plane stress

(a)

(b)   Analysis type: solid

Figure 5.11 2D and 3D post-tensioning analysis types. (a) 2D model. (b) 3D model. (Data from LUSAS®, "LUSAS Bridge/Bridge Plus Bridge Engineering Analysis," 2012, http://www.lusas.com/products/information/eurocode_pedestrian_loading .html.)

short-term composite models, respectively, are demonstrated in Figure 5.13. Many customized 2D prestressed beam computer programs are available for analysis where customization is made by dividing beams into small segments of prismatic members with tendons modeled as applied loading within each segment, which was discussed in Section 5.2.1. For a 2D beam

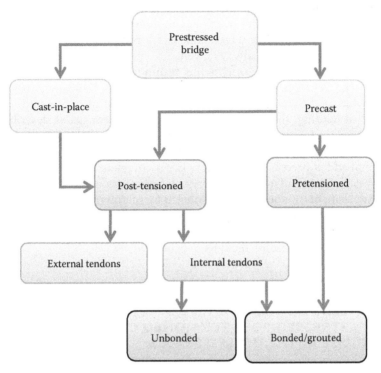

Figure 5.12 Types of prestressing analysis.

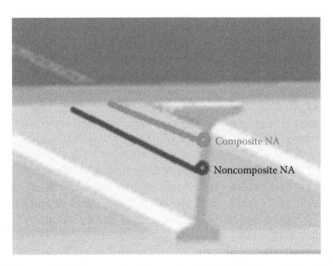

Figure 5.13 2D model with its associated neutral axis (NA) locations. (a) Framing plan.
(b) Cross section.

model, moments and shears are direct results from analysis, and there is no need to integrate stresses to get beam moments for strength limit state capacity check. No matter which code is adopted for design, stress limits for concrete and steel are always given.

On the other hand, the 3D modeling technique has become more sophisticated and more popular nowadays to understand the behavior of a bridge during different construction stages. Instead of modeling tendons as applied loading, they are modeled as resisting elements as described in Section 5.2.2. In routine bridge analyses, prestressed beams are usually modeled as beams while tendons are modeled as a series of truss elements with embedded pretensioning forces. For a complete 3D bridge model, in which deck are simulated by shell or solid elements with rigid connection to beam elements, tendons can be modeled by spatial truss elements sharing appropriate nodes with shell, solid or beam elements. An illustration of 2D modeling is described in Section 5.3, and a more detailed demonstration of 3D modeling is covered in Sections 5.4 through 5.7.

5.3 2D ILLUSTRATED EXAMPLE OF A PROTOTYPE PRESTRESSED/POST-TENSIONED CONCRETE BRIDGE IN THE UNITED STATES

Based on AASHTO specifications (2013), a design case for a concrete alternate with a continuous prestressed and then post-tensioned precast I-beam bridge is analyzed as a single beam staged from simple to continuous beams. The total length of the bridge is 198.86 m (652'-5"), with five continuous spans of 39.5 m (129'-7") each (Figure 5.14a). The clear roadway width is 13.41 m (44'), and out-to-out distance is 17.98 m (59') with 3–3.66 m (12') lanes. Five 1880-mm (74") deep precast bulb-T girders are used in the design with 3.81-m (12'-6") girder spacing (Figure 5.14b). A 200-mm (8") deck slab is used in the composite construction with another 13-mm (1/2") wearing surface.

Precast girder is formed by the semi-light weight concrete with initial concrete strength (f_{ci}') of 31 MPa (4500 psi) and final concrete strength (f_c') of 48.3 MPa (7000 psi). Concrete strength of the cast-in-place concrete is 34.5 MPa (5000 psi). All the prestressing tendons are 1862-MPa (270-ksi) stress-relieved seven-wire strands with modulus of elasticity of 1.9×10^5 MPa (28×106 psi). The prestressing steel strand's diameter is 13 mm (1/2"), and the post-tensioning steel strand's diameter is 15 mm (0.6"). Figure 5.15 shows the profile of the post-tensioning conduits (prestressing strands are not shown) and three cross sections at the end spans. Cross sections A–A and C–C (Figure 5.15b) show the thickened webs at the ends of the precast beam. The construction sequence is listed as follows:

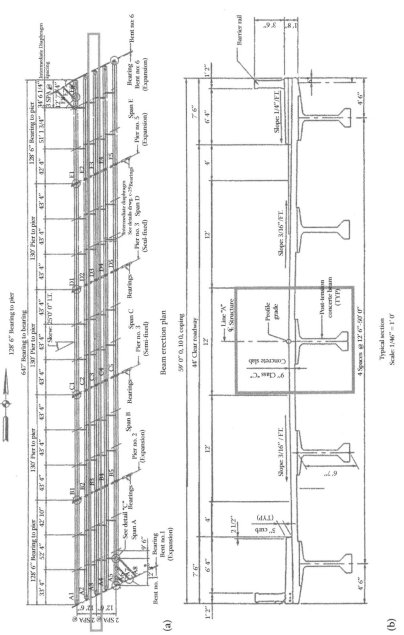

Figure 5.14 (a) Five-span precast and (b) prestressed concrete bridge made continuous with post-tensioning tendons.

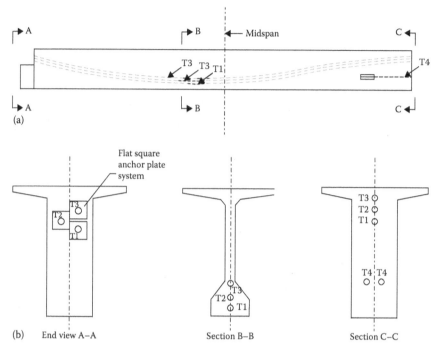

Figure 5.15 Post-tensioning (a) layout and their (b) cross sections at the end span of a continuous precast prestressed/post-tensioned concrete bridge.

1. Erect precast prestressed beams on early-made concrete abutments and supports
2. Install duct splices for post-tensioning tendons and pour beam splices and diaphragms at piers. At this stage, stress and grout tendons T1
3. Pour in-span diaphragms. At this stage, stress and grout post-tension tendons T2
4. Pour deck. At this stage, stress and grout tendons T3 for full post-tensioning
5. Construct sidewalk and barrier/railing and complete the job

In the process three 2D beam models with different section properties are built. The first noncomposite sectional model with different levels of tendon forces is used for stages 1, 2, and 3. The second *short-term* composite sectional model with full tendon forces is used for stage 4, whereas the third *long-term* composite sectional model with full tendon forces is used for stage 5. Note here that *short-term* and *long-term* composite sections are used by AASHTO to refer to the section properties of n and $3n$, respectively, where n is the modulus ratio between steel and concrete materials. For the consideration of pretensioning/post-tensioning tendon modeling and its analysis, the "tendon modeling through primary moments" as discussed in Section 5.2.1 is used in the calculation by Merlin-DASH/PBEAM, a 2D line girder program. This tedious procedure of generating primary fixed-end moments can also be employed to a generic finite element analysis package, but the process would be cumbersome. Results show that the program checks stress limits of the concrete (Figure 5.16) and the reinforcing steel under the serviceability limit states as well as ultimate moments and shears under the strength limit states.

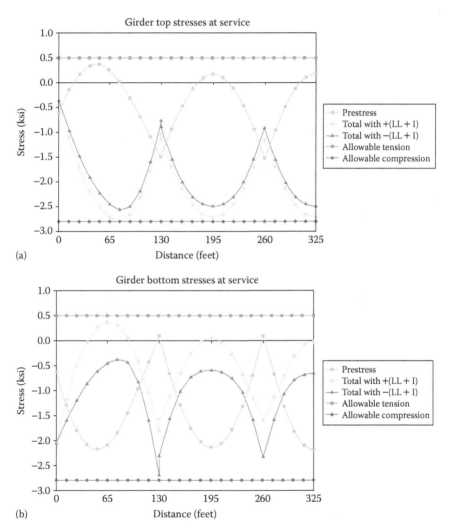

(a)

(b)

Figure 5.16 (a) Top and (b) bottom stresses of a five-span precast, prestressed concrete bridge.

5.4 3D ILLUSTRATED EXAMPLE OF A DOUBLE-CELL POST-TENSIONING CONCRETE BRIDGE—VERZASCA 2 BRIDGE, SWITZERLAND

In European practice, post-tensioning is more popular. A Swiss bridge with cast-in-place double-cell concrete beam is taking as an example in this section. The Bridge Verzasca 2, which locates on the main road between Bellinzona and Locarno, in the south of Switzerland, was built in 1990–1991 and consists of six spans between 25.24 and 39.70 m (82.8′ and 130.3′), with a total length of 203.6 m (668′). The pier supports are skewed at an angle of 28.8°, whereas the abutments are placed perpendicular to the bridge axis. The superstructure is a post-tensioned continuous girder with a cast-in-place double-cell section (Schellenberg et al. 2005).

The cross section changes in the region over the piers where negative moments are expected. In this region the three webs of the double-cell section are widened. Also, the bottom flange is thickened continuously from 200 to 300 mm (8″ to 12″) in this region.

Diaphragms are placed over each pier, providing a higher torsional rigidity. Accounting for the diaphragms as well as a cross section of the beam, a total of three cross sections can be determined. The post-tensioning tendons are anchored approximately at the section of dead load point of contraflexure, where the webs change their width, providing required spaces for the tensioning procedure.

Each tendon stretches over one span including both neighboring piers in such a way that the tendons overlap over a single pier. Their distribution over the cross section is shown in Figure 5.17.

5.4.1 Visual Bridge design system

Visual bridge design system (VBDS) is an AutoCAD-based finite element program (Wang and Fu 2005). VBDS was specially developed for the calculation of bridge structures including their construction processes. The basic idea is to define construction stages and the incremental actions of each stage that can be accumulated to obtain the final results.

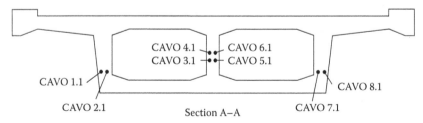

Figure 5.17 Location of the tendons in the cross section.

To build the model for VBDS, the following steps must be completed:

- Define entity geometries in AutoCAD.
- Create beam, truss, or plate elements, assigning them material properties.
- Assign section properties to the elements.
- Define construction stages and enter elements into each construction stage.
- Define boundary conditions for each stage.
- Create load cases and apply them to the construction stages.
- Define creep and shrinkage properties.

5.4.2 Verzasca 2 Bridge models

To demonstrate the analyses of Verzasca 2 Bridge at different levels of detail, five models are created:

- *Model 1*—Continuous girder with constant cross section (Figure 5.18)
- *Model 2*—Continuous girder with skew supports (Figure 5.19)
- *Model 3*—One girder built in a single stage (Figure 5.20)
- *Model 4*—Girder built with actual construction stages
- *Model 5*—Three girders skew supported (Figure 5.23, later in the chapter)

The models become increasingly more sophisticated, with Model 5 being the most complicated. This progression of complexity allows for not only a better possibility of finding errors while building the models but also a better interpretation of the results.

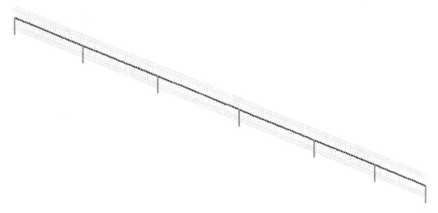

Figure 5.18 Elements of Verzasca 2 Bridge model 1.

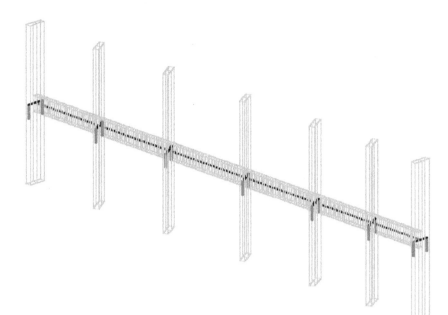

Figure 5.19 Elements of Verzasca 2 Bridge model 2.

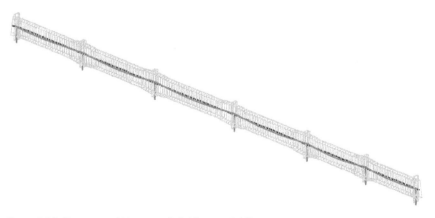

Figure 5.20 Elements of Verzasca 2 Bridge model 3.

5.4.2.1 Model 1: Continuous girder with constant cross section

The purpose of this very simple model is to have one that can be verified by hand. This model presents a good opportunity to check the results. The bridge is modeled with only one beam, which has a constant cross

Table 5.1 Spans of Verzasca 2

Span	Length (m)	Elements
1	33.57	30
2	36.26	30
3	39.69	30
4	36.51	30
5	29.40	25
6	25.24	25

section. Truss elements are placed as supports. This model is depicted in Figure 5.18. The magnitudes of moment of inertia of the beam elements are represented in gray. While all beam elements have the same properties in this example, the gray elements form a straight line.

The spans are described in Table 5.1. A uniform weight of 219.3 kN/m (15 kip/ft) is applied on each beam element. This represents the structural weight for a reinforced concrete cross section of 8.6 m² (92.6 ft²), with a density of 25.5 kN/m³ (162 lb/ft³) assumed.

5.4.2.2 Model 2: Continuous girder with skew supports

Skew supports have an influence mainly on the torsional moment of the superstructure. The skew supports would be taken into account only if they generate a change in the distribution of the vertical moments. To analyze the influences on the torsional moment, skew supports are added in Model 2, which is presented in Figure 5.19.

To model the skew supports, further elements have been created. They are aligned in the direction of the supports and have other section properties as the already-existent beam elements. To prevent deformations of these elements, their moment of inertia has been set 10 times higher than the moment of inertia of the beam elements. Notice that the beam elements need torsional stiffness to obtain the actual distribution of the internal forces.

5.4.2.3 Model 3: One girder built in a single stage

Model 3 is further developed from Model 1. There are two key differences. First, the cross section changes across the beam, with wider webs and bottom flanges in the zones of the piers. Second, the post-tensioning tendons are also included in the model.

The girder, which is modeled into 3D beam elements, is divided into three different cross sections:

- *Section 1*—In the middle of the spans
- *Section 2*—With thickened webs and a thickened bottom slab, used where negative vertical moments are expected, in the zones next to and over the piers
- *Section 3*—In the diaphragm areas

The vertical position of these elements is at the neutral axis of each cross section, which, together with the horizontal measures, defines the geometry of the beam.

To model the post-tensioning tendons, truss elements are created. Only one truss element represents all eight of the individual tendons that are distributed over the cross section, as seen in Figure 5.17. The geometry of the tendon is approximately at the middle of the actual positions. It is important that they end at the same vertical location as the beam elements, so that they can be connected with vertical rigid elements. Therefore, the line that represented the geometry of the tendon was cut with vertical lines placed at every two, three, or even four elements of the beam. Again, it was more important to have the connection at the suitable positions, rather than have intervals with the same number of elements between them. Suitable positions are (1) at the anchorage of the tendons, (2) at the middle span, (3) over the pier, and (4) where section changes occur. The tendon is divided between these points if the remaining length is longer than four elements or if a straight line between these points would fail to keep the geometry of the tendon.

Figure 5.21 shows the beam and tendon elements connected with rigid elements. The two horizontal lines are not used for this model but are used to assist in visualizing the upper and lower edges of the cross section. The shade scale changes of beam elements as shown in Figure 5.21 indicate the change of cross section. Different cross sections with different areas of steel are used for truss elements to simulate the changes of total strands in the longitudinal direction.

The boundary conditions stay the same as those for Model 1. In Figure 5.22 the right part of Model 3 is shown. Notice that the support

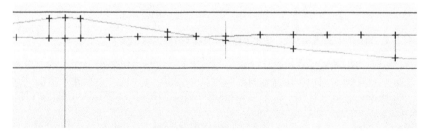

Figure 5.21 Detail of Verzasca 2 Bridge model 3.

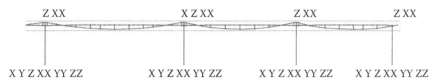

Figure 5.22 Boundary conditions of Verzasca 2 Bridge model 3.

between spans 4 and 5 is fixed in the longitudinal direction of the bridge. This point is marked with an additional X as X Z XX.

Besides the structural weight, Model 3 also takes into account the post-tensioning forces. Although this model considers the bridge built all at once, the definition of the loads already includes the incremental loads. Therefore, instead of defining the structural weight of the whole model, the weights are divided into six construction stages. The tendon forces are also defined as they are applied on the structure during construction. These forces will be discussed in Model 4 for the construction stages. For this model, it is assumed that all these loads are applied at the same time.

5.4.2.4 Model 4: Girder built with actual construction stages

Model 4 is exactly the same as Model 3, but with the added consideration of the construction stages. Span 5 is built in stage 1. These construction stages include part of the neighboring span, ending where the post-tension tendon is anchored (in the section of dead load point of contraflexure). Then, in stage 2, the rest of short span 6 is built. Spans 4 to 1 are built consecutively in stages 3 to 6.

Each stage also has a fixed sequence:

- Cast the concrete
- Stretch the tendon to 30% of the final stress after five days
- Stretch the tendon to 70% of the final stress after 14 days
- Remove the falsework and formwork
- Stretch the tendon to the final stress, as soon as the next stage's tendon is stressed to 70% of the final stress

The time sequence of the construction stages has been assumed according to the dates that the plan for each stage was checked. Table 5.2 shows the dates and the assumed construction time.

5.4.2.5 Model 5: Three girders skew supported

Model 5 (Figure 5.23) is the most complicated model of the Swiss bridge in this series. In regard to the creep and shrinkage effects, there should not be

Table 5.2 Assumption of the construction time

Building stage	Control date of plan	Construction time (weeks)
1	July 18	6
2	September 9	4
3	October 8	6
4	December 8	6
5	January 29	6
6	March 31	6
Total	34 weeks	34 weeks

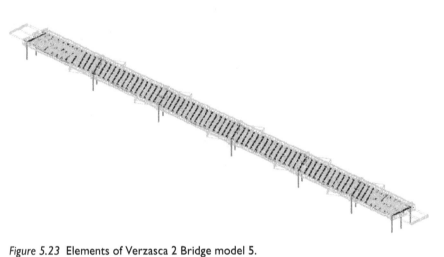

Figure 5.23 Elements of Verzasca 2 Bridge model 5.

essential differences between this model and Model 4. However, Model 5 will provide more accurate results due to its static analysis, especially with the effects of the skew supports.

The superstructure is modeled with three beams. Each represents one web of the double-cell box cross section. Taking Model 4 as a starting point, the geometry of the four middle spans can be copied and reproduced. As the abutments are placed perpendicularly to the longitudinal axis, beams in the first and sixth spans have to be extended or shortened, respectively. The geometry of the tendons is newly created according to the construction plans. The three beams are connected with transverse beam elements, which provide the transverse flexural rigidity of the section.

The cross section is divided into three by making two cuts in the middle of each cell. The new area and moment of inertia are calculated for the beam in the middle web and flanges. For both beams on the side, the remaining area and moment of inertia are divided by two. The torsional inertia for each beam is one-third of the total torsional inertia. This distribution is used for the three different cross sections along the bridge axis.

To simulate the transverse flexural rigidity, the section properties of the virtual connections are calculated according to Bakht and Jaeger (1985), where the following equations are given for cellular structures:

$$D_y = 0.5 * E_c * t * H^2 \tag{5.1}$$

$$D_{yx} = Gc * t * H^2 \tag{5.2}$$

where:
D_y is the transverse flexural rigidity
D_{yx} is the transverse torsional rigidity
t is the thickness of top and bottom flanges
H is the height between the centerline of both flanges

The acting forces are also distributed to the three beams. The structural weight is divided according to the axial areas. The tendon forces are easily distinguished, because they are located in the webs. The sequence of the construction stages and their loads are the same as those in Model 4.

5.4.3 Verzasca 2 Bridge analysis results

The vertical bending moments in the beam along the bridge axis are shown as results. All the results are given in kN-m. To simplify the discussions in this section, the spans are still counted from left to right, span 1 between abutment A and pier 1 and span 6 between pier 5 and abutment B.

5.4.3.1 Model 1: Continuous girder with constant cross section

The vertical moments of this simple model (Figure 5.24) serve as starting points for the discussion of the results of the next models. Model 1 is built in one single stage and has a uniform dead load of 219.3 kN/m acting on the entire structure. The moments are distributed according to the span lengths.

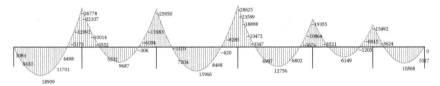

Figure 5.24 Moment distribution, Verzasca 2 Bridge model 1.

5.4.3.2 Model 2: Continuous girder with skew supports

Model 2 takes into account the skew supports. It is easy to recognize the better distribution of the negative moments by increasing the bending moment over piers 2 and 4, from 25,850 to 26,596 kN-m and from 19,355 to 22,739 kN-m, while decreasing over pier 3 from 28,625 to 27,913 kN-m. Because the abutments are placed perpendicularly to the bridge axis, the moments over piers 1 and 5 increase as well (Figure 5.25).

The torsional moments in the beam due to the skew supports are shown in Figure 5.26. While these moments are not essential in the subject of creep, they will not be taken into consideration in Models 3 and 4, but are taken into account in Model 5, as the superstructure is modeled three-dimensionally.

5.4.3.3 Model 3: One girder built in a single stage

Compared with Model 1, where the beam had a continuous cross section, the higher moment of inertia in the region of the piers causes higher negative moments (Figure 5.27).

Figure 5.28 shows the vertical moments in the beam caused by the post-tensioning procedure. In this case all tendons are also stressed at the same time. As explained in Section 5.4, the tendons overlap in the region of the piers. Thus the positive moments are much higher than the negatives, although the distance to the neutral axis is 50% larger at midspan than over the pier. The moments caused by time-dependent effects in Model 3 can be neglected as the entire bridge was cast in one single stage.

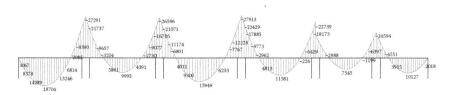

Figure 5.25 Moment distribution, Verzasca 2 Bridge model 2.

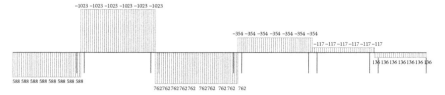

Figure 5.26 Torsional moment distribution, Verzasca 2 Bridge model 2.

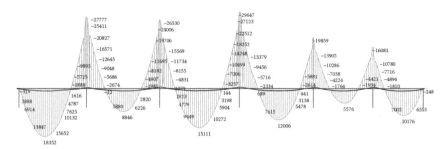

Figure 5.27 Vertical moments due to structural weight, Verzasca 2 Bridge model 3.

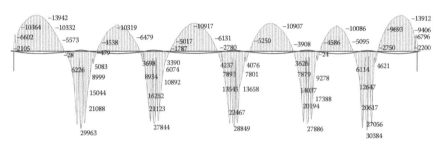

Figure 5.28 Vertical moments due to post-tensioning, Verzasca 2 Bridge model 3.

5.4.3.4 Model 4: Girder built with actual construction stages

As Model 4 takes into consideration the construction sequence and the age of the concrete in each new stage, the creep effect produces internal moments.

Figure 5.29 shows the elastic moment distribution along the beam. Compared with Model 3, all negative moments are reduced. While each span was built ending as simply supported, the negative moments over the piers are caused by only one span and the positive moments are higher. For example, the negative moments over pier 5 are about 1500 kN-m after

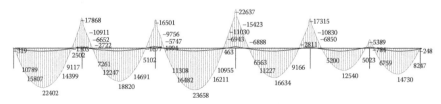

Figure 5.29 Accumulated moments due to structural weight and creep effect, Verzasca 2 Bridge model 4.

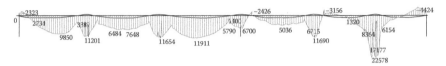

Figure 5.30 Accumulated moments due to structural weight and post-tension, Verzasca 2
Bridge model 4.

the first stage, and not 0, because the first construction stage ends 3.75 m
over the support. Once span 6 is built continuously to span 5 in the second
stage, the negative moment over pier 5 increases to around 7800 kN-m.
Due to the structural weight of span 4, the moment over pier 5 decreases to
3700 kN-m and increases again with the structural weight of span 3, and so
on. In Figure 5.30, the distribution shows the addition of all moments due
to structural weight and post-tensioning, each in its corresponding static
system. Note that the cracking moment of the beam is around 12,600 kN-m
for section 1 and 16,380 kN-m for section 2 next to the diaphragms.

5.4.3.5 Model 5: Three girders skew supported

The results of Model 5 are similar to the results of Model 4, but now the
moments are distributed to three beams, whereas they were all on the same
beam in Model 4. The moments in the middle beam are 33% higher than
those in the beams at the sides. This can be explained by the fact that the
moment of inertia in the middle beam is 33% higher.

The skew supports that are not taken into account in Model 4 also affect
the distribution of the moments in the different beams. This effect is recog-
nizable in both end spans. The front beam has larger negative moments over
pier 5, because it is nearer to abutment B. Exactly the same effect occurs
over pier 1, where the back beam receives more negative moments, due to a
shorter first span.

Creep and shrinkage not only cause a redistribution of the internal forces
but are also essential factors whenever displacements are evaluated. For the
purpose of comparison, incremental displacements of all 19 stages in the construction sequence are accumulated once for the elastic displacements and once
more for displacements due to creep and shrinkage, for AASHTO and for CEB-
FIP. Then displacements are divided into vertical and horizontal components.

From the vertical displacements shown in Figure 5.31, the construction
sequence can be reenacted. The peaks are located where the construction
stages changed. The sequence was from span 5 leftward to span 1.

The vertical displacements are mainly due to creep and the horizontal
due to shrinkage effects. The horizontal displacements due to shrinkage

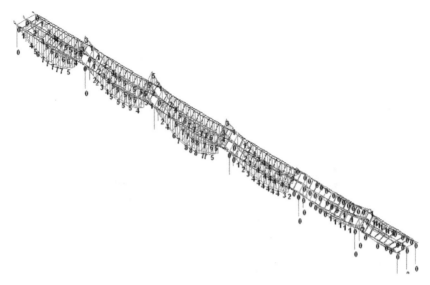

Figure 5.31 Vertical displacements due to structural weight and post-tensioning, Verzasca 2 Bridge model 5.

increase continuously from pier 4 in contrast to the displacements due to the post-tensioning. The displacements at abutment A reach 33 mm (1.54″) with AASHTO and 29 mm (1.14″) with CEB-FIP specifications and are proportional to the shrinkage coefficients at the time of five years.

The vertical displacements due to creep are more difficult to interpret because of the number of changes in the internal forces during construction. In general, CEB-FIP yields higher deformations due to creep than AASHTO.

5.5 3D ILLUSTRATED EXAMPLE OF US23043 PRECAST PRESTRESSED CONCRETE BEAM BRIDGE—MARYLAND

American practice places precast beams from pier to pier and then casts the diaphragms and the slab in the second step. The bridge US23043 was built in 2001 in the state of Maryland. It is located on Route 113 and was part of a multiphase project to create a bypass for the town of Showell. Figure 5.32 shows the perspective view of US23043 Bridge.

The 137.5-m (450′) long bridge consists of four spans, two of 38.12 m (125′) and two of 30.5 m (100′). The supports and the abutments are

Figure 5.32 US23043 Bridge, Maryland.

skewed with an angle of 30° to the bridge axis. The section consists of 11 precast and prestressed I-beams and a cast-in-place slab. The same VBDS program as in Section 5.4 is used in this analysis.

5.5.1 US23043 bridge models

Two almost identical models are created: Model 1 and Model 2. The only difference is that Model 1 has plate elements and Model 2 has beam elements to model the cast-in-place slab.

5.5.1.1 Model 1: Slab modeled with plate elements

Model 1 is a highly detailed model of the bridge US23043. As the number of elements is much higher than usual, accurate results are expected. This model has beam elements for the precast AASHTO beams and diaphragms, truss elements for the piers and the prestressing tendons, and plate elements to simulate the cast-in-place slab. Figure 5.33 shows the 3D model that contains beam, truss, and plate elements.

Although the precast AASHTO type V beams end at point "A" and are supported at point "D," the model supports the beams at point "B." For the construction periods in which the structure acts as simple spans, a joint at point "B" is added, to admit relative rotations between the beams of the two adjacent spans.

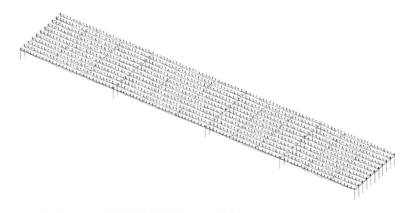

Figure 5.33 Elements of US23043 Bridge model 1.

The beam properties do not change along the bridge. The slab, which is cast in a later stage, will not change the section properties of the beam elements because the slab is modeled with additional elements.

The supports are modeled as truss elements and prevent the vertical displacements of the beam elements at these points. At the bottom end of the truss elements, all displacements and rotations are restricted. Figure 5.34 shows the restricted displacements with X, Y, or Z and the restricted rotations with XX, YY, or ZZ at the end of the elements. In all 55 supports of the four spans, 11 beams have the same boundary conditions in the first construction stage. Once the diaphragms are added, all rotations of the beam are admitted. Then the lateral displacements are restrained only at the abutments and the longitudinal displacements at one end of the bridge.

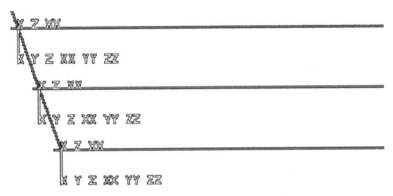

Figure 5.34 Part of US23043 Bridge model 1, boundary conditions.

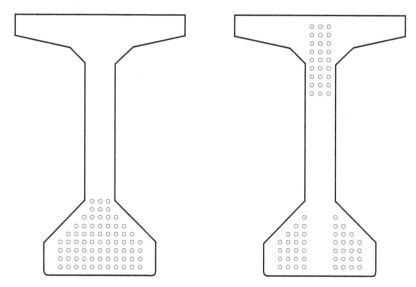

Figure 5.35 Section of the AASHTO beams with strands.

The precast and prestressed AASHTO type V beams contain different numbers of strands, depending on the span length. Figure 5.35 shows the cross sections with the reinforcement for spans 1 and 2. The I-beam section on the left is at midspan, and the one on the right is over the supports. In the first two spans, 71 12,7-mm (1/2″) diameter strands are placed; 27 of these are draped.

In the model, the prestressing truss is situated at the centroid of strands. And, while spans 3 and 4 are shorter than spans 1 and 2, the number of placed strands is smaller, which results in a different geometry of the trusses.

While spans 1 and 2 are prestressed with 71 strands and spans 3 and 4 with 43, the resulting forces applied on the truss elements are (1) $F_{1,2} = 9123.5$ kN and (2) $F_{3,4} = 5525.5$ kN.

The construction sequence of the bridge US23043 has six stages:

- *Stage 1*—Place precast I-beams from pier to pier.
- *Stage 2*—Cast midspan diaphragms, leaving a gap in the middle of the cross section.
- *Stage 3*—Pour the slab, except in the region of piers and in a gap in the middle of the cross section.
- *Stage 4*—Cast the gaps of the midspan diaphragms and the slab.
- *Stage 5*—Cast the pier diaphragms.
- *Stage 6*—Pour the slab in the regions of the piers.

Table 5.3 Construction sequence in Model I

Construction stage	Day
Stage Im	0
Stage 2m	I
Stage 3m	3
Stage 4m	8

In the model, these six stages are simplified into four. Stage 4 is included as a part of stages 2 and 3, and stages 5 and 6 are combined. To distinguish between actual and modeled sequences, the stages in the modeled sequence will be assigned both a number and the letter *m* (i.e., stage 1m, stage 2m).

The construction schedule in the model, which is shown in Table 5.3, considers the minimal possible construction time that has to be allowed between the stages. According to the instructions on the construction plans, 40 hours must be allowed between each stage in the actual sequence, except between stages 3 and 4, where only 16 hours is required. The age of the precast AASHTO beam is assumed as 60 days, which is important for the creep and shrinkage analysis.

The slab is built with plate elements to analyze the actual force distributions more accurately. Because the bridge is skewed, triangle plate elements are chosen. Between precast beams, two lines of plate elements are situated, which allow the transverse moment distribution in the slab to be obtained.

The nodes in the slab are at the same vertical location as the nodes in the beams, where elements with high rigidity connect them together. In total, 4200 plate elements are created in Model 1; 200 of them are from stage 4m.

5.5.1.2 Model 2: Slab modeled with beam elements

Model 2 is the same as Model 1, but the slab is represented by beam elements. These beam elements are located at the center of gravity of the slab, 1.1 m (3.6′) above the beam elements. Every second node of the slab elements is connected laterally with the next line of the slab elements. These connections are also simulated with beam elements and have the same torsional and bending rigidity as the plate elements in the longitudinal direction.

Creep and shrinkage properties will be assigned only to the elements in the longitudinal direction. This is the essential difference between Models 1 and 2.

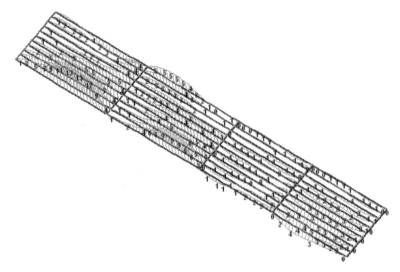

Figure 5.36 Displacements in the beam due to structural weight and prestressing, US23043 Bridge model I.

5.5.2 US23043 bridge analysis results

5.5.2.1 Model I: Slab modeled with beam elements

Model 1 will show the results of creep and shrinkage effects on bridge US23043. This model will provide the moments and longitudinal forces of the plate, beam, and prestressing tendon. The cast-in-place slab is stress free after hardening if no additional loads are applied, whereas the prestressed beam elements carry the structural weight.

The elastic displacements shown in Figure 5.36 reach 13 mm (0.5″) in one region. The displacements after five years are 148 mm (5.8″) considering creep and shrinkage with CEB-FIP and 64 mm (2.5″) with AASHTO specifications. Figure 5.37 shows the displacements due to creep and shrinkage and the total displacements after five years for both codes. Included in these results are the creep displacements that took place between the time the beams were precast and placed. These calculations, of course, depend upon how the beams were supported in this period of time. In this model, the structural weight is applied once the beam is placed at the site.

5.6 ILLUSTRATED EXAMPLE OF A THREE-SPAN PRESTRESSED BOX-GIRDER BRIDGE

A three-span single-cell haunched prestressed box girder bridge (Ketchum and Scordelis 1986) is taken as an illustrated example in this section. Figure 5.38 shows its elevation profile, and Figure 5.39 shows its cross-sectional geometry.

Figure 5.37 All displacements in the beam based on AASHTO, US23043 Bridge model I.

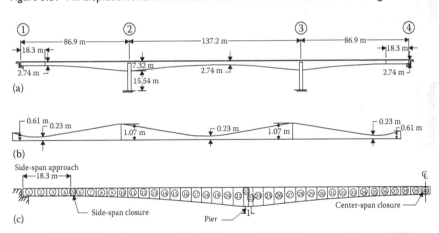

Figure 5.38 (a) Bridge elevation profile, (b) bottom slab thickness variation, and (c) segment division.

The haunched girder is cantilevered from the piers using cast-in-place segments and is later made continuous with short, conventionally erected, cast-in-place segments near the abutments and with the adjoining cantilevered girder at midspan. Each cantilever segment is post-tensioned to the previous segments with several cantilever tendons. After the closures at the abutments and at midspan, the entire bridge is prestressed with several additional continuity tendons, extending the full length of the bridge. In this example, there is no

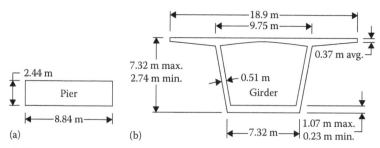

Figure 5.39 Cross section of the (a) pier and (b) main girder.

distinction between the 2D and 3D models, except that the 3D model can be used for other purposes such as wind-load and stability analyses, which will be discussed later.

The modulus of elasticity of the prestressing tendon is 1.9×10^5 MPa. The modulus of elasticity of the concrete girder is 2.86×10^4 MPa. The geometry properties for the girder cross section, pier cross section, and tendons are listed in Table 5.4.

The unit weight of the concrete of this bridge is 24.8 kN/m³. In total, 37.2 kN/m will be imposed along the deck after closure. For comparison purposes, the live loading is four lanes of AASHTO HS-20 without any multilane deduction. If the design is based on the AASHTO LRFD specifications (2013), HL-93 can be employed. The modified ACI 209 creep and shrinkage

Table 5.4 Segmental bridge section properties

Component	Moment of inertia (m⁴)/area (m²)	Component	Moment of inertia (m⁴)/area (m²)
S 22,23	173.9/20.7	S 21,24	155.1/20.2
S 20,25	131.9/19.5	S 19,26	111.6/18.8
S 18,27	94.0/18.0	S 17,28	78.8/17.3
S 16,29	65.7/16.5	S 15,30	54.6/15.9
S 14,31	45.1/15.1	S 13,32	37.1/14.4
S 12,33	30.4/13.7	S 11,34	24.7/13.0
S 10,35	20.0/12.4	S 9,36	16.1/11.7
S 8,37	13.5/11.3	S 7,38	12.2/11.2
S 6,39	11.5/11.1	S 1	15.3/13.3
S 2	12.1/11.5	S 3,4	11.2/11.1
S 5,40	11.2/11.1	T 1–16,25–28	0/0.008292
T 17–18	0/0.0166	T 19–24,29–53	0/0.004146
Pier (rigid zone)	180.0/20.9	Pier	10.7/21.6

S for girder segment; T for tendon.

model (1982) is adopted in the AASHTO LRFD specifications (2013). SFRAME (Ketchum and Scordelis 1986), developed in the University of California at Berkeley in the 1980s as a comparison with VBDS, however, adopts the original ACI 209 as its creep and shrinkage model. Although the creep models are different between two numerical models, when some parameters are taken as standards, the two creep models are very similar in nature. One great shortcoming of applying the LRFD creep model is that the maximum volume–surface area ratio used in the evaluation is limited to six, while some of the structures may require a ratio over six.

The construction sequence is modeled in 41 stages to simulate the erection and tendon prestressing of each section. It takes one week for each launching and prestressing. At day 100, the 18.3-m long girder at the side span starts to be cast, and the side spans and the center span close at day 168 and 182, respectively. All prestressing tendons are jacked at a unique stress of 1393 MPa (202 ksi), and the losses are taken as 15% of the jack stress. Unlike SFRAME, all losses are simply treated to be a constant along their path in this analysis by VBDS.

The time-dependent analysis for the 27 years following construction is performed by a smart step adjustment. The basic step is one week. It will be increased by one week whenever the differences of two adjacent analyses are less than a designated threshold or will be decreased by one week if they are above the threshold. Usually it varies between 1 and 12 weeks.

Table 5.5 shows the results and their comparison between VBDS and SFRAME. The differences between two numerical solutions are checked. Stresses of cases for maximum dual cantilever, ready to serve and 27 years later are shown. Figures 5.40 through 5.43 show some screens captured from VBDS; they show only the stress distribution on the top flange of the box girder at the maximum dual cantilever stage, after secondary dead load imposed, 27 years later, and on the stress envelop of HS-20, respectively. The jagged stress plots shown in Figures 5.40 through 5.42 are caused by the axial forces induced by the cantilever or local tendons. Jagged locations are where tendons terminate. The live load stresses show the smoothness across the whole girder. The live load analysis indicates that the live load stress along the girder may be incorrect if it is calculated by using simple girder principles based on its moment and axial force envelope. Unlike the dead load, which is already distributed over a statically determined structure before closure, the live load will cause significant axial force over the girder (–6300 kN/4000 kN at the center of the main span) because the bridge is fixed with two piers and the centroid of the girder shapes a flat arch. Therefore, the main span behaves like an arch bridge. In this case, it may not be sufficiently accurate to take the extreme moment and its correspondent axial force or the extreme axial force and its correspondent moment to calculate the stress over the girder in the main span. In VBDS, however, the

Table 5.5 Segmental bridge stresses (kN/m²) and comparisons at the center of the main span and over the pier

Stage and category	Position	VBDS	SFRAME
Maximum dual cantilever	TC	N/A	N/A
	BC	N/A	N/A
	TP	−2640	N/A
	BP	−11865	N/A
Ready to serve	TC	−4757	−4886
	BC	−8614	−10405
	TP	−3562	−3817
	BP	−12785	−13407
	MC	−14100	−17280[a]
	MP	−218900	−230500[a]
	DC	1.1 (in)	1.9 (in)[a]
27 years later	TC	−5430	−6025 (−5662[b])
	BC	−6158	−3810 (−8774[b])
	TP	−3070	−2658
	BP	−12700	−13619
	MC	−3000	1152[a]
	MP	−229300	−276500[a]
	DC	4.9 (in)	5 (in)[a]
Four lanes of HS-20	TC	−2215	−2334
	BC	2932	3248
	TP	2151	1554
	BP	−2742	−1603

[a] Measured from graphs.
[b] Recalculated based on the provided moments and section properties.
BC—bottom at the center of the main span; BP—bottom over the pier; DC— displacements at the center of the main span; MC—moment at the center span; MP—moment at the pier; TC—top flange at the center of the main span; TP—top over the pier.

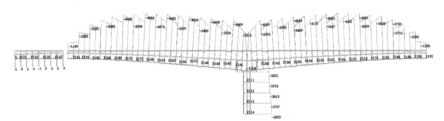

Figure 5.40 Stress (kN/m²) distributions on the top flange of the box girder at maximum dual cantilever stage.

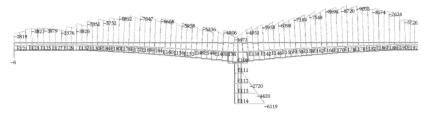

Figure 5.41 Stress (kN/m²) distributions on the top flange of the box girder after closure.

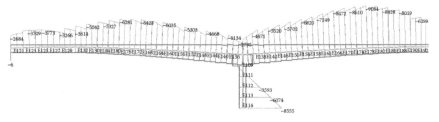

Figure 5.42 Stress (kN/m²) distributions on the top flange of the box girder after 27 years.

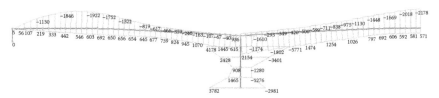

Figure 5.43 Stress (kN/m²) distributions on the top flange of the box girder due to four lanes of AASHTO live loads.

extreme stress is calculated by loading over the stress influence surfaces or influence lines, not over the simple axial force or bending moment.

5.7 ILLUSTRATED EXAMPLE OF LONG-SPAN CONCRETE CANTILEVER BRIDGES—JIANGSU, PEOPLE'S REPUBLIC OF CHINA

The long-span prestressed concrete continuous rigid-frame bridges are usually built with the balanced cantilever method. The layout of longitudinal tendons is determined according to the stress states in the cantilever stage and completion stage, and the tendons are correspondingly divided into cantilever tendons and continuity tendons (Pan et al. 2010).

- The conventional layout of longitudinal tendons is shown in Figure 5.44, including cantilever tendons in the top slabs, cantilever bent-down tendons in the webs, and continuity tendons in the bottom slabs and webs. The layout and number of tendons are mainly determined based on the envelope of the bending moment of the girder under all kinds of loads.

- From the end of 1980s, the elimination of the bent-down tendons in the webs, and instead the addition of vertical prestressing rods in the webs, was proposed, as shown in Figure 5.45. This straight layout method was well received by the construction industry because there were few ducts in the webs, which was more convenient to the construction of the box girder. However, a large number of vertical prestressing rods may have led to rising costs. Also, after more than 10 years, excessive deflections at midspan and inclined cracks in the webs appeared in many long-span concrete cantilever bridges with this design method. As a result, designers began to throw doubt on the elimination of the webs' bent-down tendons (as demonstrated in Figure 5.46).

- In the early 2000s, designers brought their attention back to the webs' bent-down tendons, and the common layout of longitudinal tendons is shown in Figure 5.46. The phenomena of the excessive deflections at midspan and the inclined cracks in the webs are seldom seen in the long-span concrete cantilever box girder bridges constructed either more than 20 years ago or more recently in the 2000s. It was therefore concluded that the elimination of the webs' bent-down tendons is one

Figure 5.44 Conventional layout of longitudinal tendons.

Figure 5.45 Straight layout of longitudinal tendons.

Figure 5.46 Current layout of longitudinal tendons with the webs' bent-down tendons.

of the main causes for inclined cracks, which are harmful to long-term deflections, and that these tendons are actually very effective in limiting the principal tensile stress.

5.7.1 The continuous rigid frame of Sutong Bridge approach spans

Deflection is mainly a result of two opposite actions. The first action is the dead loads and live loads, and the second one is the longitudinal tendons, which usually produce the counterdeflections to the dead loads and live loads. A lesson learned from the deflection problem in long-span cantilever bridges is that the deflection control is as important as conventional stress control in prestressing design. It is commonly known, that the cantilever tendons are very efficient for balancing the dead loads in the cantilever construction stage. Their effects on the deflections of the bridge would, however, be limited after the structural system transforms (like the closure of the main span). Here, the focal point is to design the tendons applied after the cantilevers are made continuous to avoid the excessive deflections.

The continuous rigid frame of Sutong Bridge approach spans (Pan et al. 2010) is a segmental, cast-in-place concrete cantilever bridge completed in 2007, and the span distribution is $140 + 268 + 140$ m ($460' + 880' + 460'$), among the longest spans in the world. Figure 5.47 shows the bridge in construction. The width of the top slab of the box girder is 16.4 m (53.8'), and the width of the bottom slab is 7.5 m (24.6'). The height of the box girder varies from 15 m (50') at the piers to 4.5 m (14.8') at midspan. The thickness of the bottom slab varies from 1700 mm (67") at the piers to 320 mm (12.6") at midspan. The web thickness varies in steps from 1000 mm (40") at the piers to 450 mm (18") at midspan. Figure 5.48 shows the arrangement of the box girder in the bridge, and Figure 5.49 shows the segments and layout of longitudinal tendons including the cantilever webs' bent-down tendons. The central span consists of 63 segments, whereas the two side spans consist of 33 segments each, and the entire span is constructed in balanced cantilevers. A similar approach as shown in Section 5.6 is used, except that MIDAS program (MIDAS 2007) is adopted for this analysis.

Figure 5.47 Construction of the continuous rigid frame of Sutong Bridge approach spans.

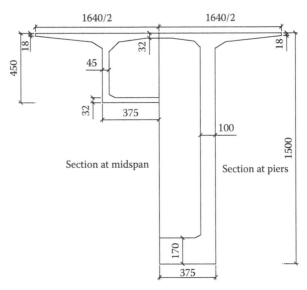

Figure 5.48 Typical section of the box girder (cm).

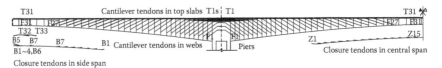

Figure 5.49 Segments and layout of tendons in the continuous rigid frame of Sutong Bridge approach spans.

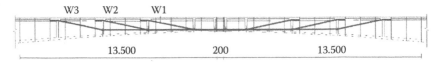

Figure 5.50 Layout of the preparatory external tendons in the continuous rigid frame of Sutong Bridge approach spans.

Two sets of additional tendons are designed to avoid excessive deflections due to uncertainties including material properties, concrete creep and shrinkage, and prestress losses. The first set is the internal tendons preset in the bottom slabs, and the other set is the external preparatory tendons. Both sets would be applied after closure or during service if necessary. As shown in Figure 5.49, there are a total of 15 internal tendons (Z1–Z15) in the bottom slab in the main span. Z1 through Z5, Z7 through Z9, Z11 through Z13, and Z15 are applied immediately after closure, and the rest are anticipated to be applied one year after the bridge is in service. As shown in Figure 5.50, there are three external tendons (W1–W3) in the middle of the main span, with each consisting of $25\Phi^j15.24$.

5.7.2 Results of webs' bent-down tendons

If the cantilever bent-down tendons in the webs as shown in Figure 5.49 are changed into the straight layout of tendons (shown in Figure 5.45), the shear force provided by the cantilever tendons can be calculated, and the comparison with the straight layout of tendons is shown in Figure 5.51.

Figure 5.51 shows that the webs' bent-down tendons can provide more shear force, which can balance the shear force induced by the dead loads. Therefore, the shear stress in the web will be reduced, and the principal tensile stress can be effectively limited.

5.7.3 Results of two approaches on deflections

Obviously, the preset additional internal tendons in the bottom slabs can effectively improve the stress deflection. Using the CEB-FIP78 creep and shrinkage prediction models, which are adopted in the previous bridge code (JTG D62-85 1985), the increments of deflections of the bridge after the completion were analyzed, and the increments with and without the preset-ting internal tendons in the bottom slabs are shown in Figures 5.52 and 5.53, respectively. Figure 5.52 shows that tensioning the presetting tendons would induce a camber of about 3 cm at midspan.

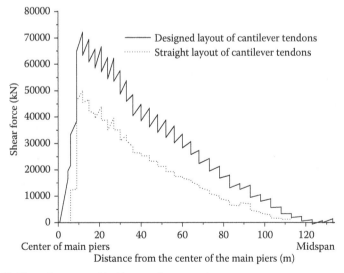

Figure 5.51 Shear force provided by cantilever tendons.

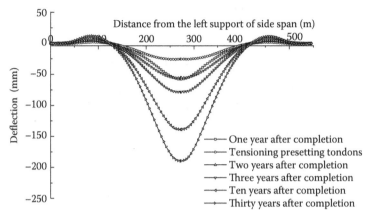

Figure 5.52 Increment of deflections of the bridge after completion with the presetting internal tendons.

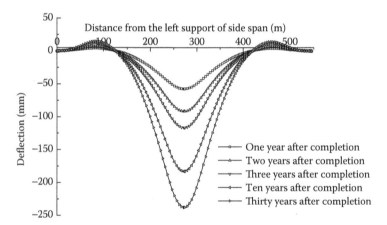

Figure 5.53 Increment of deflections of the bridge after completion without the presetting internal tendons.

With respect to the preparatory external tendons, according to the field observation, the external tendons were designed to be tensioned when the deflection or stress status was about to exceed the prediction value. After tensioning the three couples of external tendons, the upward deflection at midspan will be 14 mm. Also, it will provide an increment of 3.5 MPa normal compressive stress in the bottom slab and a 0.4 MPa normal compressive stress in the top slab.

Chapter 6

Curved concrete bridges

6.1 BASICS OF CURVED CONCRETE BRIDGES

6.1.1 Introduction

Due to urban development, more curved alignments, longer spans, more skewed supports, and more segmental construction for concrete bridges are expected. Construction methods can be cast in place with shoring or precast, curved, spliced "U" girders with a cast-in-place deck. Based on survey (Nutt and Valentine 2008) in the NCHRP report 620, except the western United States, most states are tending toward segmental construction (cantilever and span by span using both precast and cast-in-place concrete) to avoid conflict with traffic. A common application of curved structures is in freeway curved alignment or interchanges. Cross sections of curved box girders may consist of single-cell, multicell, or spread box beams, as shown in Figure 6.1. In the United States, only a very few spread box beams are used for curved concrete bridges. As for the requirement of a more refined analysis, many U.S. states use an 800-foot (244-m) radius as the trigger where designers should consider three-dimensional (3D) analysis, such as a grillage or finite element analysis (FEA) described in Chapter 5.

Sennah and Kennedy (2002) present highlights of references pertaining to straight and curved box girder bridges in the form of single-cell, multiple-spine, and multicell cross sections. The elastic analysis techniques discussed include the following:

1. Orthotropic plate theory method
2. Grillage analogy method
3. Folded plate method
4. Finite strip method
5. Finite element method (FEM)

The orthotropic plate method lumps the stiffness of the deck, webs, soffit, and diaphragms into an equivalent orthotropic plate. In the grillage

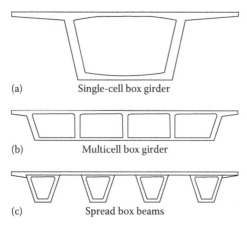

Figure 6.1 (a–c) Types of curved concrete bridge cross sections.

analogy method, the multicellular structure is idealized as a grillage of beams. Special attention should be paid to the modeling of shear lag and the torsional stiffness of closed cells. If properly done, the grillage model by this method yields results that compare well with finite element techniques. The folded plate method uses plates to represent the deck, webs, and soffit of box girders. Diaphragms are not modeled. The plates are connected along their longitudinal edges, and loads are applied as harmonic load functions. The finite strip method is essentially a special case of the FEM but requires considerably less computational effort because a limited number of finite strips connected along their length are used. Its drawback is that it is limited to simply support bridges with line supports and thus not applicable as a general analysis tool for production design.

With the advent of powerful personal computers and computer programs, the FEM has become the method of choice for complex structural problems. Many curved box girder bridges were analyzed by this technique. The versatility of this method has allowed users to investigate several aspects of bridge behavior, including dynamics, creep, shrinkage, and temperature changes.

6.1.2 Stresses of curved concrete box under torsion

Curved bridges behave quite different from straight bridges. The curvature results in off-center placement of loads and, subsequently, induces torsion into the superstructure. The torsion, in turn, causes the shear stresses to increase and plays an important role in a curved structure's behavior. Also, the curved geometry of the bridge will result in the development of transverse moments, which can increase the normal stresses on the outside edges of the bridge and can result in higher tension and/or compression stresses (Fu and Yang 1996). Post-tensioned bridges also have an additional equivalent transverse load, which can result in significant tension on the inside of the curve and compression on the outside edge (Fu and Tang 2001). The magnitudes of such effects depend on the radius of curvature, span configuration, cross-sectional geometry, and load patterns among other parameters. The global structural analysis is required to capture such effects.

In the early development by Hsu (1994), a set of equations is given for solving single-cell torsion. A reinforced concrete prismatic member is subjected to an external torque T as shown in Figure 6.2a. The external torque is resisted by an internal torque formed by the circulatory shear flow q along the periphery of the cross section. The shear flow q occupies a zone, called the shear flow zone, which has a thickness denoted t_d. This thickness t_d, or an equivalent thickness for a uniform shear stress, is a variable determined from the equilibrium and compatibility conditions. It is not the same as the given wall thickness h of a hollow member. Element A in the shear flow zone (Figure 6.2a) is subjected to a shear stress $\tau_{lt} = q/t_d$ as shown in Figure 6.2b.

In bridge engineering, many reinforced concrete bridges consist of multicell boxes. Therefore, a set of simultaneous equations to analyze structural torsion for multicell boxes is needed (Fu and Yang 1996; Fu and Tang 2001). In this chapter, equations for single- and multicell box are listed.

6.1.2.1 Equations for multiple cells

Assume a structural section has N cells (Figure 6.3). According to restraint condition $\theta = \theta_1 = \theta_2 = \dots = \theta_N$, a set of simultaneous equations for cell i can be obtained.

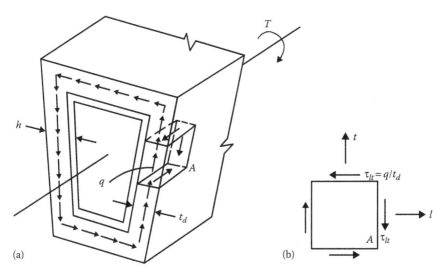

Figure 6.2 Hollow box subjected to torsion. (a) Shear flow in an element. (b) Shear stress on element A.

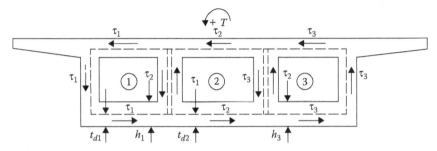

Figure 6.3 Shear stresses in a multicell section.

6.1.2.2 Equilibrium equations

A prestressed concrete element, as shown in Figure 6.4a, is reinforced orthogonally with longitudinal and transverse (prestressing or nonprestressing) steel reinforcements. The applied stresses on the element have three stress components, σ_l, σ_t, and τ_{lt}. The longitudinal steels are arranged in the l-direction (horizontal axis) with a uniform spacing of s_l. The transverse steels are arranged in the t-direction (vertical axis) with a uniform spacing of s as shown in Figure 6.4a. After cracking, the concrete is separated by diagonal cracks into a series of concrete struts, as shown in Figure 6.4b. The cracks are oriented at an angle α with respect to the l-axis. The principal stresses on the concrete strut itself are denoted as σ_d and σ_r. According to the unified theory (Hsu 1993), after transformation, the governing equations for equilibrium condition are shown as follows:

$$\sigma_l = \sigma_d \cos^2 \alpha + \sigma_r \sin^2 \alpha + \rho_l f_l + \rho_{lp} f_{lp} \tag{6.1}$$

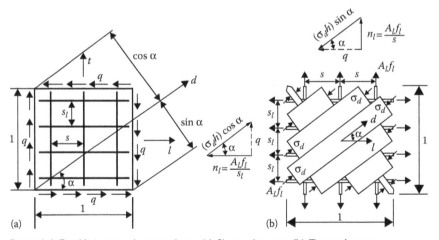

Figure 6.4 Equilibrium in element shear. (a) Shear element. (b) Truss element.

$$\sigma_t = \sigma_d \sin^2 \alpha + \sigma_r \cos^2 \alpha + \rho_t f_t + \rho_{tp} f_{tp} \tag{6.2}$$

$$\tau_{lt} = (-\sigma_d + \sigma_r) \sin \alpha \cos \alpha \tag{6.3}$$

$$T = \tau_{lt} (2 A_0 t_d) \tag{6.4}$$

where:

σ_l, σ_t, and τ_{lt} are the three homogenized stress components of the composite element (Figure 6.4a)

σ_d and σ_r are the concrete stresses in d- and r-directions, respectively (Figure 6.4b, where r-direction is perpendicular to d-direction and not shown)

α is the angle between l and d axes

f_l and f_t are the stresses in steel in the l- and t-directions, respectively

f_{lp} and f_{tp} are the stresses in the prestressing steel in the l- and t-directions, respectively

ρ_l and ρ_t are the steel ratio in the l- and t-directions, respectively

ρ_{lp} and ρ_{tp} are the prestressing steel ratio in the l- and t-directions, respectively

T is the external torque

A_0 is the cross-sectional area bounded by the centerline of the shear flow zone

t_d is the shear flow zone thickness

It should be noted that, for a multicell box under pure torsion, $\sigma_l = \sigma_t = \sigma_r = 0$ and, assuming a structural section has N cells (Figure 6.3), a set of simultaneous equations for cell i can be obtained.

$$\tau_{lti} = -\sigma_{di} \sin \alpha_i \cos \alpha_i \tag{6.5}$$

$$T_i = \tau_{lti} (2 A_{0i} t_{di}) \tag{6.6}$$

6.1.2.3 Compatibility equations

Similarly, the governing equations for compatibility condition were based on the unified theory (Hsu 1993) and later extended by Fu and Yang (1996) and Fu and Tang (2001). It should be noted that for a multicell box under pure torsion, a set of simultaneous equations for cell i was simplified as

$$\frac{\gamma_{lti}}{2} = (-\varepsilon_{di} + \varepsilon_{ri}) \sin \alpha_i \cos \alpha_i \tag{6.7}$$

$$\theta = \theta_i = \frac{p_{0i}}{2 A_{0i}} \gamma_{lti} \tag{6.8}$$

$$\psi_i = \theta \sin 2\alpha_i \tag{6.9}$$

6.1.2.4 Constitutive laws of materials

The softening concrete stress–strain curve proposed by Hsu (1993, 1994) is adopted here. General expressions for the constitutive laws of concrete and steel for a multicell box are as follows:

Concrete struts:

$$\sigma_{di} = k_{1i}\zeta_i f_{c'} \tag{6.10}$$

$$k_{1i} = \xi_1(\varepsilon_{dsi}, \zeta_i) \tag{6.11}$$

$$\zeta_i = \xi_2(\varepsilon_{di}, \varepsilon_{ri}) \tag{6.12}$$

Steel:

$$f_{li} = \xi_3(\varepsilon_{li}) \tag{6.13}$$

$$f_{ti} = \xi_4(\varepsilon_{ti}) \tag{6.14}$$

Combining governing equations for compatibility condition based on the unified theory (Hsu 1993) with selected constitutive equations, in this case, the softening concrete stress–strain curve, solution can be derived. For details of the solution of a single cell, refer to Hsu (1993, 1994), Fu and Yang (1996), and Fu and Tang (2001).

6.1.3 Construction geometry control

Curved bridges can be built segmentally or nonsegmentally where segmental bridges may adopt precast or cast-in-place construction of bridge members. The short-line match-cast joint method of precasting concrete segments has proved to be the most versatile and reliable way of building precast segmental bridges. The geometry control of segments casting in yard is a unique issue of precast segmental bridges, and its application is critical to reproduce the designed bridge curves after assembling. This long-standing topic is always a part of the design and construction of segmental bridges, especially for curved segmental bridges. More details about this topic will be discussed in Section 18.6.

6.2 PRINCIPLE AND MODELING OF CURVED CONCRETE BRIDGES

A variety of modeling approaches can be applied when analyzing horizontal curved bridges. Among these methods, plane frame analysis, spine beam analysis, and 3D FEM are the most popular methods that are used in practice. Plane frame analysis is acceptable for curved bridges that have a central angle less than 12°. For bridges that have a central angle greater than 12°, curve geometry should be considered in the analysis model, and 3D spine frame analysis is required when a curved bridge is modeled as a series of straight (or curved) frame elements in the centerline. Otherwise, specialized curved beam elements are preferred. On the other hand, the 3D FEAs are less vulnerable to applicability and modeling scope. Although this analysis is still an approximate method, a closer to actual bridge behavior can be generated by creating a more complex bridge model. With today's advanced technology with mesh-generating power, a bridge-designated FEA program can build a finite element bridge model to ensure the correctness of the model, and it is more frequently used in practice.

6.2.1 Modeling of curved concrete bridges

Curved concrete bridges, based on their level of required accuracy can be modeled into different types, from spine model to grid model to 3D finite element model.[*] Also, based on their emphases, bridges can be analyzed as decoupled super- or substructural model or a global bridge model. A global bridge, which includes the entire bridge with all frames and connecting structure, may be needed for certain circumstances, especially for earthquake analysis as discussed in Chapter 17. The three types of modeling are described briefly as follows:

1. Spine model. Spine models as shown in Figure 6.5a simplify the whole cross section, no matter single- or multicell boxes. The 3D frame element considers six degrees of freedom at both ends of the element and is modeled at their neutral axis. In this model, prestressing can be considered as equivalent loads with axial, vertical, and translational equivalent forces, or prestress tendons can even be included in the model as truss elements, as described in Chapter 5. Figure 6.6a demonstrates a single-box sitting on two bearings by connecting the neutral axis by rigid element. Different types of bearings, such as polytetrafluoroethylene (PTFE), stainless steel sliders, rocker bearings, or elastomeric bearings, may be used, and they should be modeled accordingly with directional restraints or springs. For bearing-supported connections, only up to three translational degrees of freedom are restrained, but the rotational degrees of freedom are free. Three 3D rigid truss elements can be used to simulate the three translational restraints. Figure 6.6b, adopted from Priestley et al. (1996), illustrates a typical

---

[*] 3D FEM model in this chapter refers to a finite element model in 3D that differs from spine and grillage models. Usually a 3D finite element model contains plane shell elements and other types of elements.

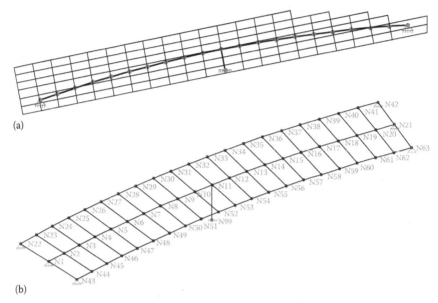

Figure 6.5 Modeling of curved concrete box bridges. (Data from Nutt, R. and Valentine, O., "NCHRP Report 620—Development of Design Specifications and Commentary for Horizontally Curved Concrete Box-girder Bridges," Transportation Research Board, Washington, DC, 2008.) (a) Typical spine beam model. (b) Typical grillage model.

monolithically single-column bent where the super- and substructures are tied together. This model uses frame elements, effective bending stiffness, cap with large torsional and transverse bending stiffness to capture superstructure (Caltran 2012). The calculation of bending and torsional stiffness can be found in Chapter 2.

2. Grillage model. However, spine model cannot capture the super-structure carrying wide-roadway, high-skewed bridges. In these cases grillage model as shown in Figure 6.5b is recommended (Caltran 2012). Grillage models are used regularly for modeling steel composite deck superstructures. For complicated concrete structures where superstructures cannot be considered stiff such as very long and narrow bridges and interchange connectors, grillage models can be used. This analysis approach requires the structure to be modeled as a 3D grid of frame elements in which the superstructure is comprised of both longitudinal and transverse beams located at the vertical center of gravity of the superstructure. Section properties are based on the box section with equivalent effective width as shown in Chapter 2.

For bridges with single- or multicell box (or spread multiple boxes) as shown in Figure 6.1, properties can be calculated as shown in

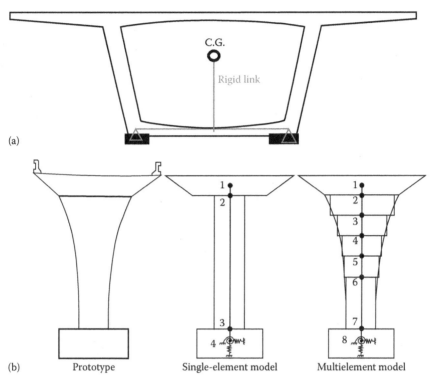

(a)

(b)   Prototype   Single-element model   Multielement model

Figure 6.6 Super- and substructure connection. (a) Bearing-supported connection. (b) Mono-lithic connection. (Data from CalTran, "Structural Modeling and Analysis," LRFD Bridge Design Practice, August 2012, http://www.dot.ca.gov/hq/esc/techpubs/manual/bridgemanuals/bridge-design-practice/pdf/bdp_4.pdf.)

Figure 6.7 provided by VBDS (Wang and Fu 2005). For multicell box bridge, either a spine beam model with multicell properties is used or a grillage model with each beam line associated with its respective web is adopted. The section properties for longitudinal frame elements are modeled as shown in Figure 6.8 (Nutt and Valentine 2008). A_x is considered as the tributary cross-sectional area of longitudinal segment as shown in the figure. A_y for vertical shear counts on the area of web only, and A_z for transverse shear is considering the area of tributary deck and soffit slabs in the same figure. I_{zz} and I_{yy} represent the tributary moments of inertia with respect to horizontal and vertical axes, respectively. J is estimated by using the total torsional moment of inertia divided by the number of webs to assume equally divided. For multicell boxes, transverse section properties can be assumed as combined transverse deck and bottom slab properties with respect to the box neutral axis.

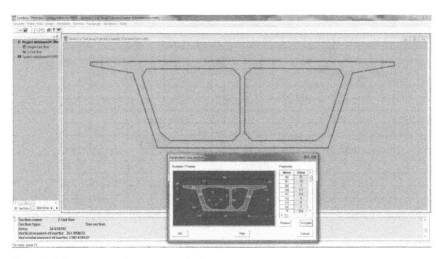

Figure 6.7 Box sectional property calculation.

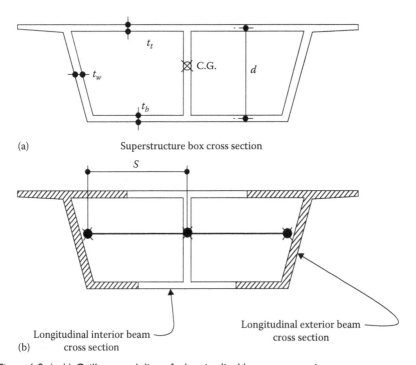

(a)   Superstructure box cross section

(b)   Longitudinal interior beam
            cross section

Figure 6.8 (a, b) Grillage modeling of a longitudinal box cross section.

3. 3D finite element model. 3D FEM is considered the most sophisticated method among the three models. Earlier, it was most used in refined local analysis as shown in the NCHRP Report (Nutt and Valentine 2008). However, with today's advanced technology and meshing capability, it becomes a powerful tool in detailed local stress investigation to make sure all is within the allowables. An illustrated example is provided for the demonstration of this type of modeling.

6.2.2 Modeling of material properties

The mathematical properties of structural components are usually assumed according to the codes issued by the responsible authority. These properties for static loading, including stress–strain relationship, concrete-cracking effect, yield, and ultimate strength of steel and concrete, were discussed in Chapter 4 for RC bridges.

For nonstatic loading, which could affect bridge member stiffness, the nonlinear properties of concrete are required in modeling, and they will be covered in Chapter 17—Dynamic/Earthquake Analysis.

6.2.3 Modeling of live loads

When all of the girders in a span are parallel and the span is contained entirely within the limits of a vertical and/or horizontal curve, the profile effect is simply the sum of the vertical curve effect and the horizontal curve effect.

$$\Delta_{total} = \Delta_{vertical\ effect} + \Delta_{horizontal\ effect} \tag{6.15}$$

When analyzing concrete curved bridges using FEA, it is crucial to model the live load value and position along the longitudinal direction of the bridge to yield proper live load response. Also vehicular effects, especially centrifugal forces, should be considered. The horizontally curved bridge is applied by a lateral load due to the centrifugal force from traffic. According to the AASHTO LRFD bridge design specification, the centrifugal force is defined as the product of design truck weight and a C factor, which is defined as

$$C = f\frac{v^2}{Rg} \tag{6.16}$$

where:
  g is the gravitational acceleration
  R is the bridge curvature radius
  v is the design lane speed
  f is equal to 4/3 (to 1.5) for all limit state other than fatigue

In the case of a multilane bridge, a multilane presence factor should be included.

6.2.4 Modeling of lateral restraint and movement

Bearings in a horizontal curved bridge may be restraint in lateral to prevent movement due to the centrifugal forces from traffic load, thermal movement, and prestress shortening. The bearing should be so modeled to reflect its actual movement and restraints in all directions.

6.3 SPINE MODEL ILLUSTRATED EXAMPLES OF PENGPO INTERCHANGE, HENAN, PEOPLE'S REPUBLIC OF CHINA

This illustrated example is a ramp bridge located in Pengpo Interchange, one of the major transportation hubs in Henan, China. It has a total length of 343.465 m, with a radius of 130 m for the first 150 m, a left transition curve for the next 50 m, and a radius of 400 m for the rest of the ramp bridge. The bridge is shown in Figure 6.9. The bridge is designed as cast-in-place prestressed concrete continuous box girder bridge, with a bridge roadway width of 7.5 m + 2 × 0.5 m. The span layout of this bridge contains two 6 continuous spans with the first one 6 × 30 m and the second one 6 × 25.88 m (Figure 6.9). Typical cross section of the bridge is shown in Figure 6.10. The bridge uses elastomeric bearing pads with various sizes ranging from 500×87 mm^2 to 900×115 mm^2.

The purpose of this analysis is to find the reason of damage to bearings and substructure pier columns due to lateral movement. Pengpo Bridge support arrangement plan is shown in Figure 6.11a, and its movement sketch is shown in Figure 6.11b. During the inspection, it was found that though the bridge superstructure's performance meets the original design requirement, large shear deformation and transversal displacement occurred throughout the bearings in the ramp bridge, ranging from 10 mm at bearing #0 to 90 mm at bearing #6.

To understand the cause of the damage, a 3D model of the first 150-m ramp bridge was generated by CSIBridge, as shown in Figure 6.12. The elastomeric bearing pads were modeled by introducing the stiffness values in vertical, horizontal, and lateral directions provided by bearing pads. A lateral load of moving vehicle centrifugal force was introduced following the AASHTO instructions, which is triggered by 20 tons of vehicles traveling at about 60 km/hour. The deflection is shown in Figure 6.13. The CSIBridge model results show that displacements among the supports are different from 7.25 mm at bearing #0 to 56.3 mm at bearing #6, which is reasonably close to the field-inspection results. It is detected from the results that the centrifugal force from traffic may be the main reason that this example bridge got damaged.

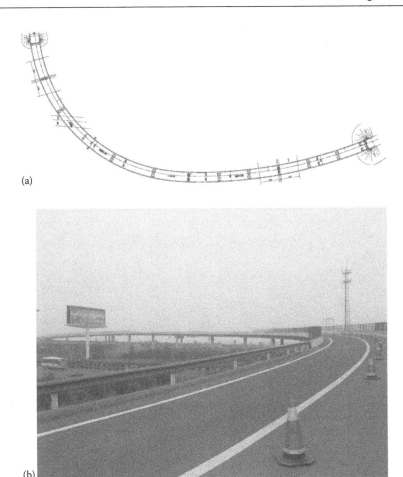

(a)

(b)

Figure 6.9 (a) Sketch. (b) View of Pengpo Bridge, Henan, China.

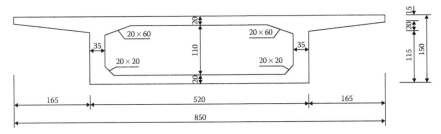

Figure 6.10 Typical cross section (mm) of Pengpo Bridge, Henan, China.

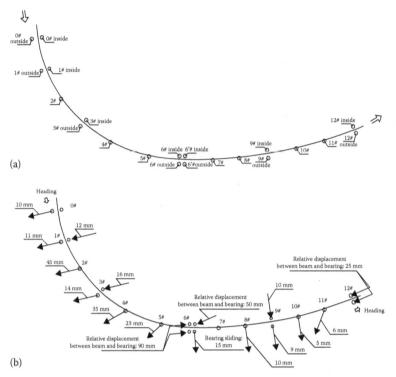

(a)

(b)

Figure 6.11 Pengpo Bridge: (a) support arrangement plan and (b) its movement sketch.

Figure 6.12 CSIBridge FEA model.

Figure 6.13 CSIBridge model after centrifugal load has been applied.

6.4 GRILLAGE MODEL ILLUSTRATED EXAMPLES—FHWA BRIDGE NO. 4

This skewed bridge example as shown in Figure 6.14a is used to illustrate the modeling technique. This monolithic concrete bridge is one of the FHWA examples series (Mast et al. 1996) and is also used as an illustrated example in Chapter 17. It consists of three spans. The total length is 97.5 m (320′), with span lengths of 30.5, 36.6, and 30.5 m (100′, 120′, and 100′), respectively. In the longitudinal direction, the intermediate bent columns are assumed to resist the entire longitudinal force, whereas the seat-type abutments provide vertical but no longitudinal restraint. As shown in Figure 6.14a, all substructure elements are oriented at a 30° skew from a line perpendicular to a straight bridge centerline alignment. The superstructure is a cast-in-place concrete box girder with two interior webs. The intermediate bents have a cross-beam integral with the box girder and two round columns that are pinned at the top of spread footing foundations. Because this model was used for

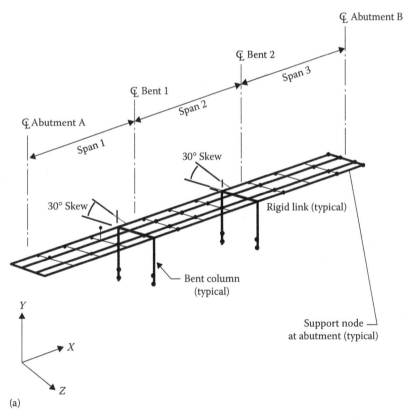

(a)

Figure 6.14 Details of super- and substructural elements. (a) Grid model. (Continued)

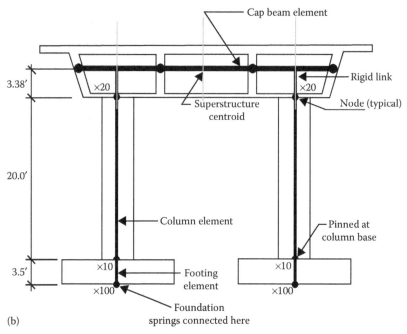

Figure 6.14 (Continued) Details of super- and substructural elements. (b) Details of bent elements. (Data from FHWA 1996.)

earthquake analysis, the intermediate bent foundations were modeled with equivalent spring stiffness for the spread footing to capture the soil effect. In this grillage model, section properties, A_x, A_y, A_z, I_{xx}, I_{yy}, and J, are calculated as described in the early section and later in Chapter 17 for verification purpose. The superstructure has been modeled with four elements per span, and the element axes are located along the centroid of the superstructure. The bents are modeled with 3D frame elements that represent the cap beams and individual columns. As columns are pinned to the column bases, two elements were used to model each column between the top of footing and the soffit of the box girder superstructure. A rigid link was used to model the connection in between. The final model is shown in Figure 6.14a. Note that unlike what is demonstrated in Chapter 17, no plastic hinge is modeled here.

6.5 3D FINITE ELEMENT MODEL ILLUSTRATED EXAMPLES—NCHRP CASE STUDY BRIDGE

To demonstrate the 3D finite element model of curved concrete bridge, the bridge example B-1 in the NCHRP study by Nutt and Valentine (2008) is adopted. This fictitious bridge is a cast-in-site curve bridge with a curve

R of 122 m (400′) and spans of 61 m + 91 m + 61 m (200′ + 300′ + 200′). The box section has two cells as shown in Figure 6.15. In this model, box girder is modeled instead by 3D plane shell elements, bents are modeled by 3D frame elements, and bearings are modeled by rigid truss elements. As shown in Figures 6.16 and 6.17, flanges and outer webs are meshed into three-node triangular plane shell elements so as to incorporate curvature, whereas the middle web is meshed into four-node rectangle plane shell elements. In the longitudinal direction, the box girder is meshed in every 1.2 m (4′) and around 0.6′ (2′) in the transverse direction. The entire bridge is modeled into 14,700 plane shell elements as well as 18 truss and frame elements as support.

In addition to structural weight as dead load, as described in the NCHRP Report (2008), a concentrated load of 100 kip (445 kN) is applied in the middle of the midspan with three different locations, on the top of the outer web, middle web, and inner web. Figure 6.18 shows the major principal stress distribution on the top of flange due to structural weight. Figure 6.19 shows the longitudinal stress distribution in the transverse direction at the pier due to structural weight. Figure 6.20 shows the longitudinal stress

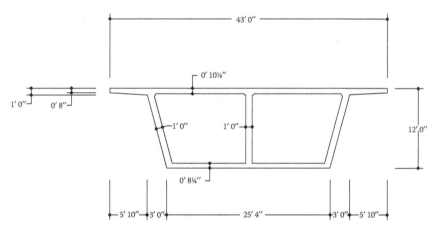

Figure 6.15 Typical cross section of a box girder. (Example B-1, Data from Nutt, R. and Valentine, O., "NCHRP Report 620—Development of Design Specifications and Commentary for Horizontally Curved Concrete Box-Girder Bridges," Transportation Research Board, Washington, DC, 2008.)

Figure 6.16 3D view of finite element model.

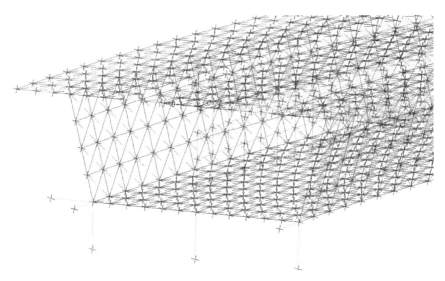

Figure 6.17 Meshes in 3D finite element model.

216   −1

Figure 6.18 Major principal stress on top flange due to structural weight (kip/sf).

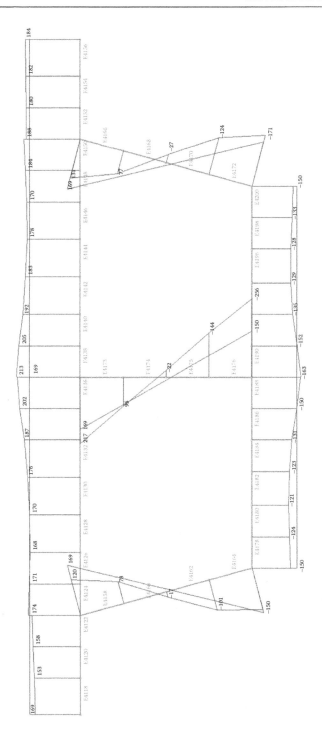

$F_x = 55.4799; F_z = -47.3194; F_z = -224.561$
$M_x = -3395.21; M_y = 173.336; M_z = 81054.3$

Figure 6.19 Longitudinal stress distribution at pier due to structural weight (kip/sf, straight lines are stresses by the beam theory) (87490).

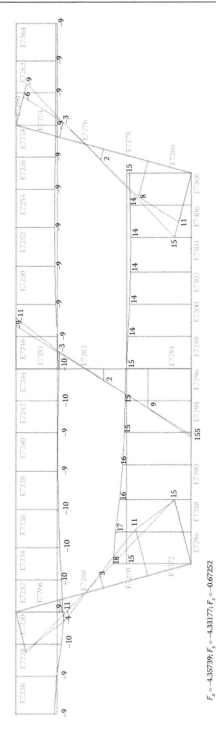

$F_x = -4.35739; F_y = -4.33177; F_z = -0.67252$
$M_x = 44.1402; M_y = -27.7553; M_z = -4217.52$

Figure 6.20 Longitudinal stress distribution in the middle of middle span due to 100-kip concentrated load on the top of outer web (kip/sf, straight lines are stresses by the beam theory) (4875).

Figure 6.21 3D view of the grillage model.

Figure 6.22 Girder moment distribution due to structural weight (kip-ft).

Figure 6.23 Girder moment distribution due to 100-kip concentrated load on the top of outer web (kip-ft).

distribution in the transverse direction in the middle of midspan due to a concentrated load on the top of outer web. Straight lines in both Figures 6.19 and 6.20 are obtained by the beam theory from moment calculated by stress integration over the entire cross section.

Alternatively, the same bridge is modeled using grillage model as shown in Figure 6.21. The box girder is modeled by three beams at web locations with a longitudinal mesh of 1.2 m (4′). The transverse beams are also meshed at a space of 1.2 m (4′) in the longitudinal direction. The total number of 3D frame elements for box girder is 2275, and the total number of elements in the model is 2309. Figures 6.22 and 6.23 show the moment distribution on the girder due to structural weight and 100-kip (445 kN) concentrated load on the top of outer web, respectively. In comparison with the finite element model, the total moments on three beams at pier location due to structural weight and in the middle span due to 100-kip (445 kN) concentrated load on the top of outer web are −87,490 kip-ft (−118,615 kN-m) and 4,875 kip-ft (6,609 kN-m), respectively, whereas the integrated moments from plane shell elements on these locations are −81,054 kip-ft (−109,889 kN-m) and 4,217 kip-ft (5,717 kN-m) accordingly, which are very close in this example.

Chapter 7

Straight and curved steel I-girder bridges

7.1 BEHAVIOR OF STEEL I-GIRDER BRIDGES

7.1.1 Composite bridge sections under different load levels

A composite steel I-girder bridge can be considered as a series of I-girders with their concrete deck acting compositely with the steel girders (Figure 7.1). Figure 7.1c shows three different noncomposite or composite steel sections and their respective stress diagrams. For steel girder bridge analysis, the respective section properties are used at different load stages. For steel multigirder bridges with cast-in-place concrete decks, there are four general loading stages in the construction sequence:

- *Stage 1*—Erection of structural steel framing (girders and cross frames)
- *Stage 2*—Placement of the structural deck slab (wet concrete)
- *Stage 3*—Placement of appurtenances (e.g., barriers, railings, overlays) representing the long-term (LT) loading
- *Stage 4*—Bridge in-service condition (e.g., carrying live loads; vehicular, rail, pedestrian) representing the short-term (ST) loading

The normal stress distribution $\sigma(x)$ in the concrete slab of a composite beam does not have a constant value but varies where the maximum flexural normal stress occurs at the junction point of the slab and steel girder web, as illustrated in Figure 7.2. This phenomenon is caused by the lag of shear strain at the top of the concrete slab and is referred to as shear lag effect. Effective width of a cross section at a given location depends on the structural layout and loads. For design purposes, it is convenient to define the effective width for the concrete slab. The effective (b_e) and transformed widths (b_{tr}) are illustrated in Figure 7.2. Results of a recent study as shown in NCHRP Report (Chen et al. 2005),

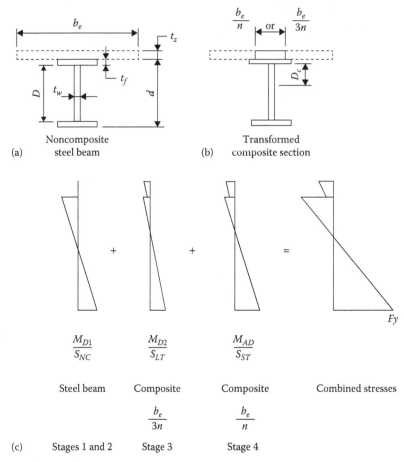

Figure 7.1 (a–c) Steel section properties and their respective stress diagrams.

which was later adopted by AASHTO load and resistance factor design (LRFD) specifications (2013), recommend that the full slab width half-way between adjacent girders can be counted as the effective width for the concrete slab. Calculation of the effective widths and their respective section properties will be shown in the example of Section 7.3. For each of the loading stages described earlier in this section, a distinct set of section properties exists and must be used in their respective analyses to properly ascertain design forces and deflections to evaluate strength and serviceability criteria.

Steel I-girder can be made of rolled beams or welded plate girders. For typical bridges, fabricators usually prefer rolled beams. Generally, rolled

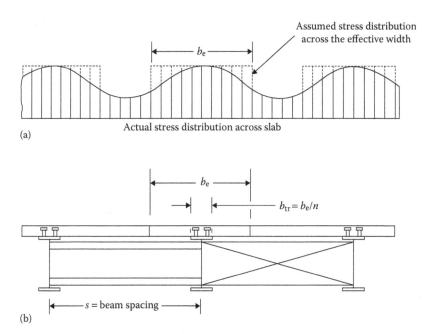

Figure 7.2 Definition of effective width and transformed width of a steel composite section. (a) Actual stress distribution and (b) effective width and transformed width.

beams are more economical than welded girders. Other considerations are delivery or specific requirements, such as camber and curvature. Typical diaphragms or cross frames as shown in Figure 7.2 are designed for

- Lateral loads transferring.
- Stability of the bottom flange for all loads when it is in compression.
- Stability of the top flange in compression prior to curing of the deck.
- Live loads distribution.

Diaphragms or cross frames can be specified as either of the following:

- *Permanent*—if they are required in the bridge's final condition
- *Temporary*—if they are required only during construction

The difference between diaphragms and cross frames is that diaphragms consist of a transverse flexural component, whereas cross frames consist of a transverse truss framework; both carry vertical shear and moment from one beam to the others. For straight bridges, the general recommendation is to place cross frames either parallel to skewed supports or normal to the girders if the deflection between girders is constant at cross-frame connections and the skew angle is equal to or less than 20°. Otherwise place cross frames normal to the girders.

The behavior of the steel girder bridges may be grouped as either (1) straight and nonskewed or (2) curved and/or skewed bridge. According to G13.1 by the AASHTO/NSBA Steel Bridge Collaboration (2011), the behavior of curved and skewed steel girder bridges can be broadly divided into two categories:

Basics—Curved or skewed steel girder bridges, or both, experience the same effects of gravity loading (dead load and live load) as straight girder bridges.

Curvature and skew effects—Torsional and warping stresses, flange lateral bending, load shifting and warping, and twisting deformations.

In Section 7.1.2, different effects will be characterized as effects of curvature.

7.1.2 Various stress effects

Early steel bridges are primarily straight and simple-span bridges and can be analyzed by hand. The advent of computers can easily handle indeterminate structures, such as continuous span bridges, but are still mainly straight bridges subjected to major-axis shear and bending moment effects of the main girders. A curved girder and/or skewed girder bridge, in addition to the basic vertical shear and bending effects, will be subjected to torsional effects (Nakai and Yoo 1988). Torsion in steel girders causes both normal stresses and shear stresses. Because I-shaped girders are in opened sections and thus have low St. Venant torsional stiffness, they carry torsion primarily by means of warping. The total normal stress in an I-shaped girder is a combination of any axial stress, major-axis bending stress, lateral bending stress, and warping normal stress (Figure 7.3). The total shear stress is the sum of vertical shear stress, horizontal shear stress, St. Venant torsional shear stress (generally relatively small), and warping shear stress (Figure 7.4). For nonskewed straight steel bridge analysis, only the major-axis bending stress (second term on Figure 7.3) and the vertical shear stress (first term on Figure 7.4) are dominant, and the rest of the terms can be ignored in the design phase, but have to be included in other load combinations for code checking.

The relatively low St. Venant torsional stiffness of I-shaped girders is a result of their open cross-sectional geometry. The St. Venant torsional shear flow around the perimeter of the cross section can develop only relatively small force couples. Without significant force couples, compared to the close section (described in Chapter 8 for steel box girder bridges), the ability of I-shaped girders to carry torque through St. Venant torsional response is low.

I-girders carry torsion through the combination of pure torsion and restrained warping. Diaphragms and/or cross frames provide lateral

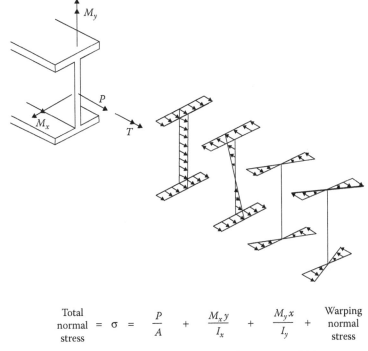

$$\begin{array}{ccccccccc}
\text{Total} & & & & & & & & \text{Warping} \\
\text{normal} & = & \sigma & = & \dfrac{P}{A} & + & \dfrac{M_x y}{I_x} & + & \dfrac{M_y x}{I_y} & + & \text{normal} \\
\text{stress} & & & & & & & & \text{stress}
\end{array}$$

Figure 7.3 Illustration of the general I-girder normal stresses, which can occur in a curved or skewed I-shaped girder.

restraint to girders. The total torsional resistance is the sum of nonrestrained torsion (pure torsion) and restrained torsion (warping), as expressed in Equation 7.1.

$$T = GJ\theta' - EC_w\theta''' \tag{7.1}$$

where:

G is the shear modulus of elasticity of steel
J is the torsional constant of cross section and can be approximated using Equation 7.2 for rolled and built-up I shapes
E is the modulus of elasticity of steel
C_w is the warping constant of cross section and can be approximated as $I_y h^2/4$ for rolled and built-up I shapes
I_y is the lateral moment of inertia about y-axis
h is the distance between centerlines of top and bottom flanges
θ is the rotation angle of cross section along the girder axis

For the calculation of section properties, including C_w, refer to AISC, *Design Guide 9: Torsional Analysis of Structural Steel Members* (2003).

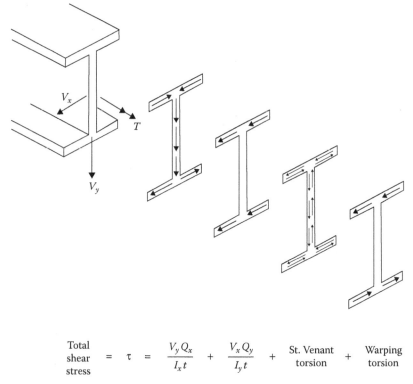

$$\begin{array}{c}\text{Total} \\ \text{shear} \\ \text{stress}\end{array} = \tau = \frac{V_y Q_x}{I_x t} + \frac{V_x Q_y}{I_y t} + \begin{array}{c}\text{St. Venant} \\ \text{torsion}\end{array} + \begin{array}{c}\text{Warping} \\ \text{torsion}\end{array}$$

Figure 7.4 Illustration of the general I-girder shear stresses, which can occur in a curved or skewed I-shaped girder.

The normal warping stresses for I-girders caused by torsion represent one source of what are called flange lateral bending stresses. These are an important part of the design equations for flange stresses in I-girders. In that case, all terms on Figures 7.3 and 7.4 have to be counted for.

7.1.3 Section property in the grid modeling considerations

During the development of any structural analysis, section properties are assigned to the members. The distribution of forces through the system is highly dependent on member stiffness parameters such as EI_x, EI_y, GJ, and EC_w. EC_w, the warping stiffness parameter, is not used in a generic structural analysis method based on the beam theory with six degrees of freedom (DOFs) per node. For special analyses, cross-sectional warping deflection, the seventh DOF can be included to consider the warping of thin-wall cross sections. Thus, the additional warping stiffness is required. For curved structure, EC_w is often the dominant contributor to the individual girder torsional stiffness. Without consideration of EC_w, the local twisting responses of the girders cannot be modeled accurately. On the other hand, a full three-dimensional (3D) finite element analysis (FEA), in which the thin-wall sections are modeled by plane shell elements, bypasses the need for the modeling of warping stiffness within the single beam element used to model the girder in two-dimensional (2D) grid analysis approaches.

A rigorous solution of grid analysis to take care of the warping problem of a thin-wall beam requires the warping deflection as an additional DOF. Several researchers (e.g., Hsu and Fu 1990; Fu and Hsu 1995) have included the warping deflection as the seventh DOF, in addition to the regular six DOFs, at each node for the curved beam analysis to consider the warping effect. For the case of partial warping restrained, an effective torsional constant, K_{eff}, was proposed by Fu and Hsu (1994) and later improved by Elhelbawey and Fu (1998) to consider warping effects in a regular six DOFs analysis. A simple, easy-to-apply effective torsional constant for the rotational stiffness of a restrained open section was developed to take both the pure torsion and the warping torsion into account. This effective (equivalent) torsional constant, K_{te}, can be easily calculated and used for any generic finite element structural analysis program.

The original torsional constant for most common structural shapes, J, can be approximated by Equation 7.2.

$$J = \sum bt^3/3 \tag{7.2}$$

where b and t are the width and thickness of the thin-wall elements, respectively. The effective (equivalent) torsional constant, K_{te}, developed by Fu and Hsu (1994), can be expressed as

$$K_{te} = J \cosh\frac{\lambda}{2} / \left\{ \cosh\frac{\lambda}{2} - 1.0 \right\} C \tag{7.3}$$

where:
λ^2 is the GJ/EC_w, ($\lambda = l/a$, where a is used in AISC documents)
C is the correction factor that equals $\{1.0/[1.0 + 2.95\ (b/l)^2]\}$
l is the unbraced length
b is the flange width

Once the effective torsional constant is determined, the stiffness matrix for a grid structure can be derived by using the traditional straight beam method with three DOFs (torsional rotation, bending rotation, and deflection) per node. The stiffness matrix of an element in a grid model with warping partially restrained is as follows:

$$[K_e] = \begin{bmatrix} \dfrac{GK_{te}}{l} & 0 & 0 & \dfrac{-GK_{te}}{l} & 0 & 0 \\ 0 & \dfrac{4EI_y}{l} & \dfrac{-6EI_y}{l^2} & 0 & \dfrac{2EI_y}{l} & \dfrac{6EI_y}{l^2} \\ 0 & \dfrac{-6EI_y}{l^2} & \dfrac{12EI_y}{l^3} & 0 & \dfrac{-6EI_y}{l^2} & \dfrac{-12EI_y}{l^3} \\ \dfrac{-GK_{te}}{l} & 0 & 0 & \dfrac{GK_{te}}{l} & 0 & 0 \\ 0 & \dfrac{2EI_y}{l} & \dfrac{-6EI_y}{l^2} & 0 & \dfrac{4EI_y}{l} & \dfrac{6EI_y}{l^2} \\ 0 & \dfrac{6EI_y}{l^2} & \dfrac{-12EI_y}{l^3} & 0 & \dfrac{6EI_y}{l^2} & \dfrac{12EI_y}{l^3} \end{bmatrix} \quad (7.4)$$

A similar study was done years later by the NCHRP Project 12-79 Report 725 (White et al. 2012) with two equivalent equations with warping fixity at each end of a given unbraced length L_b (Equation 7.5a) and warping fixity at one end and warping free boundary conditions (Equation 7.5b), where J_{eq} is equivalent to K_{te} in Equation 7.3.

$$J_{eq(fx-fx)} = J\left\{1 - \frac{\sinh(pL_b)}{pL_b} + \frac{[\cosh(pL_b)-1]^2}{pL_b\sinh(pL_b)}\right\}^{-1} \quad (7.5a)$$

$$J_{eq(s-fx)} = J\left[1 - \frac{\sinh(pL_b)}{pL_b\cosh(pL_b)}\right]^{-1} \quad (7.5b)$$

A cross frame between girders for a grid analysis can be formed by steel beam, X-type, and K-type cross frames. For the 3D-modeling purpose, at least four or five nodes are needed for the definition of a cross frame as seen on the right side of Figure 7.2. For a 2D grid model, idealization in beam solutions is used to simulate the *exact* equivalent beam stiffness of this cross frame. In the NCHRP Project 12-79 Report 725 (White et al. 2012), it is called *Timoshenko beam element*.

This approach simply involves the calculation of an equivalent moment of inertia, I_{eq}, as well as an equivalent shear area As_{eq} (as shown in Equations 7.6 and 7.7) for a shear-deformable (Timoshenko) beam element representation of the cross frame.

1. The equivalent moment of inertia is determined first based on pure flexural deformation of the cross frame (zero shear). The cross frame is supported as a cantilever at one end and is subjected to a force couple applied at the corner joints at the other end (Figure 7.5).

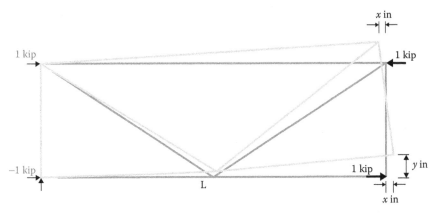

Figure 7.5 Calculation of equivalent moment of inertia based on pure bending. (Data from White, D.W. et al. "Guidelines for Analysis Methods and Construction Engineering of Curved and Skewed Steel Girder Bridges," NCHRP Project 12-79 Report 725, TRB, Washington, DC, 2012.)

$$\theta = \frac{2x}{L}, \quad I_{eq} = \frac{ML}{E\theta} \tag{7.6}$$

2. Using an equivalent Timoshenko beam element rather than an Euler–Bernoulli element, the cross frame is still supported as a cantilever but is subjected to a unit transverse shear at its tip (Figure 7.6).

$$A_{seq} = \frac{VL}{G\left[\Delta - (VL^3/3EI_{eq})\right]} \tag{7.7}$$

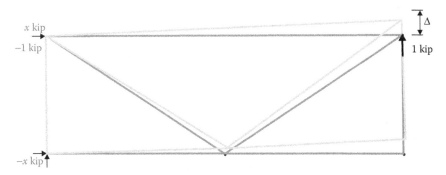

Figure 7.6 Calculation of equivalent shear area based on tip loading of the cross frame supported as a cantilever. (Data from White, D.W. et al. "Guidelines for Analysis Methods and Construction Engineering of Curved and Skewed Steel Girder Bridges," NCHRP Project 12-79 Report 725, TRB, Washington, DC, 2012.)

7.2 PRINCIPLE AND MODELING OF STEEL I-GIRDER BRIDGES

7.2.1 Analysis methods

It is always recommended to perform some kind of simplified verification of the results of more complex analysis models by means of simpler analysis models or hand calculation, or both. These types of checks are extremely valuable to allow the designer an opportunity for better understanding the behavior of the structure and validating the correctness of the more complicated analysis. It is also advised to perform a number of simple check calculations directly based on the analysis results. For instance, the simplest check when performing an analysis is to see whether the summation of dead load reactions equals the summation of the applied dead loads and whether the distribution of dead load reactions among the various support points matches the anticipated internal load distribution in the structure.

Depending on the complexity of the steel framing, the level of analysis required can range from simple hand calculations to 3D finite element modeling, which are briefly discussed here:

1. *Beam charts.* In the United States, there are a number of standard beam design charts and other design aids that can be of use to the designer. The AISC Manual includes a table of beam shear, moment, deflection, and reaction graphs and formulas for the cases of uniform load and point load. Although these patterns of loading are typically too simplified to be of direct benefit to bridge engineers, these design aids can serve a valuable purpose by providing a handy resource for finding approximate analysis methods for use in the preliminary design or in the checking of more complicated analyses.

2. *Line girder analysis method.* This method is referred to as *approximate* method in the AASHTO LRFD specifications (2013). The line girder analysis method uses load distribution factors to isolate a single girder from the rest of the superstructure system and evaluates that girder individually. When modeling, beam elements are lined up with the neutral axis. For composite sections, there are four stages, as described earlier, where their neutral axes and sections may change accordingly. Figure 7.7 shows the perspective view of a composite section with its associated neutral axis locations. The live load distribution factors can be simply determined by some approximate formulas for both straight bridges and curved bridges (AASHTO 2013).

3. *Grid analysis method.* This method is also referred to as plane grid or 2D grillage analysis method. In this method the structure is divided into plane grid elements (as shown in Figure 7.8) with three DOFs at each node (vertical displacement, rotation angles about the

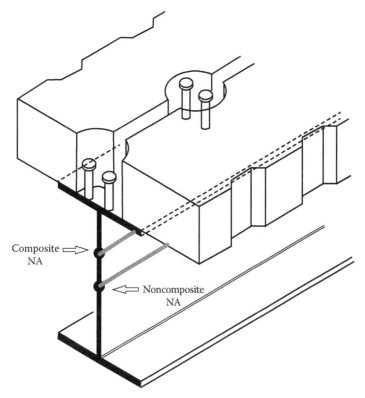

Figure 7.7 Line girder model with its associated neutral axis locations.

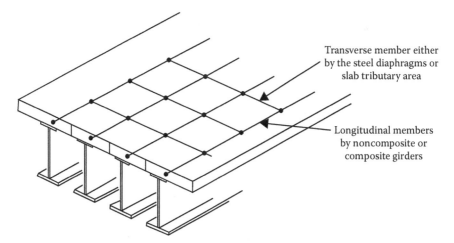

Figure 7.8 Grillage model representing the concrete deck on a steel I-girder bridge.

longitudinal and transverse axes). This method is most often used in steel bridge design and analysis.

4. *Plate and eccentric beam analysis methods.* This method is an advancement of a 2D grid/grillage analysis model. The deck is modeled using plane shell elements, whereas the girders and cross frames are modeled using beam elements offset from the plane shell elements to represent the offset of the neutral axis of the girder or cross frame from the neutral axis of the deck.

The offset length is typically equal to the distance between the centroids of the girder and deck sections. This method is more refined than the traditional 2D grid method. For this modeling approach, beam element internal forces obtained from this method need to be eccentrically transformed to obtain the composite girder internal forces (bending moment and shear) used in the bridge design. More details and sketch of the model are further discussed in the next point, 3D FEA methods.

5. *3D FEA methods.* The 3D FEA method is meant to encompass any analysis or design method that includes a computerized structural analysis model where the superstructure is modeled fully in three dimensions: modeling of girder flanges using line or beam elements or plate-, shell-, or solid-type elements; modeling of girder webs using plate-, shell-, or solid-type elements; modeling of cross frames or diaphragms using line or beam, truss, or plate-, shell-, or solid-type elements (as appropriate); and modeling of the deck using plate-, shell-, or solid-type elements. This method is fairly time consuming and complicated and is arguably deemed to be most appropriate for use for complicated bridges (e.g., bridges with severe curvature or skew or both, unusual framing plans, unusual support/substructure conditions, or other complicating features). 3D analysis methods are useful for performing refined local stress analysis of complex structural details (AASHTO/NSBA 2011).

However, there are some complications associated with 3D analysis methods. For instance, in a 3D analysis, generally used girder moments and shears are not directly calculated. Instead, the model reports stresses in flanges, webs, and deck elements. If the designer wishes to consider girder moments and shears, a postprocessor with some kind of conversion or integration of the stresses over the depth of the girder cross section will be required. A demonstration of this kind of conversion is shown in Figure 7.9. In this figure force results of a steel girder section are shown where top and bottom flanges are modeled by beam elements (element numbers 30 and 90) and web by two shell elements (element numbers 838 and 898). Neutral axis is in the middle for a symmetric section. Resultants for beam elements are shown in F_x (axial force = 649.72 kip or 2890 kN), F_y (transverse force = 9.95 kip or 44.3 kN for the top flange), F_z (vertical force = 0.28 kip or 1.2 kN for the top flange),

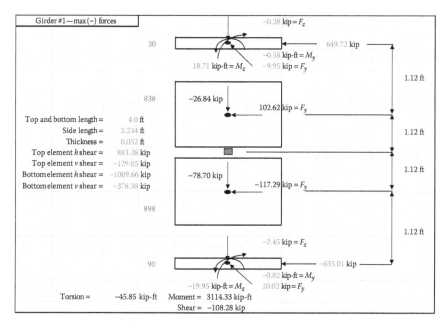

Figure 7.9 Conversion of FEM stress resultants to beam moments and shears.

M_y (transverse moment = 0.58 kip-ft or 0.8 kN-m for the top flange), and M_z (vertical moment = 18.71 kip-ft or 25.4 kN-m). Resultants for shell elements are shown in F_y (horizontal force = 102.62 kip or 456.5 kN for the web top element) and F_z (vertical force = 26.84 kip or 119.4 kN for the web top element). By integrating all resultants of these four elements, moment, shear, and torsion can be obtained at the central or any location of the cross section. This process can be a significant undertaking, particularly with regard to proper proportioning of deck stresses and deck section properties to individual girders.

When and how to use a refined 3D FEA for engineering design is a controversial issue, and in the United States such an approach has not been fully incorporated into the AASHTO specifications to date (2013). The typical AASHTO methodology for design is generally based on the assessment of nominal (average) stresses calculated by simplified methods, such as P/A or Mc/I, and not localized peak stresses obtained by shell- or solid-based finite element models. Refined analysis can provide substantially more detailed and accurate information about the stress state of the structure. This could allow for more cost-effective and reliable design but often comes with increased engineering effort and increased potential for error. The results are often more sensitive to the input parameters and the mathematical assumptions that are employed by the software. For instance, a given element will have a unique formulation, interpolation, integration, and software implementation, all of which will affect results. However, if properly modeled, in the forensic or load test cases, such refined analysis is commonly adopted due to its refinement and accuracy. 3D FEA models are described here:

a. *In-plane shell–beam model.* Hays Jr. et al. (1986) and Mabsout et al. (1997) modeled the deck slab using quadrilateral shell elements in plane with five DOFs per node and the steel girders using 3D beam elements with six DOFs per node (Figure 7.10). The bridge deck slab and steel girders shared nodes where the steel girder is present. This model is essentially a 2D FEA, and it is not capable of capturing the effect of the offset between the center of gravity of the steel girder and that of the deck slab. Furthermore, it cannot capture the system's actual boundary conditions, that is, the supports in the actual system are located at the bottom of the steel girder rather than at the center of gravity of the deck slab.

b. *3D brick–shell model.* Tarhini and Frederick (1992), Eamon and Nowak (2001), Baskar et al. (2002), and Queiroz et al. (2007) used eight-node linear solid brick elements with three displacement DOFs in each node to model the concrete deck. The girders were modeled using quadrilateral shell elements, which contain three displacement and two rotational DOFs per node (Figure 7.11). The cross frames were modeled using 3D two-node truss elements with three displacement DOFs per node. Tarhini and Frederick

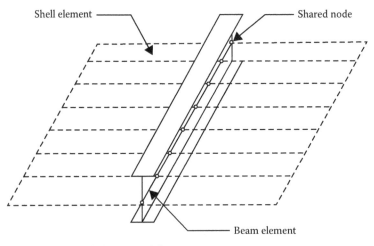

Figure 7.10 In-plane shell–beam model.

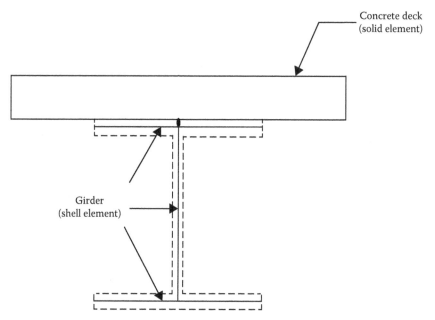

Figure 7.11 3D brick–shell model.

(1992) modeled full composite action by imposing no release at the interface nodes at the concrete deck and girders. Queiroz et al. (2007) vmodeled the longitudinal and transverse reinforcement in the deck slab as a smeared layer of equivalent area in solid brick elements.

c. *3D shell–beam model.* Tabsh and Tabatabai (2001) and Issa et al. (2000) modeled deck slab using four-node rectangular shell elements with five DOFs per node. Each component of the steel girder, that is, top and bottom flange and web, was modeled separately. Top and bottom flanges were idealized as two-node beam elements with six DOFs per node. The steel web was idealized using four-node rectangular shell elements and the cross frames were idealized using two-node beam elements. Rigid beam elements were used to model the full composite action between the concrete deck and steel girders as shown in Figure 7.12.

d. *3D shell–shell model.* Fu and Lu (2003) idealized the bridge deck with isoparametric quadrilateral shell elements, and the reinforcement was modeled as a smeared 2D membrane layer with isoparametric plane stress element. The steel girder flanges were modeled using eight-node plane shell elements and web by eight-node plane stress elements. This modeling selection clearly generates an

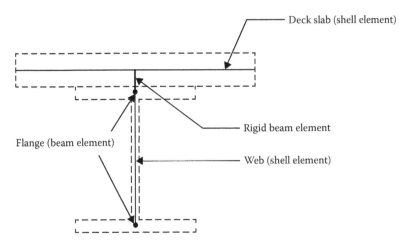

Figure 7.12 3D shell–beam model.

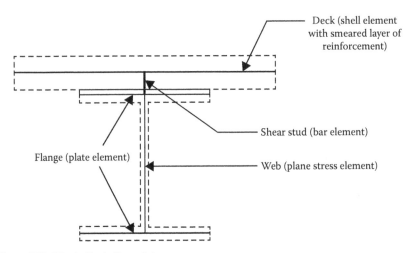

Figure 7.13 3D shell–shell model.

incompatibility at the flange and web connection. However, the authors did not discuss this issue or its potential effect on the results. The shear studs were modeled using bar elements (Figure 7.13).

e. *3D brick–beam model.* Ebeido and Kennedy (1996), Barr et al. (2001), Chen (1999), and Sebastian and McConnel (2000) used eccentric beam model as shown in Figure 7.14 to idealize the bridge superstructure in which the bridge deck was modeled using four-node plane shell elements. The longitudinal steel girders and cross frames were idealized using 3D two-node beam elements with six DOFs for each node.

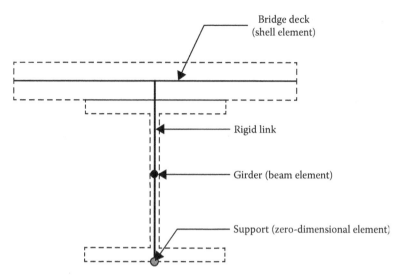

Bridge deck
(shell element)

Rigid link

Girder (beam element)

Support (zero-dimensional element)

Figure 7.14 3D brick–beam model.

Chung and Sotelino (2006) used four different techniques to model an I-girder bridge superstructure. In their approach the bridge deck was modeled using shear flexible shell elements (S8R in the commercial software ABAQUS, 2007) and the steel girders were modeled by four different models, named G1, G2, G3, and G4, to assess the suitability of each technique. In the G1 model, the girder flanges and webs were modeled using shell elements. The shell elements used to model the flanges were placed at the mid-surface of the flanges using the offset option in ABAQUS to obtain the correct moment of inertia of steel girders. The only difference between the G1 and G2 models is that in the latter the flanges were modeled using beam elements placed at the location coinciding with the center of the flange. The use of beam elements reduced the computational cost as compared to G1 model. In the G3 model, the web was modeled using a beam element and both flanges were modeled using shell elements. This model was considered further to investigate the incompatibility that possibly exists between model G1 where the in-plane rotational DOF of the flange shell and drilling rotational DOF of the web shell are shared at the flange and web joint. Rigid links were used to connect the shell and beam elements to ensure full composite action. In the G4 model either Euler beam elements or shear flexible Timoshenko beam elements were used to model the steel girder. All four models were evaluated numerically by looking at the maximum deflection due to concentrated load applied at the center of a simply supported I-shaped beam (Figure 7.15). It should be noticed that the analytical solution to this problem is readily available from the theory of elasticity. G1 and G2 models required significant mesh refinement to converge to the analytical solution as compared to G3 and G4 models.

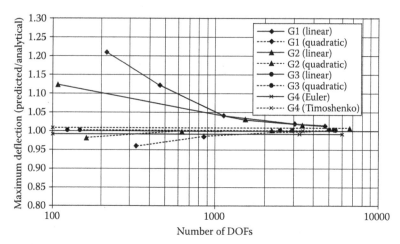

Figure 7.15 Convergence of finite element girder models. (Data from Chung, W. and Sotelino, E.D., *Engineering Structures*, 28, 63–71, 2006.)

7.2.2 Modeling in specific regions

One phenomenon rarely considered in the modeling process in the United States is the effect of curb and parapet. Bridge investigation shows that, if a slab deck effectively acts with the parapets, it is better able to carry a load near the edges. However, the practice in the United States is not to count on the contribution of the edge stiffening in the design process. If the existing bridge needs to be load rated or the serviceability, such as deflection, is the concern, the edge stiffening may be considered in the bridge model.

The resistance offered by steel girders to different loads on the bridge depends on the amount of composite action between the deck and steel girders. To model the partial composite action in the FEA, different methods are proposed. Tarhini and Frederick (1992) modeled the partial composite action with three linear spring elements with properties based on the amount of expected slip.

Baskar et al. (2002) used two different techniques to model the composite action in ultimate strength. In the first method, the surface interaction technique was used to model the composite action. This technique allows incompatible strains and slip between the nodes in two different sets. More specifically, the bond strength at steel and concrete interface and the strength of shear stud were combined and modeled as shear between two surfaces. A bilinear curve similar to shear force versus slip curve for shear stud was used to model the slip. The surface behavior option in ABAQUS was used to model the vertical tensile strength of the stud. This method was unable to capture local effects such as slab failure and stud connector failure. In the second method, general beam elements were used to model the shear studs and the area of the beam element was modified to account for the strength of the embedded shear stud in concrete. Both techniques were evaluated by comparing their results with the experimental load versus deflection plot for a cantilever beam subjected to a point load at the tip. The results obtained using the surface interaction technique were found to be in good agreement with the experimental data. However, even though the second technique was able to closely match the results in the initial stage of the deformation, it had a slower convergence with the post peak load versus deflection plot, and more mesh refinement than in the first method was required for the results to converge to the experimental results. This kind of nonlinear analysis technique is mostly used for research purpose.

A steel I-girder bridge may be designed and built composite or non-composite. Negative moment region could be complicated as it might be considered as noncomposite with or without shear connectors, even the bridge was designed and built composite. For analyzing a continuous composite steel girder bridge, no matter using line girder analysis method or grid analysis method, assumption has to be made in the negative moment region. AASHTO LRFD specifications (2013) state that stiffness characteristics of beam–slab-type bridges may be based on the full participation of concrete decks due to the fact that crack does not mean ineffective until total concrete failure. Figure 7.16a models the noncomposite sections, whereas Figure 7.16b demonstrates a case that if shear connectors with steel reinforcements are considered in the negative moment region, full composite sections are used throughout the analysis. However, when calculating stresses in the negative moment area, steel section properties with steel reinforcements in the slab, instead of the full composite section, are conservatively used.

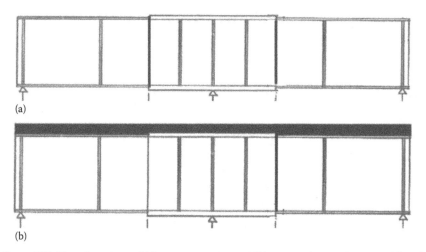

Figure 7.16 Elevation view of (a) noncomposite and (b) composite sections considered in the analysis and design.

7.2.3 Live load application

For steel girder analysis, the following girder influence surfaces should be analyzed in terms of

- Moment (M)
- Shear (V)
- Torsion (left of the joint)
- Torsion (right of the joint)
- Deflection (D)
- Reaction (R)
- Diaphragm internal forces

For grid analysis, any influence surface as shown in Figure 7.17 can be considered as a series of influence lines. For instance, Figures 7.18 and 7.19 are considered as a set of moment influence surface for Joint 3 of girder 2. Vehicular loading and lane loading move laterally within the traffic lane or outside the lane as long as the distance between vehicles are maintained to yield the highest reaction (distributed force) for the concerned (primary) girder. Ordinates

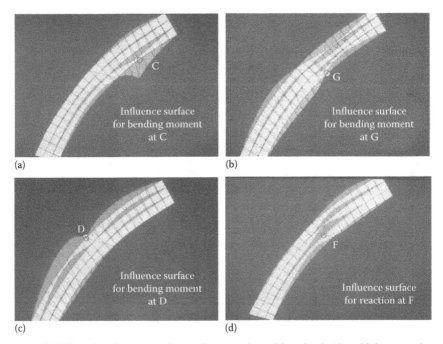

(a)   (b)

(c)   (d)

Figure 7.17 Sample influence surfaces of a curved steel I-girder bridge. (a) Inner girder in-span bending moment at C. (b) Inner girder interior support bending moment at G. (c) Outer girder interior support bending moment at D. (d) Second interior girder interior support reaction at F.

of their respective influence lines are used to multiply the fraction of vehicular loading to that girder, and areas under their respective influence lines are used to multiply the fraction of lane loading to their respective girders.

To determine the extreme live load locations and corresponding extreme values, influence lines are obtained from the influence surface for each girder. Figure 7.17 shows the 3D perspective views for exemplar positive and negative moment influence surfaces. Their exemplar 2D views for all four girders from the DESCUS-I program (2012) are shown in Figures 7.18 and 7.19. Placement of live loads on one influence surface is shown in

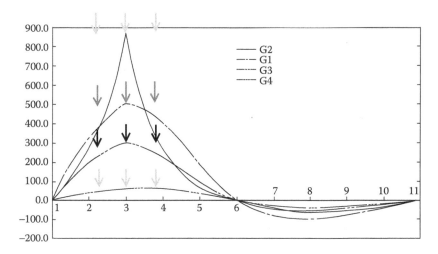

Figure 7.18 DESCUS-I example—positive moment influence lines of girder 2.

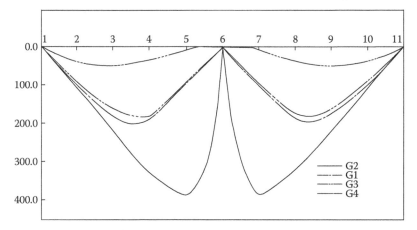

Figure 7.19 DESCUS-I example—negative moment influence lines of girder 2.

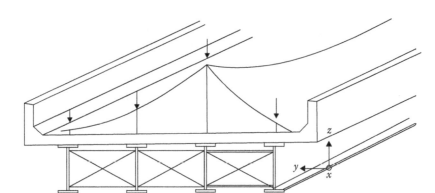

Figure 7.20 Placement of live load on influence surface.

Figure 7.18 for a 2D plot or in Figure 7.20 in 3D perspective view where each downward force can be considered a fraction of a truck axle.

7.2.4 Girder–substringer systems

A typical girder–substringer configuration is illustrated in Figure 7.21. In this figure, three main girders are shown. At intervals associated with typical brace points of the main girders, floor beams span between them. Within each girder bay, a series of smaller, more closely spaced substringers are carried on the floor beams and support the deck.

Typical 2D and advanced grid analysis methods (methods 2 and 3 as described in Section 7.2.1) can readily capture the distributions of dead loads and live loads in such a system. The tradeoffs are the increased complexity in live load application; the proliferation of load placement options, which comes with the consideration of transverse location in addition to longitudinal, the proliferation of output, and potentially the postprocessing demands of assembling composite section forces from disparate model elements.

An analysis approach that may be considered is the use of a basic 2D grid model to explore only the load distribution properties of the system. A manageable regime of unit line load placements, unit area load placements of one lane in width, and full-deck area load can provide moment, shear, and reaction results for the stringers and girders. By comparing such results to cases

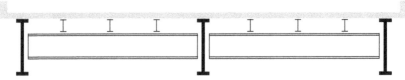

Figure 7.21 Cross-sectional view of a girder–substringer system.

in which similar loadings are applied directly to isolated models of a stringer or girder, effective live load distribution factors can be extracted. These factors could then be used in line girder analyses, provided the capacity for the override of typical AASHTO distribution factors is available. The stringers in such models would be supported at floor beam locations and the girders at pier locations. The flexibility of the stringer supports at floor beam locations is arguably built in by virtue of the method used to construct distribution factors. The reaction results from the stringer model would provide input to floor beam analyses.

Usually, 2D grid model is assuming hinge or roller at a support location. Considering the flexibility of the bent, the vertical stiffness offered by the long-span steel straddle bent will be less than that offered by the concrete hammerhead bent because the straddle bent cap possesses significant vertical flexibility, whereas the concrete hammerhead is essentially rigid in the vertical direction. If several supports of a multispan continuous steel girder bridge are concrete hammerhead bents, with one support being a long-span steel hammerhead bent, the response of the girders to vertical loading will be different from that of the girders supported by all concrete hammerhead bents (Figure 7.22).

For example, consider a bridge with a relatively wide, multigirder cross section, supported at one or more bents by a steel straddle bent (Figure 7.23). In this case, the vertical stiffness offered by the support for the leftmost girder in the cross section will be different from that offered by the support for the rightmost girder in the cross section.

7.2.5 Steel I-girder bridge during construction

The sequence of erection, as well as the number of girders in place and connected by cross frames during erection, will affect the response of the girders to loading. In the United States, many owner agencies require that contract

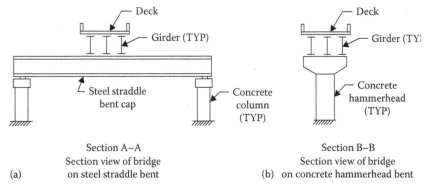

Section A–A
Section view of bridge
(a)   on steel straddle bent

Section B–B
Section view of bridge
(b)   on concrete hammerhead bent

Figure 7.22 Section views of (a) a bridge with girders sitting on a steel straddle bent versus (b) girders sitting on a concrete hammerhead.

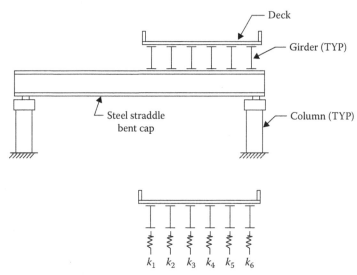

Figure 7.23 Straddle bent cap modeled by support stiffness. (a) Girders supported by straddle bent cap and (b) support stiffness.

plans clearly indicate the assumed erection sequence and designers should be ready to assess different erection sequences during shop drawing review if the contractor chooses to erect the girders in a different way. Depending on the complexity of the steel framing and the proposed erection sequence, the level of analysis required can range from simple hand calculations to 3D finite element modeling. In general, for a simple framing plan such as a simple-span bridge with no skew, hand calculations may be sufficient. On the other hand, for a large curved steel I-girder bridge where vertical and lateral displacements may be of concern to ensure proper fit-up or where lateral bending stresses at certain stages of erection may be of concern, a full 3D FEA may be warranted (White et al. 2012).

The 2D or 3D model can be created for the completed steel framing and then reconstructed stage by stage in accordance with the proposed erection sequence. For the analysis of the steel erection sequence, dead loads and construction loads need to be determined and applied to the appropriate elements in the model. Dead loads typically include the self-weight of the structural members and detail attachments. Wind loads must be considered in the analysis of the steel erection sequence.

Increasingly, engineers are required to evaluate the stability of steel members under partial stages of completion, for instance, the behavior of a beam suspended by a crane or spreader beams during lifting or the behavior of partly completed spans during erection with beams cantilevered or partly suspended by holding cranes. Prior to the casting of deck concrete, uneven solar heating may cause the misalignment of girders and other construction issues.

Therefore, analysis based on such temperature changes during erection may be required as well. The erection of the steel framing, whether the bridge is straight or curved, is one of the most critical stages with regard to ensuring stability, and these factors may need to be considered in the models during construction.

Deck placement effects must be considered in the design of steel bridges. When a portion of the deck slab is pouring, deck concrete casted in previous stages may be cured enough to form a composite action. Therefore, the moment of inertia in the previously poured sections has to be so adjusted to reflect the stiffness changes. The deck placement sequence also has an effect on other aspects of bridge behavior including uplift, deflections, and bearing rotations. Staging analysis process due to deck placement based on ACI209 (2008) are shown here.

1. Creep coefficient ($\varphi[t,t_0]$): The general form of the creep equation is

$$\varphi(t,t_0) = \frac{(t-t_0)^\psi}{d + (t-t_0)^\psi} \varphi_u \tag{7.8}$$

where:
$(t - t_0)$ is the time since application of load
ψ and d (in days) are constants
φ_u is the ultimate creep coefficient

$$\varphi_u = (\varphi_u)_{avg} \cdot \gamma_c \tag{7.9}$$

where:
$(\varphi_u)_{avg} = 2.35$
γ_c is the cumulative product of six applicable correction factors for loading age, relative humidity, volume–surface ratio, and concrete composition (slump, aggregate, and air content)

2. Strength at age t (f_{cmt}). The general form of the strength equation is

$$f_{cmt} = \frac{t}{a+bt} f_{cm28} \tag{7.10}$$

where:
f_{cm28} (in MPa or psi) is the strength of concrete at an age of 28 days
a (in days) and b are constants depending on the concrete type

3. Modulus of elasticity at loading age t_0 and age t (E_{mcto} and E_{cmt}):

$$E_{mcto} = 33\gamma_c^{1.5}\sqrt{f_{cmto}} \text{ (psi) or } E_{mcto} = 0.043\gamma_c^{1.5}\sqrt{f_{cmto}} \text{ (MPa)} \tag{7.11a}$$

$$E_{cmt} = \frac{E_{cmto}}{1 + \varphi(t,t_0)} \tag{7.11b}$$

where:

E_{cmt} (MPa or psi) is the effective modulus and is used to compute the modulus ratio between concrete and steel

γ_c (kg/m³ or lb/ft³) is the unit weight of concrete

7.3 2D AND 3D ILLUSTRATED EXAMPLE OF A HAUNCHED STEEL I-GIRDER BRIDGE—MD140 BRIDGE, MARYLAND

Bridge No. 6032 is on MD140 over Maryland Midland Railroad, MD27, and West Branch in Carroll County, Maryland. It was originally built in 1952. However, due to severe deterioration over years, the superstructures of this bridge including all girders and deck were rebuilt in 2005. Description of the structure is given in Table 7.1. The rebuilt of this bridge was partially funded by the Federal Highway Administration's Innovative Bridge Research & Construction (FHWA-IBRC) Program as an application of a high-performance steel (HPS) bridge.

In the design process, AASHTO 2D line girder method was adopted. As mentioned in Section 7.2.1, for the line girder approximate method, certain conditions have to be met, such as

Table 7.1 Description of the MD140 Bridge structure

Item	Description
Structure identification	Bridge #06032
Location	MD140 over MD27—in Carroll County
Structure type	15-Steel-girder bridge
Span length(s)	148'–152' (45.11–46.33 m) two-span bridge
Girder web depth	Varied from 45" to 81" (1143–2057 mm)
Roadway width	61' (18.6 m) clear roadway width with 5' (1.5 m) sidewalk Northbound 6' (1.8 m) median and 50' (15.2 m) clear roadway width with 5' (1.5 m) sidewalk Southbound
Structure width	130' (39.6 m) out-to-out superstructure
Cross-frame type	K-type intermediate cross frames; channel-end diaphragms
Girder spacing	5@10'–0" (3.05 m), 2@7'–3" (2.21 m), median 5'–0" (1.52 m), 2@7'–3" (2.21 m), 4@9'–9" (2.97 m)
Structural steel	F_y = 70 ksi (483 MPa)
Abutments	Concrete abutment
Construction phases	Three phases; (1) girders 6–11, (2) girders 1–5, (3) girders 12–15
Pouring sequence	Pouring sequence nos. 1–3 for phase 1, sequence nos. 4–6 for phase 2, sequence no. 7 for phase 2 closure, sequence nos. 8–10 for phase 3, sequence no. 11 for phase 3 closure

Figure 7.24 MD140 haunched 15-steel girder system.

- Assumed constant deck width, parallel beams with about the same stiffness
- Use of *design* trucks
- Designed within the bound for that structural type
- Limited ranges of applicability, such as applicable for straight bridge and for constant girder spacing only (When exceeded, the AASHTO LRFD specifications mandate refined analysis.)

This bridge fits all the conditions, and 2D line girder method was adopted for the analysis, and the girder sections were designed accordingly. Figure 7.24 shows the parabolic-haunched 15-steel girder system with one interior girder isolated for the analysis. The prospective view of this line girder is shown in Figure 7.7. The calculation of the section properties in three stages, noncomposite (N = infinity), LT composite ($N = 3n$), and ST composite ($N = n$) sections, are shown in Figure 7.25. This table lists the 2D beam section properties at the interior pier location used in the analysis where S_{top}, S_{bot}, and S_{topc} refer to section moduli at the top and bottom of the steel girder and the top of concrete slab, respectively. Also, Q_{slab} is the first moment of inertia of the slab, and I_x is the moment of inertia of the composite section wherein AASHTO Q_{slab}/I_x is used for the calculation of the shear connector fatigue requirement. Herein, n is the modulus ratio of the steel girder to concrete deck and N is the actual modulus ratio used in that stage where N is the infinity mean steel section only. For 4000 psi (27.6 MPa) normal concrete, $n = 8$ is used.

As this bridge was the first few applications of HPS in the state of Maryland, full bridge testing, including 3D FEA, in all phases and stages was conducted. As described in Table 7.1, this 15-girder system was reconstructed in three phases: (1) the first phase—girders 6–11, (2) the second phase—girders 1–5, and (3) the third phase—girders 12–15. Each phase has three pouring sequences. 3D FEAs by ANSYS were conducted (Figure 7.26).

Bridge Input Parameters

Span L =	152	ft
Slab t =	10	in
Girder S =	10	ft
Haunch =	2	in
Top flange	14	0.75
Web	0.75	x 45
Bot flange	16	x 2
Total steel D		

(Top flange 14 × 2, Web 0.75 × 45, Bot flange 16 × 2)

Effective Width Calculation

$L_{eff}/4$ =	342	
$12t + \max(b_f/2, t_w)$ =	127	
S =	120	
	22	in
	2	in
	45	in
	2	in
	49	in

Full Girder Width

full S (in) =	120	

(a)

Section	A	d_b	Ad_b	d	Ad^2	I_0
Top flange	28	48	1344	24.50	16810.66	9.333333
Web	33.75	24.5	826.875	1.00	33.93	5695.313
Bot flange	32	1	32	−22.50	16196.16	10.66667
Sum	93.75		2202.875		33040.749	5715.313
y_{bot} =	23.50	in	y_{top} =	25.50	in	
$I_x = I_0 + Ad^2$ =	38756.06	in^4				
S_{top} =	1519.69	in^3	S_{bot} =	1649.38	in^3	

(b)

Figure 7.25 (a) Demonstration of the 2D beam section properties at the interior pier location. (b) Section properties (N = Infinity). Highlighted areas are input items. (Continued)

Section	A	d_b	Ad_b	d	Ad^2	I_0
Slab	50	54	2700	19.82	19640.98	416.67
Haunch	0.666667	50	33.33333333	15.82	166.84	0.00
Top flange	28	48	1344	13.82	5347.54	9.33
Web	33.75	24.5	826.875	−9.68	3162.67	5695.31
Bot flange	32	1	32	−33.18	35229.88	10.67
Sum	144.4167		4936.208333		63547.91	6131.98
$y_{bot} =$	34.18	in		$y_{top} =$	14.82	in
$I_x = I_0 + Ad^2 =$	69679.89	in⁴		$S_{topc} =$	2598.09	in³
$S_{top} =$	4701.85	in³		$S_{bot} =$	2038.60	in³

(c)

Section	A	d_b	Ad_b	d	Ad^2	I_0
Slab	150	54	8100	11.67	20424.38	1250.00
Haunch	2	50	100	7.67	117.62	0.00
Top flange	28	48	1344	5.67	899.81	9.33
Web	33.75	24.5	826.875	−17.83	10730.78	5695.31
Bot flange	32	1	32	−41.33	54664.39	10.67
Sum	245.75		10402.875		86836.99	6965.31
$y_{bot} =$	42.33	in		$y_{top} =$	6.67	in
$I_x = I_0 + Ad^2 =$	93802.3	in⁴		$S_{topc} =$	5024.53	in³
$S_{top} =$	14065.70	in³		$S_{bot} =$	2215.92	in³
$Q_{slab} =$	1750.33	in³				

(d)

Figure 7.25 (Continued) (c) Section properties ($N = 3n = 24$). (d) Section properties ($N = n = 8$).

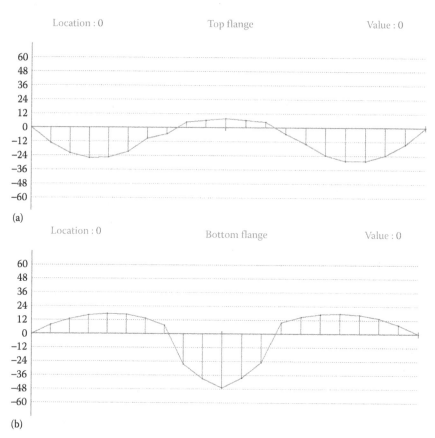

Figure 7.26 Stress diagrams of MD I 40 Bridge (in ksi). (a) Top-flange stress diagram. (b) Bottom-flange stress diagram.

Bridge testing results of these nine stages were collected and compared with the FEA results for all phases of construction. For the FEA, steel girders were represented by 2D shell elements joined together in an I shape and the concrete deck was represented by 3D solid elements as shown in Figure 7.11—3D brick–shell model. When analyzing fresh concrete pouring as dead loads, steel-only cross sections were used, rather than composite sections used in analyses in later phases.

The graphic results shown for the 3D finite element staging analysis are for the third construction phase where girders 1 through 11 have been constructed and concrete deck was poured in the first and second phases. The third-phase construction is for girders 12 through 15, and the pouring sequence is starting from the north-side 46.33-m (152′) span (Figure 7.27), the second pour on the south-side 45.11-m (148′) span (Figure 7.28), and then the third closure pour in three consecutive days. The

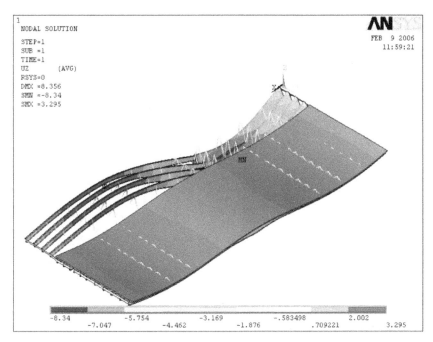

Figure 7.27 Total deflection after deck pouring on girders 12 to 15 in north span.

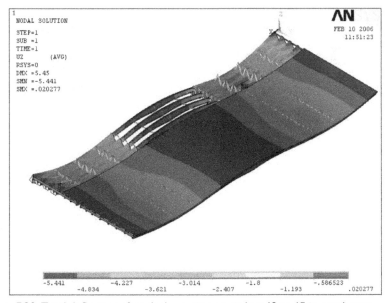

Figure 7.28 Total deflection after deck pouring on girders 12 to 15 in south span.

study was made with the consideration of creep and shrinkage that the deck being constructed still matched up in elevation to the deck poured in the previous phase, not causing noticeable misalignments in the deck.

7.4 2D AND 3D ILLUSTRATED EXAMPLE OF A CURVED STEEL I-GIRDER BRIDGE—ROCK CREEK TRAIL PEDESTRIAN BRIDGE, MARYLAND

The Rock Creek Trail Pedestrian Bridge serves as a connection for the Rock Creek Hiker–Biker Trail and provides a safe crossway for pedestrians and cyclists. The bridge spans over Veirs Mill Road (MD-586) near Rock Creek Park in Montgomery County, Maryland. Table 7.2 lists the parameters of the curved portion of the bridge. At the erection stage of the construction process, the curved inner girder at the supports of Piers 1 and 3 was uplifted after the temporary supports were released. Due to the aberrant response of the curved spans during construction, the spans were modeled and studied for two stages: (1) bridge under construction with dead load only and (2) bridge under live load. This two-span, two-girder

Table 7.2 Description of the Rock Creek Trail Pedestrian Bridge structure

Description	Variable
Number of girders	2
Number of spans	2
Radius of curvature of girder 1 (inner)	220.0′ (67 m)
Radius of curvature of girder 2 (outer)	230.0′ (70 m)
Span lengths of girder 1	2@161.33′ (49.2 m)
Span lengths of girder 2	2@168.67′ (51.4 m)
Spacing between girders	10.0′ (3 m)
Roadway width	29.0′ (8.8 m)
Overhang width, left and right	2.33′ (0.7 m)
Curb width, left and right	1.33′ (0.4 m)
Design slab depth (excluding integral wearing surface)	9.0″ (229 mm)
Integral wearing surface	0.5″ (13 mm)
Haunch depth and width	5.0″ and 24.0″ (127 and 610 mm)
Type of concentration	Composite
Ultimate strength of concrete	4.0 ksi (27.6 MPa)
Yield strength of steel	50 ksi (A992) (345 MPa)
Live loading	H-10 and pedestrian loading

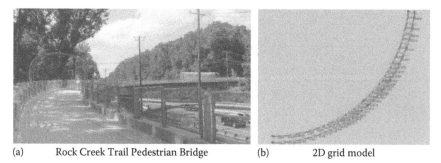

(a)   Rock Creek Trail Pedestrian Bridge   (b)   2D grid model

Figure 7.29 Rock Creek Trail Pedestrian Bridge and its 2D grid model.

bridge is used in this section as a demonstration for 2D grid model and 3D finite element model.

In 2D grid model, DESCUS-I program (2011) was used in the design phase. To establish the curved grid model, torsional constant J_{eq} (Equation 7.5a and b, or K_{te} in Equation 7.3) for the curved steel section and moment of inertia I_{eq} for the cross frame as described in Section 7.1.3 were adopted by the program. The bridge and its grid model are shown in Figure 7.29a and b, respectively.

A 3D finite element model of the same bridge was established by Bridge Analysis Generator in SAP2000 (2007) (later version named CSiBridge) without considering the vertical altitude difference and the superelevation. The concrete deck and two steel I-girders are simulated with shell elements, and diaphragms are simulated with frame elements, as shown in Figure 7.30. Table 7.3 lists the comparison of reactions for these two models.

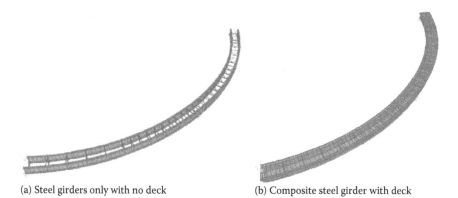

(a) Steel girders only with no deck   (b) Composite steel girder with deck

Figure 7.30 3D finite element models of Rock Creek Trail Pedestrian Bridge by SAP2000.

Table 7.3 Reaction comparison

	Grid model		Shell element (kip)		Frame element (kip)	
	Inner reactions	*Outer reactions*	*Inner reactions*	*Outer reactions*	*Inner reactions*	*Outer reactions*
Steel girders	7.9	56.4	6.73	62.48	8.82	57.65
	273.5	82.5	289.17	55.61	318.47	31.78
	7.9	56.3	6.00	63.82	8.07	59.01
			Sum = 483.81		Sum = 483.80	
Steel girders with concrete deck			11.70	137.57	20.24	134.07
			650.72	115.34	716.59	51.43
			9.94	140.83	18.44	137.39
			Sum = 1066.10		Sum = 1078.16	

7.5 2D AND 3D ILLUSTRATED EXAMPLE OF A SKEWED AND KINKED STEEL I-GIRDER BRIDGE WITH STRADDLE BENT

In early 1960s in the United States, before heat curving on steel girders was made popular, girders were made kinked at field splice locations to accommodate complex (curved or flaring) framing. It is recommended to provide close cross frame(s) with the splice to help resist lateral loads on the girder due to the kink.

In this illustrated example, 2D and 3D models were generated by DESCUS-I and CSiBridge (2011), respectively. This two-span concrete–steel composite bridge consists of five I-girders with three different sections. The 2D grillage model is shown in Figure 7.31. The bridge layout line is comprised of three segment lines kinked with three different slopes as shown in Figure 7.32 for the perspective view and Figure 7.33 for the plan view of the 3D finite element model. The bridge diaphragms are inverted K-type braces with top and bottom chords. Most of them are normal to the layout line, except at those kinked locations. The bent is comprised of a concrete cap

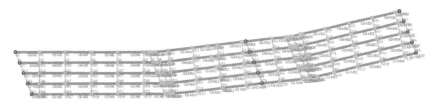

Figure 7.31 2D grillage model.

Figure 7.32 Perspective view of the 3D finite element model.

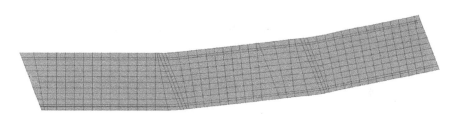

Figure 7.33 Plan view of the 3D finite element model.

beam and two concrete columns, supporting the superstructure, as shown in Figure 7.34. In the 2D grillage model, all supports can be considered fixed at the vertical direction.

As illustrated in Figure 7.35, concrete deck was modeled by shell elements, whereas the five I-girders were modeled by 3D frame elements. The connections between concrete deck and steel girders are simulated by displacement constraints with corresponding nodes constrained together. To better simulate the bearing, foundation springs at the start and end abutments are fixed at vertical and horizontal directions as shown in Figure 7.36.

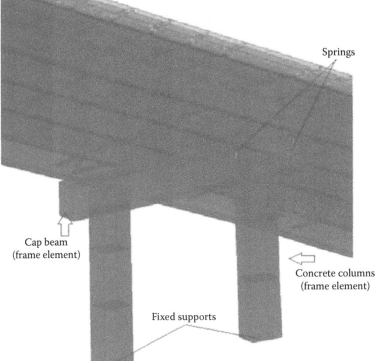

Springs

Cap beam
(frame element)

Concrete columns
(frame element)

Fixed supports

Figure 7.34 Bent region modeling detail of the 3D finite element model.

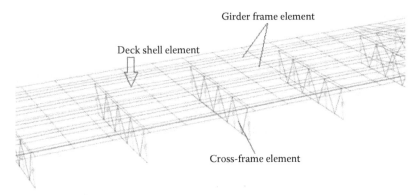

Girder frame element

Deck shell element

Cross-frame element

Figure 7.35 Superstructure modeling detail of the 3D finite element model.

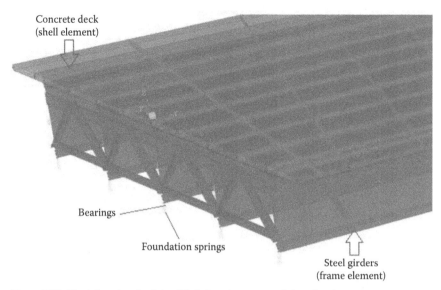

Figure 7.36 Modeling detail of the 3D finite element model in the abutment area.

7.6 2D AND 3D ILLUSTRATED EXAMPLE OF A GLOBAL AND LOCAL MODELING OF A SIMPLE-SPAN STEEL I-GIRDER BRIDGE—I-270 MIDDLEBROOK ROAD BRIDGE, GERMANTOWN, MARYLAND

MD Bridge No. 1504200 I-270 over Middlebrook Road is a simple-span composite steel I-girder bridge with a span length of 42.7 m (140′). Located at I-270 over Middlebrook Road near Germantown, Maryland, it carries three traffic lanes in the southbound roadway and five traffic lanes in the northbound roadway. This bridge has a 76° parallel skew of its bearing lines. The bridge diaphragms are inverted K-type braces with top and bottom chords. All of them are parallel to the bearing lines. A research project sponsored by the U.S. Department of Transportation's Research and Innovative Technology Administration (RITA), under the Commercial Remote Sensing and Spatial Information (CRS&SI) Technologies Program required a pilot testing bridge to develop and perform a LT field monitoring test of a wireless Integrated Structural Health Monitoring (ISHM) system. The Middlebrook Road Bridge with active fatigue cracks at the connection plates of the K-type bracing was selected. Complete pilot testing was performed using acoustic emission (AE), accelerometer, deflection, and strain sensors for bridge information collection. To simulate the bridge behavior under traffic load, global and local models were built, of which the global model was used to monitor the global behavior and stresses near the crack area, whereas the local model was used to find the stress concentration factor (SCF) at the exact crack location, called hot spot.

Stress concentration factor. For steel bridges, weld toes are usually the critical fatigue damage regions. However, the monitoring sensors installed in bridges are not necessarily located in these areas. To know the hot spot stress near the weld toe, it is necessary to convert the nominal stress obtained from the monitoring sensors into the corresponding hot spot stress near the weld toe for fatigue life evaluation. The SCF is defined as the ratio of the hot spot stress value to that of the nominal stress and can be calculated from stress values obtained from the global model of coarse mesh and the local model of refined mesh. With the SCF value obtained, the hot spot stress can be obtained by multiplication of the nominal stress with the SCF value. The SCF value of a welded joint is commonly obtained by experiment or numerical finite element method.

Global model. A 3D model of the Southbound consisting of eight I-girders as shown in Figure 7.37 was generated by CSiBridge (2011). The concrete deck, eight I-girders, and connection plates, which connect diaphragms and girder webs, were modeled by shell elements, whereas all the diaphragms were modeled by truss elements. The translations of x-, y-, and z-directions are fixed at the abutments. To locate the hot spot, a global model refined mesh around the hot spots was built for analysis, and a detailed view of this global model is presented in Figure 7.38.

Figure 7.37 Global model of I-270 Middlebrook Bridge.

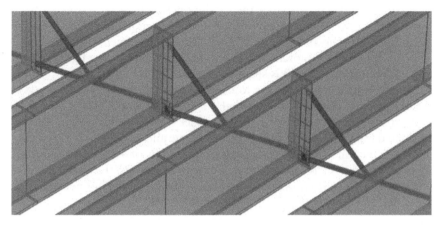

Figure 7.38 Zoom-in view of the global model.

Local model. To study the behavior of the bridge, the entire superstructure was first analyzed by a large, coarse finite element model. The global model contains only the main components of the bridge and is mainly for modal analysis, displacement output of the whole bridge, critical fatigue location determination, and so on. However, to investigate the stresses or strains of a certain area or a certain joint, it is necessary to employ a series

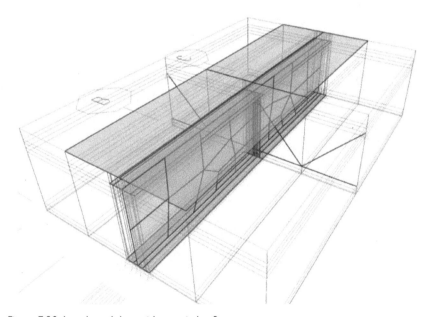

Figure 7.39 Local model at midspan girder 3.

of small, refined submodels, or the local models, which are extracted from the global model. When extracting local models, enough buffer zones surrounding the focus area should be included in the refined local models so that the effect of the notch stress concentration may be negligible. Local model at midspan girder 3 is shown in Figure 7.39.

The interest of this project is to determine the location of fatigue crack, crack path, and crack rate; the global model of the whole bridge cannot be more refined, and therefore the local model of this critical region can be facilitated for this purpose. It is a common issue and also crucial in modeling a local refined model that the boundary conditions of a local model are set up correctly to truly reflect its mechanical connections to the global model.

Boundary conditions of the local model—To set up the boundary conditions, the following guidelines should be followed:

1. The boundary nodes should apply the same displacements obtained from the global model.
2. The boundary nodes should apply the same external forces obtained from the global model as internal forces.

In the local model to investigate the hot spot at the connection plates of the K-type bracing, equivalent forces were applied at the other ends of the K-type bracing. Results obtained from the local model can be used for the calculation of SCF for fatigue study of the hot spots.

Chapter 8

Straight and curved steel box girder bridges

8.1 BEHAVIOR OF STEEL BOX GIRDER BRIDGES

The steel box girder may be defined as a longitudinal structural member with four steel plates, two webs, and two flanges, arranged to form a closed box section as shown in Figure 8.1a. For modern highway structures, the more common arrangement for the box girder is an open top, which is usually referred to as the tub girder. In this case, two steel webs with narrow top flanges similar to those of the plate girders are joined together by a full-width bottom flange as shown in Figure 8.1b. Due to buckling, the thin steel plates' resistance to compression is reduced in comparison to their strengths. An economic design may be achieved when longitudinal and/or transverse stiffeners are provided. Such stiffeners may be of open or torsionally rigid closed sections, as shown in Figure 8.1c for web/bottom flange and top flange, respectively.

During fabrication and erection, the section may be completely open at the top, or it may be braced by a top lateral-bracing system connected to the top flanges (Figure 8.2). A composite box girder bridge may take the form of single box, multibox also called twin box, or multicellular box (Figure 8.3). To close the top opening and complete the box, a reinforced concrete deck slab is added, which acts compositely with the steel section by means of shear connectors to ensure full interaction between them. During construction, the steel girders are subjected to the wet concrete load and other construction loads without the composite action that results from the hardened concrete deck.

During the construction stage, the open box girder behavior may be more complicated than it is closed after the deck concrete is cast. The usual practice of assuming the system to be noncomposite during construction requires substantial top-flange bracing to form a quasiclosed box section. The noncomposite steel section must support both the fresh concrete and the entire construction loads, hence steel box girders are at their critical stage during construction. The open section of the bath-tub girder is a major concern because of its relatively low torsional stiffness. A lateral-bracing system is usually installed to increase

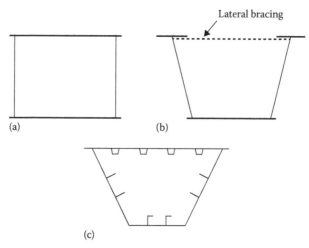

Figure 8.1 Steel box girders. (a) Unstiffened closed box girder. (b) Unstiffened tub girder with lateral bracing. (c) Stiffened closed box girder.

Figure 8.2 Twin-box girder bridge. (Courtesy of TXDOT.)

the torsional stiffness. Bracing systems commonly consist of a horizontal truss attached to the girder near its top flange to increase its torsional stiffness. Consideration of distortional effects may be limited to local regions between internal intermediate diaphragms. The distortion of the cross section can be reduced by using closer internal cross frames and diaphragms. External bracing between girders may be necessary in curved bridges to control the deflections and rotations of the girders, thereby facilitating the placement of the concrete roadway deck. The box girder cross section possesses a high torsional stiffness after the concrete deck gains its full strength because the cross section is considered as a fully closed section. However, internal intermediate diaphragms and top-flange lateral bracing may still be needed as a box girder is an unstable open section with very little torsional stability before the concrete is hardened.

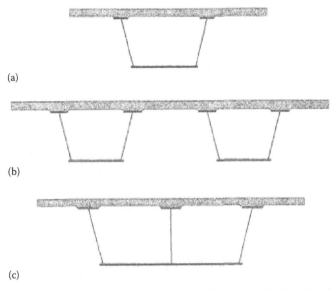

(a)

(b)

(c)

Figure 8.3 Steel/concrete composite box girders. (a) Single box. (b) Multibox (twin-box). (c) Multicellular box.

Horizontally curved box girders applicable for both simple and continuous spans are used for grade separation and elevated bridges where the structure must coincide with the curved roadway alignment. This condition occurs frequently at urban crossings and interchanges and also at rural intersections where the structure must conform to the geometric requirements of the highway. Horizontally curved bridges will undergo bending and associated shear stresses as well as torsional stresses due to the horizontal curvature even if they are subjected only to their own gravitational load. The bridge can be treated as a series of interconnected beams where the beam theory can be used for the behavior of the individual elements. Figure 8.4 shows the general behavior of an open box section under gravity load showing separate load effects. An arbitrary uniform load on a simple-span box girder (Figure 8.4a) contains bending and torsional load components that have corresponding bending and torsional effects, which will be described further in Sections 8.1.1 and 8.1.2.

8.1.1 Bending effects

The bending load (Figure 8.4b), causes the section to

1. Deflect rigidly (longitudinal bending)
2. Deform (bending distortion)

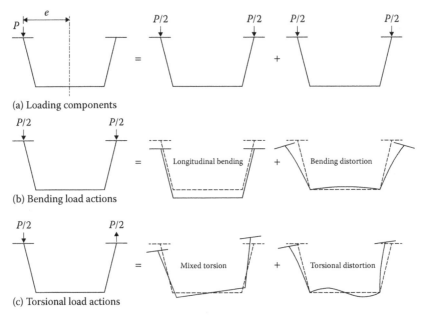

(a) Loading components

(b) Bending load actions

(c) Torsional load actions

Figure 8.4 General behavior of an open box section under gravity load showing separate effect.

8.1.1.1 Longitudinal bending

A survey conducted by ASCE Task Committee on horizontally curved steel box girder bridges revealed that box girders in the United States typically have an average span-to-depth ratio of 23 for single spans and 25 for continuous girder spans (Heins 1978). For girders with such a large span-to-depth ratio, any vertical load may cause significant longitudinal bending and thus longitudinal bending stresses in the girder.

Assuming elastic behavior, normal stresses due to longitudinal bending, f is given according to the beam theory as

$$f = \frac{M}{S} \tag{8.1}$$

where:
   M is the bending moment
   S is the section modulus

Shear stresses associated with the moment gradient also occur and are calculated by

$$f_v = \frac{VQ}{It} \tag{8.2}$$

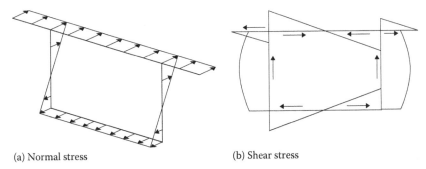

(a) Normal stress   (b) Shear stress

Figure 8.5 (a) Normal and (b) shear stress components of longitudinal bending stress.

where:
V is the shear force
I is the moment of inertia of the section
Q is the first moment of area under consideration
t is the width of the section where shear stress is considered

Normal and shear components of longitudinal bending stress are illustrated in Figure 8.5a and b, respectively.

8.1.1.2 Bending distortion

When any vertical load is applied on a box girder, bending distortion in transverse direction, or local transverse bending, occurs at the same time as longitudinal bending. This local bending effect could be significant before a box girder is closed on top. The AASHTO guide specifications (2003) state that if the box girder does not have a full-width steel top flange, the girder must be treated as an open section. In open box girders, this distortion causes outward bending of the webs, upward bending of the bottom flange, and in-plane bending of the top flange (Figure 8.4b). The transverse bending could cause the cross section to change shape. Therefore, to prevent bending distortion, the top bracing (ties and struts) as shown in Figure 8.2 is usually placed between top flanges.

8.1.2 Torsional effects

In their studies (Hsu 1989; Hsu et al. 1990; Fu and Hsu 1995), Hsu and Fu modified Vlasov's theory on curved thin-walled beams (Vlasov 1965; originally developed for open sections such as I-girders shown in Equation 7.1) to represent the behavior of both open and closed sections for box girder analysis. The torsional load (Figure 8.4c) causes the section to

1. Rotate rigidly (mixed torsion)
2. Deform (torsional distortion)

8.1.2.1 Mixed torsion

In curved box girder bridges, a vertical load may cause the girder twisted about its longitudinal axis because of the bridge curvature. Uniform torsion occurs if the rate of change of the twist angle is constant along the girder and longitudinal warping displacement is not restrained and maintaining a constant. St. Venant analyzed this problem and found that the St. Venant shear stresses occur in the cross section (Figure 8.6). If there is a variation of torque or if warping is prevented or altered along the girder, longitudinal torsional warping stresses develop.

In general, both St. Venant torsion and the warping torsion are developed when thin-walled members are twisted. Box girders are usually dominated by St. Venant torsion because the closed cross section has a high torsional stiffness. Box girders have large St. Venant stiffness, which may be 100–1000 times larger than that of a comparable I-section. The longitudinal normal stresses resulting from the restrained warping in closed box sections are usually negligible (Kollbrunner and Basler 1969).

St. Venant stiffness of the box section is a function of the shear modulus of the steel (G) and the torsional constant J (or K_t), which is related to the cross-sectional geometry. In curved box girder bridges, St. Venant torsion provides most of the resistance that is given by

$$T = GJ \frac{d\theta}{dz} \tag{8.3}$$

where:
  T is the torque on the cross section of the member
  θ is the twist angle of the cross section
  z is the longitudinal axis of the member

For box sections, as shear stress flows are formed in closed cells, equivalent torsional constant for open sections as shown in Chapter 7 is no longer applicable. The torsional constant J in Equation 8.3 for a single-cell box girder is given by

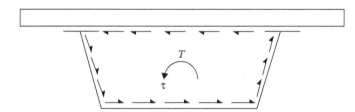

Figure 8.6 St. Venant torsion in a closed section.

$$J = \frac{4A^2}{\sum (b/t)}$$ (8.4)

where:

  A is the enclosed area of the box section
  b is the width of the individual plate element in the box
  t is the thickness of the plate element in the box

For the approximation of the torsional constant of a multicell box, the intermediate webs can be ignored as shear stress flows over these webs are negligible due to countereffects from two adjacent cells. Therefore, Equation 8.4 is still applicable as if the intermediate webs were removed. When calculating torsional constant using Equation 8.4, open-section segments such as cantilevered flanges can be ignored as resistance to torsion from these segments is not comparable to that from closed cells, or simply sum the torsional constant of these open segments (Equation 7.2) and that of closed cells (Equation 8.4) as the total of the entire section.

For analysis purposes, top lateral bracing, as shown in Figure 8.2, may be transformed to an equivalent thickness of plate t_{eq} by

$$t_{eq} = \left(\frac{E}{G} \right)\left(\frac{2A_d}{b} \right)(\cos^2 \alpha \sin \alpha)$$ (8.5)

where:

  E is the steel modulus of elasticity
  G is the steel shearing modulus of elasticity
  A_d is the area of the lateral-bracing diagonal
  b is the clear box width between top flanges
  α is the angle of lateral-bracing diagonal with respect to transverse
    direction

Kollbrunner and Basler (1966) provide a more complete list of equivalent thickness for quasibox girder as shown in Table 8.1.

To properly close the section and minimize warping stresses, the cross-sectional area of the lateral-bracing diagonal A_d should be at least $0.03b$.

The internal stresses produced by St. Venant torsion in a closed section are shearing stresses around the perimeter, as shown in Figure 8.6, and defined by

$$\tau = \frac{T}{2At}$$ (8.6)

Table 8.1 Equivalent thickness of the top bracing for the quasiclosed box

Type no.	Type of lateral bracing	Equivalent thinness (t_{eq})
1	2_b, d, A_d, A_f, λ, $S_d = qd$	$\dfrac{E}{G}\ \dfrac{2\lambda b}{d^3/A_d + 2\lambda^3/(3A_f)}$
2	2_b, d, A_d, A_f, A_v, λ, $S_d = qd$	$\dfrac{E}{G}\ \dfrac{2\lambda b}{2d^3/A_d + 4b^3/A_v + \lambda^3/(6A_f)}$
3	2_b, d, A_d, A_v, A_f, λ, $S_d = \dfrac{qd}{2}$	$\dfrac{E}{G}\ \dfrac{2\lambda b}{d^3/(2A_d) + \lambda^3/(6A_f)}$
4	2_b, d, A_d, A_v, λ, $S_d = \dfrac{qd}{2}$	$\dfrac{E}{G}\ \dfrac{2\lambda b}{d^3/A_d + 8d^3/A_v + \lambda^3/(6A_f)}$

where:

τ is the St. Venant shear stress in any plate

T is the internal torque

A is the enclosed area within the box girder

t is the thickness of the plate

8.1.2.2 Torsional distortion

Torsional load causes the cross section to deform through bending of the walls (Figure 8.4c). Normal stresses as shown in Figure 8.7 result from warping torsion restraint and from distortion of the cross section. If the box girder has no cross frames or diaphragms, the distortion is restrained only by the transverse stiffness of the plate elements. In an open box girder cross section, due to the lack of distortional stiffness, the torsional distortion can be prevented through the use of internal cross frames (Figure 8.8) connecting top and bottom flanges. Figure 8.9 illustrates the general box girder normal stresses, which can occur in a curved or skewed box-shaped girder.

Closed box sections, on the other hand, are extremely efficient at carrying torsion by means of St. Venant torsional shear flow because the shear flow around the circumference of the box has relatively large force couple distances (Figure 8.10). For this reason, a box-shaped girder can carry relatively

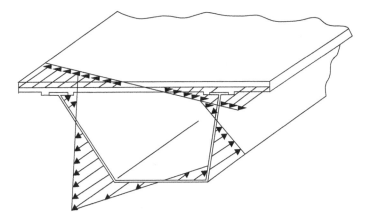

Figure 8.7 Warping stresses in a box girder.

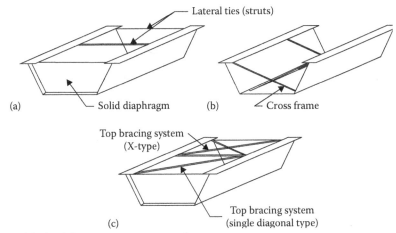

Figure 8.8 (a–c) Bracing system terminology.

large torques with relatively low shear flows. The shear flow around the circumference of the box follows a consistent direction (clockwise or counterclockwise) at any given location along the length of the girder. As a result, when combined with vertical shear in the webs, this shear flow is always subtractive in one web and additive in the other.

In addition, box girders are subjected to cross-sectional distortion when subjected to eccentric loading such as overhang loads and eccentrically applied live loads. This cross-sectional distortion results in out-of-plane (transverse) bending stresses and longitudinal warping normal stresses in the webs and full-width flanges of the box cross section. These bending stresses may be estimated by using a beam-on-elastic-foundation (BEF) analogy method (Wright et al. 1968), which was further improved

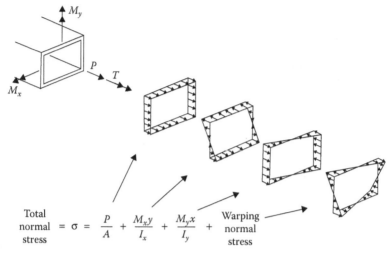

$$\text{Total normal stress} = \sigma = \frac{P}{A} + \frac{M_x y}{I_x} + \frac{M_y x}{I_y} + \text{Warping normal stress}$$

Figure 8.9 Illustration of the general box girder normal stresses, which can occur in a curved or skewed box-shaped girder.

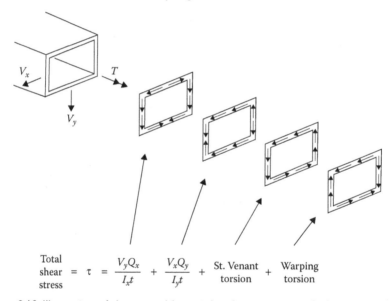

$$\text{Total shear stress} = \tau = \frac{V_y Q_x}{I_x t} + \frac{V_x Q_y}{I_y t} + \text{St. Venant torsion} + \text{Warping torsion}$$

Figure 8.10 Illustration of the general box girder shear stresses, which can occur in a curved or skewed box-shaped girder.

by implementing a equivalent BEF (EBEF) analogy into the beam element model as the supplement (Hsu et al. 1995; Fu and Hsu 1995; Hsu and Fu 2002). As stated in AASHTO (2013), the effects of cross-sectional distortion are typically controlled by providing adequately spaced internal intermediate diaphragms. Cross-sectional distortion, the resulting stress effects, and the design of internal intermediate diaphragms are discussed in detail in Fan and Helwig (2002).

It should be noted that all box girders, even straight box girders, are subjected to torsional loading. Because of the curvature, any vertical load applied on a curved girder bridge, such as structural weight, and centric or eccentric concentrated load will cause torsion. For a straight girder bridge, torsion is caused by eccentric loads such as construction loads or live loads.

8.1.3 Plate behavior and design

Box girders are formed by plates to resist in-plane and out-of-plane loading. For a closed box as shown in Figure 8.11a, all four sides can be treated as plates. For tub girder bridges most frequently used in the United States, the bottom flange is treated as the plate, which can be unstiffened as shown in Figure 8.11b or stiffened as shown in Figure 8.11c. The important geometric parameters are thickness t, width b, and length a, as seen in Figure 8.11b. The ratio b/t, often called the plate slenderness, influences the local buckling of the plate panel; the aspect ratio a/b may also influence buckling patterns and may have a significant influence on strength (SCI 2000; ESDEP course). In a tub girder bridge, the longitudinal length a and the transverse width b can be assumed as the junctions between web and bottom plate and the

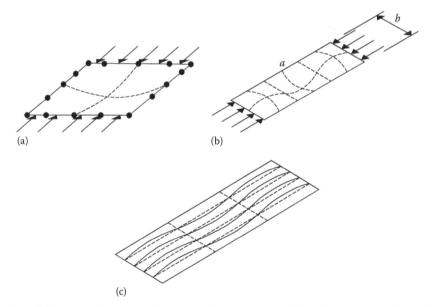

(a)   (b)

(c)

Figure 8.11 Box girder bottom flange under in-plane action. (a) Unstiffened plate with small aspect ratio a/b. (b) Unstiffened plate with large aspect ratio a/b. (c) Stiffened plate.

locations of the vertical bracing, respectively. If the aspect ratio a/b is relatively small, the postbuckling mode appears as shown in Figure 8.11a. As the aspect ratio increases, the critical mode changes, tending toward the multimode situation, all depending on the a/b aspect ratio. In this case, the stiffened bottom flange (as shown in Figure 8.11c) is recommended to assure higher buckling mode for higher strength capacity.

8.2　PRINCIPLE AND MODELING OF STEEL BOX GIRDER BRIDGES

There are many methods available for analyzing curved bridges. Of all the available analysis methods, the finite element method (FEM) is considered to be the most powerful, versatile, and flexible method (FHWA/NSBA/HDR 2012). Among the refined methods allowed by AASHTO LRFD specifications (2013) the three-dimensional (3D) FEM is probably the most involved and time consuming method, and it is the most general and comprehensive technique for static and dynamic analyses capturing all aspects affecting the structural response. The other methods proved to be adequate but are limited in scope and applicability. Due to the recent development in computer technology, the 3D FEM has become an important part of engineering analysis and design. FEA packages are used practically in all branches of engineering nowadays. A complex geometry, such as that of continuous curved steel box girder bridges, can be readily modeled using the finite element technique, in which steel plates and concrete deck of a box girder may be modeled as plane shell elements. The method is also capable of dealing with different material properties, relationships between structural components, boundary conditions, as well as statically or dynamically applied loads. The linear and nonlinear structural response of such bridges can be analyzed with good accuracy using this method. Live load application is the same as that shown in Chapter 7 (Section 7.2.3) where girder influence surfaces are generated to obtain the maximum effects due to live load.

8.2.1　2D and 3D finite element method

In a two-dimensional (2D) grid analysis, the entire tub girder section with concrete slab, steel top and bottom flanges, webs (with or without longitudinal stiffeners), and top-flange lateral bracing is modeled as a beam. The stiffness of the beam can be calculated from the whole cross section of the girder or empirical estimates as shown in the illustrated example. When calculating sectional properties, internal vertical diaphragms or cross frames can be ignored. A relatively torsionally stiff beam element along the centerline of each box (i.e., the shear center) is used to connect the slab at the web positions. This can be done with short *dummy* transverse slab beams modeled with either no stiffness before the hardening of concrete during construction or assigning transverse slab bending stiffness. This form of 2D grid model for a twin-box bridge with cantilevers is illustrated in Figure 8.12.

For a box girder bridge, a 3D finite element model can be used to more accurately simulate each part of the section and bridge component. As shown in Figure 8.13, web and flange plates of a box girder bridge are modeled by plane shell elements, whereas bracing or diaphragm components are modeled by beam or truss elements. As far as finite element modeling is concerned, the same five modeling techniques described in Chapter 7 (Figures 7.10 through 7.14) can be adopted for box girders. Among the five, 3D brick–shell model and 3D shell–shell model are more suited for box sections where the bottom flange is modeled by using shell elements and longitudinal stiffeners by eccentric beam elements to correctly quantify the lateral and torsional stiffness of the cross section. Girder flanges can be modeled by beam or, more commonly, shell elements; webs are modeled by using shell elements (at least two to capture the parabolic-curved shear); and cross frames and bracing are modeled by using truss/beam elements with their respective proper areas and bracing configuration. The deck typically can be modeled by using eight-node solid elements (Figure 8.14) or four-node plane shell elements (Figure 8.15).

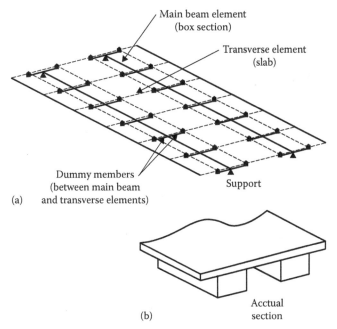

Figure 8.12 (a, b) 2D grillage model for a twin-box girder bridge.

Figure 8.13 Finite element model of a box girder bridge with internal bracing system.

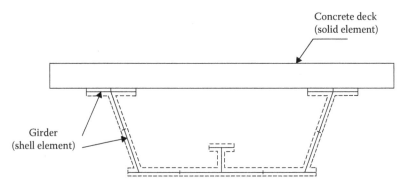

Figure 8.14 3D brick–shell model.

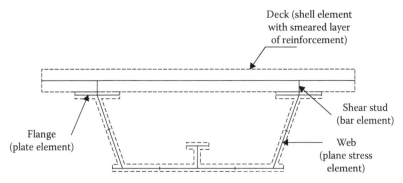

Figure 8.15 3D shell–shell model.

8.2.2 Consideration of modeling steel box girder bridges

8.2.2.1 Design considerations

Steel box girders are at their critical stage during construction because the noncomposite steel section must support both the fresh concrete and the entire construction loads. A comparison study using different modeling techniques was made recently (Begum 2010). Curved and straight box girders in the study are two span bridges that have the same general construction, consisting of a bottom flange, two sloped webs, and top flanges attached to the concrete deck with shear connectors. The negative bending region, where the bottom flange is in compression, is stiffened by longitudinal stiffeners. There are internal diaphragms or cross frames at regular intervals along the span and lateral bracing at top flange. The cross frames maintain the shape of the cross section and are spaced at regular intervals to keep the transverse distortional stresses and lateral bending stresses in flanges at acceptable levels.

8.2.2.2 Construction

From a designer's point of view, the most critical stage is during construction when the box is quasiclosed and the casting sequence of the concrete may affect girder stresses and deflections. Most steel box girder bridges are using disk, pot, or spherical bearing (Figure 8.16), although elastomeric bearing pads have been successfully employed in some applications. Collectively, these bearings are known as high-load multirotational bearings and suited for curved steel box girder bridges. Of the three bearing systems, spherical bearings have the greatest rotation capacity and most trouble-free maintenance record. Pot bearings have been troublesome; disk bearings, on the other hand, have fewer documented failures than pot bearings. The main purpose of these bearings is to allow the girders to expand and contract to accommodate daily and annual thermal changes that the bridge undergoes as well as accommodating construction and live load rotations.

Free or fix of bearings should be correctly simulated in the superstructure analysis model to accurately analyze the response of the structure to various loading conditions. The bearing orientations must be reproduced and modeled correctly, especially for curved bridges, not only for thermal load analysis but also for dead load (DL), live load, and centrifugal force analyses.

Depending on the specific configuration of a structure, improper modeling of bearing conditions (boundary conditions) could have a significant impact on the correctness of the analysis results. Boundary conditions should be carefully modeled, and, in cases where the support stiffness is not known with certainty (e.g., with integral abutments), it may be advisable to run more than one analysis with different assumptions to assess the sensitivity of the structural response to the different boundary condition assumptions,

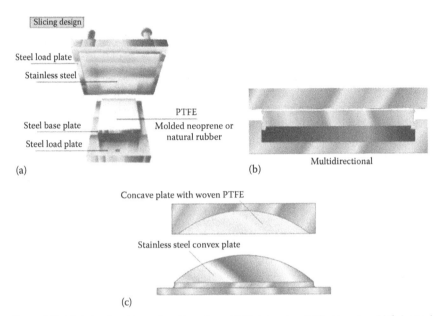

Figure 8.16 High-load multirotational bearings. (a) Disk bearing. (b) Pot bearing. (c) Spherical bearing.

with consideration given to designing for the resulting force and deflection envelopes. Note also that during bridge erection, bearing points may be temporarily blocked (partially fixed), so the construction cases may not have guided or nonguided (free) bearing points. This may be a consideration if significant thermal movements are anticipated at partially erected structural conditions.

Compared to I-shape girder, a steel box girder is stiff and difficult to adjust in the field. NCHRP Report 12-79 (White et al. 2012) advises to detail tub girders for no-load fit or steel DL fit (with consideration given to possible temporary shoring or hold cranes; if sufficient shoring or temporary support is provided, detailing for no-load fit may be more appropriate). It should be noted that almost all structural analyses are based on the assumption that the structure is under initial no-load (undeformed, unstrained) geometry. The stresses and forces in the system are based on the deformations from this configuration, including any lack-of-fit effects (White et al. 2012).

8.2.2.3 Description of the noncomposite bridge models

A two-span noncomposite, single steel box girder bridge as shown in Figure 8.17 is used in this chapter as an example to illustrate different modeling methods. The total span length of this bridge is 97 m (320′). A lateral-bracing system is installed at the top-flange level in the open-top

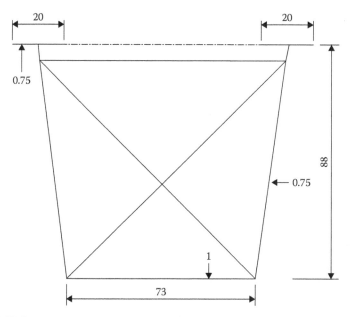

Figure 8.17 Cross-sectional dimensions (in) of the box.

box girder to form a quasiclosed box, thereby increasing the torsional stiffness. Crossed diagonal bracing systems are considered part of lateral-bracing systems. Internal transverse bracing or internal cross frames are provided at regular intervals in the box. In the negative bending region, longitudinal stiffener is provided in the bottom flange. The cross-sectional dimensions are shown in Figure 8.17.

To compare the differences resulting from curvature, the same bridge is modeled as both straight and curved bridges. There are four different types of models, which are as follows:

1. Straight box shell model (M1)
2. Curved box shell model (M2)
3. Straight box beam model (M3)
4. Curved box beam model (M4)

8.3 2D AND 3D ILLUSTRATED EXAMPLES OF A STRAIGHT BOX GIRDER BRIDGE

The finite element modeling and analysis performed in this example for a straight bridge and in the next example for a curved box girder bridge is done using a general purpose, multidiscipline finite element program, ANSYS. ANSYS has an extensive library of truss, beam, shell, and solid elements. Shell elements were used to model the structural components of the box girder bridges (webs, bottom flange, top flanges, and the solid diaphragms), whereas truss and beam elements were used to model top bracing trusses and cross frames:

1. Shell 63 (elastic shell). It is a four-node element that has both bending and membrane capabilities. The element has five degrees of freedom at each node; translations in x, y, and z directions; and rotations about y and z axes (of the element's local coordinate system). Large deflection capabilities are included in the element. This type of element can produce good results for a curved shell surface provided that each flat element does not extend over more than a 15° arc.
2. Link 8 (3D spar). It is a two-node, 3D truss element. It is a uniaxial tension–compression element with three translational degrees of freedom at each node. The element used for bracing is a pin-jointed structure with no bending capabilities. Plasticity and large deflection capabilities are included. The required inputs for this element are material properties and cross-sectional area.
3. Beam 188 (3D linear finite strain beam). It is a 3D linear (two-node) or quadratic beam element. Beam 188 has six or seven degrees of freedom at each node. These include three translations and three rotations in x, y, and z directions (of the element's local coordinate system). A seventh degree of freedom (warping displacement) can also be considered. This element is well-suited for linear, large rotation and/or large strain nonlinear applications. The beam elements are one-3D line elements.
4. Beam 4 (3D elastic beam). It is a uniaxial element with tension, compression, torsion, and bending capabilities. The element used for longitudinal stiffeners has six degrees of freedom at each node; translations in x, y, and z directions; and rotations about x, y, and z axes (of the element's local coordinate system). Stress stiffening and large deflection capabilities are included. The required inputs for this element are cross-sectional properties such as moment of inertia, area, and torsional constant.

Results from the ANSYS finite element models can be used in understanding the box bridge behavior. In addition, they can be used to compare the stress profiles. Therefore, creating the same general construction of straight and curved bridge models with same boundary conditions is required.

8.3.1 Straight box shell model (M1)

Straight box bridge model is made using Shell 63 elements for webs, the top flange, and the bottom flange. Shell 63 elements are used as well to model longitudinal and transverse stiffeners and solid diaphragms at the

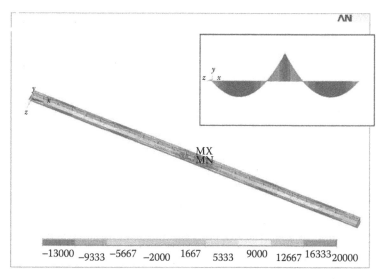

Figure 8.18 Stress contour of straight box shell model (insertion shows moment distribution).

support location. The plate thicknesses and the material properties are required inputs for Shell 63. Link 8 elements were used to model the top bracing truss and the cross frames. The stress contour of the straight box is shown in Figure 8.18.

8.3.2 Straight box beam model (M3)

The straight box beam model is made using Beam 188. Beam 4 (with a hinge at an end or a truss element) is used at supports to provide bearing support and apply boundary conditions. In the beam element model, the bracing and stiffener effects are not considered. Two cases are modeled: (1) two bearings are provided at all supports and (2) two bearings are provided in the middle support (at pier) and single bearing is provided at end supports. The insertion in Figure 8.18 shows the bending moment throughout the span.

8.3.3 Comparison results

To better understand the structural behavior, the same box girder bridge is also analyzed with DESCUS-II, a dedicated design and analysis system for straight or curved box girder bridges by using beam models. Tables 8.2 and 8.3 compared analysis results from these two systems. Table 8.2 compares support reactions, moments, and bending stresses for 2D and 3D models. In the model, twin bearings are supplied to all supports. Not like the curved model shown in Section 8.4, twin bearings at all supports behave no different from twin bearings at support 2 only. Table 8.3 compares the

Table 8.2 Comparison of a straight box girder in beam model (twin bearings at all supports)

	Support reactions in kip (kN)		Moment due to DL in lb-in (kN-m)			Bending stress in kip/in² (MPa)		
Location	DESCUS	ANSYS	Location	DESCUS	ANSYS	Location	DESCUS	ANSYS
At support 1	205.3 (913)	217.07 (966)	M at 4	7.4E+07 (8361)	7.7E+07 (8700)	Top DL stress at 4	−14.24 (−198.18)	−14.16 (−97.63)
At support 2	684.4 (3044)	715.71 (3183)				Bottom DL stress at 4	9.55 (65.84)	10.266 (70.78)
At support 3	205.3 (913)	217.06 (965)	M at 10	−1.3E+08 (−14688)	−1.37E+08 (−15479)	Top DL stress at 10	25.43 (37.44)	25.07 (172.85)
						Bottom DL stress at 10	−17.05 (−117.56)	−18.17 (−125.28)

Table 8.3 Stress comparison of straight steel box girder

	Bending stress in kip/in² (MPa)	
Location	Shell model (ANSYS)	Beam model (ANSYS)
Top DL stress at 4th pt	−14.63 (−100.87)	−14.162 (−97.64)
Bottom DL stress at 4th pt	10.53 (72.6)	10.266 (73.5)
Top DL stress at 10th pt	25.9 (178.57)	25.07 (172.85)
Bottom DL stress at 10th pt	−14.9 (−102.73)	−18.173 (−125.3)

stresses. Except the bottom stress at the interior support, other stresses are very close between the 2D and 3D FEM models.

8.4 2D AND 3D ILLUSTRATED EXAMPLES OF A CURVED BOX GIRDER BRIDGE—METRO BRIDGE OVER 1495, WASHINGTON, DC

8.4.1 Curved box shell model (M2)

This curved box having a radius of 91 m (300′) is modeled using Shell 63 similar to that of the straight box shell model. This model was the same as model M1, except that it is horizontally curved and two spans of 160 ft (48.8 m) each. Both M1 and M2 have the same general construction as mentioned in the description of the noncomposite bridge model. The stress contour of the curved box is shown in Figure 8.19.

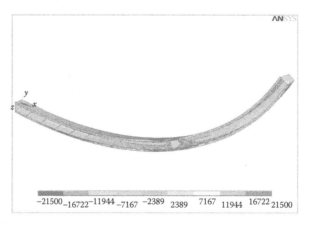

Figure 8.19 Stress contour of curved box shell model.

8.4.2 Curved box beam model (M4)

In this model, Beam 188 is also used to model the curved box girder. The model is similar to that of the straight box except the curvature. Similar to the straight box beam model, Beam 4 (with a hinge at an end or a truss element) is used at supports to model boundary conditions. Two cases regarding boundary conditions are modeled: (1) Two bearings are provided at all supports and (2) two bearings are provided at the middle support (at pier) and single bearing is provided at end supports. Figure 8.20 shows the bending moment diagram throughout the span.

The stability of the single box girder under the maximum overturning combination of DLs, wind load, and live load with its centrifugal effects is also analyzed. Figure 8.21 shows the reaction forces where the insertion shows the twin-bearing intermediate support at the pier location. The maximum bearing conditions under various load combinations are checked. The beam element shown in solid circle in the insertion of Figure 8.21 representing the shear center is connected with rigid links supported by two hinges at bearings. Tables 8.4 through 8.6 compare beam and shell models for curved box girder bridges. Tables 8.4 and 8.5 compare support reactions, moments, and bending stresses for 2D and 3D models.

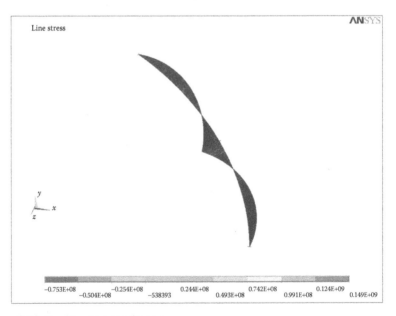

Figure 8.20 Bending moment diagram.

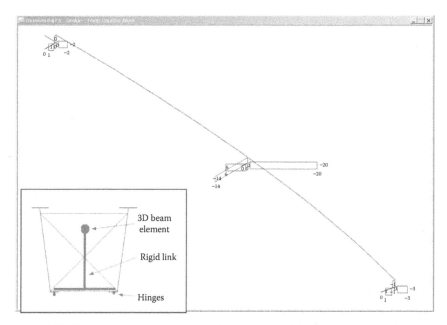

Figure 8.21 3D beam element model with boundary conditions and their reactions.

Table 8.4 is for twin bearings at all supports, and Table 8.5 is for twin bearings at interior support only. Slight shifts on support reactions, moments, and bending stresses between twin bearings at all supports and at support 2 only are noticed. Table 8.6 compares the stresses. All bending stresses for beam model are higher (or more conservative) than those by 3D FEM model.

8.5 2D AND 3D ILLUSTRATED EXAMPLES OF THREE-SPAN CURVED BOX GIRDER BRIDGE—ESTERO PARKWAY BRIDGE, LEE COUNTY, FLORIDA

This three-span steel box girder bridge project is part of Estero Parkway, Lee County, Florida. The bridge was designed as a three-span continuous bridge during construction (Figure 8.22), but opened to traffic as a two-span continuous bridge (Figure 8.23) with spans of approximately 97.5–70.1 m (320′–230′) with a 1036.3 m (3400′) radius. Top view of the bridge under construction is shown in Figure 8.24. Table 8.7 lists parameters

Table 8.4 Comparison of a curved box girder in beam model (twin bearings at all supports)

Support reactions in kip (kN)			Moment due to DL in lb-in (kN-m)			Bending stress in kip/in² (MPa)		
Location	DESCUS	ANSYS	Location	DESCUS	ANSYS	Location	DESCUS	ANSYS
At support 1	202.2 (899)	211.973 (943)	M at 4	7.2E+07 (8135)	7.5E+07 (8474)	Top DL stress at 4	-13.98 (-96.39)	-13.69 (-94.39)
At support 2	680.8 (3028)	729.61 (3245)				Bottom DL stress at 4	9.37 (64.6)	9.926 (68.44)
At support 3	211.9 (943)	208.255 (926)	M at 10	-1.4E+08 (-15818)	-1.51E+08 (17061)	Top DL stress at 10	26.67 (183.89)	27.616 (190.41)
						Bottom DL stress at 10	-17.88 (-123.28)	-20.018 (-138.02)

Table 8.5 Comparison of a curved box girder in beam model (twin bearings at support 2 only)

Support reactions in kip (kN)			Moment due to DL in lb-in (kN-m)			Bending stress in kip/in² (MPa)		
Location	DESCUS	ANSYS	Location	DESCUS	ANSYS	Location	DESCUS	ANSYS
At support 1	202.2 (899)	206.09 (917)	M at 4	6.5E+07 (7344)	6.8E+07 (7683)	Top DL stress at 4	-12.66 (-87.29)	-12.43 (-85.7)
At support 2	680.8 (3028)	740.15 (3292)				Bottom DL stress at 4	8.49 (58.54)	9.011 (62.13)
At support 3	211.9 (943)	203.6 (906)	M at 10	-1.5E+08 (-16948)	-1.67E+08 (-18869)	Top DL stress at 10	29.84 (205.74)	30.643 (211.28)
						Bottom DL stress at 10	-20 (-137.9)	-22.21 (-153.13)

Table 8.6 Stress comparison of curved steel box girder

	Bending stress in kip/in² (MPa)	
Location	Shell model (ANSYS)	Beam model (ANSYS)
Top DL stress at 4th pt	−9.124 (62.91)	−13.693 (−94.41)
Bottom DL stress at 4th pt	5.73 (39.51)	9.926 (68.44)
Top DL stress at 10th pt	25.68 (177.06)	27.616 (190.41)
Bottom DL stress at 10th pt	−17.96 (−123.83)	−20.018 (−138.02)

Figure 8.22 Estero Parkway Bridge constructed as a three-span continuous bridge.

of the Estero Parkway Bridge. Calculation of the box section properties (at midspan location) in three stages, noncomposite ($N = $ infinity), long-term composite ($N = 3n$), and short-term composite ($N = n$) sections, are shown in Figure 8.25. Equivalent thickness of top lateral-bracing plate t_{eq} is calculated using Equation 8.5. The steel section during construction is considered a quasiclosed section enclosed by top lateral bracing, webs, and the bottom flange, whereas the composite box sections for long-term and short-term consideration are enclosed by the top slab and the steel

Figure 8.23 Estero Parkway Bridge opened to traffic as a two-span continuous bridge.

Figure 8.24 Top view of the Estero Parkway Bridge under construction.

Table 8.7 Description of the Estero Parkway Bridge

Description	Variable
Number of girders	4
Number of spans	2
Radius of curvature of girder 1 (inner)	3386.5′ (1032.2 m)
Radius of curvature of girder 4 (outer)	3476.5′ (1059.6 m)
Span lengths of girder 1	322.15′–229.74′ (98.2–70.0 m)
Span lengths of girder 4	340.93′–225.61′ (103.9–68.8 m)
Spacing between girders	30.0′ (9.1 m)
Roadway width	96.0′ (29.3 m)
Overhang width, left and right	8.58′ (2.6 m)
Curb width, left and right	8.58′ (2.6 m)
Design slab depth (excluding integral wearing surface)	9.5″ (241 mm)
Integral wearing surface	0.5″ (13 mm)
Haunch depth and width	2.5″ and 24.0″ (64 and 610 mm)
Type of connection	Composite
Ultimate strength of concrete	4.5 ksi (31 MPa)
Yield strength of steel	50 ksi (A992) (345 MPa)
Live loading	HL-93 by AASHTO LRFD

section itself. Equation 8.4 is used for the calculation of torsional constant J (or K_t). Also, Q_{slab} is the first moment of inertia of the slab and I_x is the moment of inertia of the composite section where Q_{slab}/I_x is used for the calculation of the shear connectors. Herein, n is the modulus ratio of steel girder to concrete deck and N is the actual modulus ratio used in that stage where $N = \infty$ means steel section only. For 4000-psi (27.6 MPa) normal concrete, $n = 8$ is used.

Bridge was modeled with 3D FEM for detailed design. To investigate the bridge behavior under construction, DESCUS-II was used to build several grid models. Figure 8.26 shows a 3D rendering of the girders only. Thickened sections can be seen in the negative moment area with spans of approximately 97.5–70.1 m (320′–230′). Due to heavy box sections and traffic control, three-span arrangement was proposed and constructed as shown in Figure 8.27 with spans of 39.6–57.9–70.1 m (130′–190′–230′) while the two-span finished model is shown in Figure 8.28. Design has to make sure that the negative moment areas near the temporary supports may carry steel and concrete DLs and construction loads. Analyses demonstrated that the strength capacities and displacements are adequate during all construction stages.

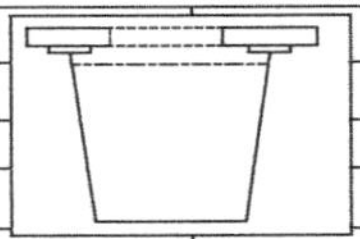

(a)

Bridge Input Parameters		Effective Width Calculation		Full Girder Width	
Span L =	334.64 ft	$L_{eff}/4$	752.94		
Slab t =	9.5 in	$12t+\max(b_f/2,t_w)=$	128	Eff. width = Total use	330
Girder S =	30 ft	$S=$	360		
Top Lat Brac =	0.1 in		10 in (distance from top)	175.84 in	
Haunch =	2.5 in		28 in		
Top flange	28	$W_t \times t_t$	1.5 in	180.82 in	
Web	0.938	$t_w \times D_w$	132 in	Angle (deg) 14°	0.970296
Bot flange	116	$W_b \times t_b$	1.5 in		
Total steel D			135 in		

(b)

Section	A	d_b	Ad_b	d	Ad^2	l_0	b/t	$J_t = bt^3/3$
Top lat brac							1758.36	63.00
Top flange	84.00	134.25	11277.00	78.29	514919.98	15.75	268.09	6.02
Web	255.21	67.5	17226.87	11.54	34012.53	370569.13	76.67	
Bot flange	172.50	0.75	129.38	−55.21	525722.67	32.34	2103.12	69.02
Sum Σ =	511.71		28633.25		1074655.18	370617.23		
Enclosed A_0	17850.06 in²		$K_t = 4A_0^2/\Sigma(b/t)+l_t$	606073.3 in⁴				
y_{bot} =	55.96 in		y_{top} =	79.04 in				
$I_x = l_0+Ad^2$ =	1445272.4 in⁴							
S_{bot} =	25828.88 in³		S_{top} =	18284.3 in³				

Figure 8.25 (a) Box beam section properties at the midspan location. (b) Section properties (N = Infinity). Highlighted areas are input items. S_{top}, S_{bot}, and S_{topc} refer to section moduli at the top and bottom of the steel girder and the top of concrete slab, respectively. (Continued)

Section	A	d_b	Ad_b	d	Ad^2	I_0	b/t	$J_t = bt^3/3$
Slab	130.63	140.75	18385.47	67.33	592080.98	982.41	456.81	3.08
Haunch	2.33	135.5	316.17	62.08	8991.09	0.19	2.14	0.39
Top flange	84.00	134.25	11277.00	60.83	310774.82	15.75	0.05	31.50
Web	255.21	67.5	17226.87	−5.92	8958.92	370569.13	290.07	
Bot flange	172.50	0.75	129.38	−72.67	911081.54	32.34	76.67	
Sum Σ =	644.67		47334.88		1831887.36	371599.83	825.74	34.97
Enclosed A_0	19820.11 in²		$K_t = 4A_0^2\Sigma(b/t) + I_t$	1902982 in⁴				
y_{bot} =	73.42 in		y_{top} =	61.58 in				
$I_x = I_0 + Ad^2$ =	2203487.2 in⁴		S_{topc} =	29948.79 in³				
S_{bot} =	35785.33 in³		S_{top} =	35785.33 in³				

(c)

Section	A	d_b	Ad_b	d	Ad^2	I_0	b/t	$J_t = bt^3/3$
Slab	391.88	140.75	55156.41	47.69	891302.99	2947.23	152.27	83.27
Haunch	7.00	135.5	948.50	42.44	12608.83	0.58	0.71	5.50
Top flange	84.00	134.25	11277.00	41.19	142524.55	15.75	0.05	31.50
Web	255.21	67.5	17226.87	−25.56	166717.47	370569.13	290.07	
Bot flange	172.50	0.75	129.38	−92.31	1469855.48	32.34	76.67	
Sum Σ =	910.59		84738.15		2683009.32	373565.04	519.77	120.27
Enclosed A_0	19820.11 in²		$K_t = 4A_0^2\Sigma(b/t) + I_t$	3023266 in⁴				
y_{bot} =	93.06 in		y_{top} =	41.94 in				
$I_x = I_0 + Ad^2$ =	3056574.4 in⁴		S_{topc} =	56664.86 in³				
S_{bot} =	32845.65 in³		S_{top} =	72877.48 in³				
Q_{slab} =	17905.27 in³							

(d)

Figure 8.25 (Continued) (c) Section properties ($N = 3n = 24$). (d) Section properties ($N = n = 8$).

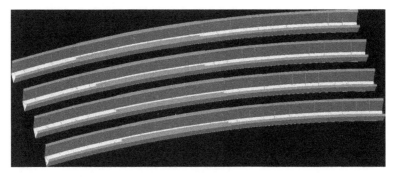

Figure 8.26 Rendering of the Estero Parkway Bridge.

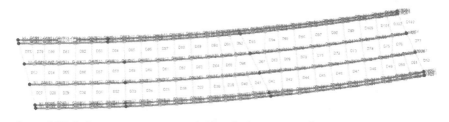

Figure 8.27 A three-span continuous bridge during construction.

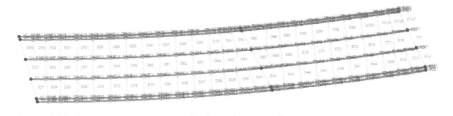

Figure 8.28 A two-span continuous bridge when complete.

Chapter 9

Arch bridges

9.1 INTRODUCTION

The arch bridge is one of the oldest types of brides and has been in existence in the world since more than 2000 years (Brown 2005). The Romans were the first to take the advantages of the arch in building bridges. There are more than 900 ancient Roman bridges found in Europe; most of them are arch bridges. Applying arch into bridges and buildings has a long history also in the East. The Anji Bridge, the oldest open-spandrel segmental stone arch bridge with a central span of 37 m, was built in AD 605 in Hebei, China (Figure 9.1). The use of cast iron as dovetails to interlock stone segments and open spandrels so as to reduce structural weight and to increase water flow during flooding made it a milestone in the long history of arch bridges. Its survival of at least eight wars, ten major floods, and numerous earthquakes, especially the 7.2-richter-magnitude earthquake in 1966, Xingtai (40 km away from the site) demonstrates the strength and advantage of the arch bridge.

Arch is sometimes defined as a curved structural member spanning an opening and serving as a support for the loads above the opening. This definition omits a description of what type of structural element; a bending and/or an axial force element makes up the arch. Nomenclatures used to describe the arch bridges are outlined in Figure 9.2. A true or perfect arch, theoretically, is one in which only a compressive force acts at the centroid of each element of the arch. The shape of the true arch can be thought of as the inverse of a hanging chain between abutments. It is practically impossible to have a true arch bridge, except for one loading condition. However, an arch is usually subjected to multiple loadings, which will produce bending stresses in the arch rib that are generally small compared with the axial compressive stress.

Arch bridges have great natural strength. In addition to pushing straight down, the weight of an arch bridge is carried outward along the curve of the arch to the supports at each end. These supports carry the load and keep the ends of the bridge from spreading out. When supporting its own weight and the weight of crossing traffic, every part of the arch is under

Figure 9.1 The Anji Bridge, Hebei, China, built in AD 605 and still in use.

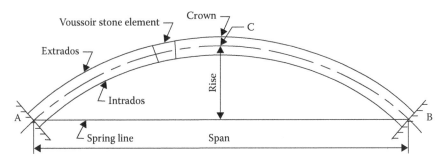

Figure 9.2 Nomenclatures used in the arch bridge.

compression. For this reason, arch bridges must be made of materials that are strong under compression. Most ancient arch bridges were merely built by stones that stay together by the sheer force of their own weight. Today, materials like steel and prestressed concrete have made it possible to build longer and more elegant arches. In the first decade of the twenty-first century, many arch bridges with main spans ranging from 400 to 550 m were built in China, which demonstrate its competitiveness against cable-stayed bridges in situations where the foundation is favorable. Table 9.1 lists the top 10 longest arch bridges in the world.

Table 9.1 Top 10 longest arch bridges in the world

No.	Name	Main span (m)	Year of built	Location
1	Chaotianmen Bridge	552	2009	Chongqing, China
2	Lupu Bridge	550	2003	Shanghai, China
3	Bosideng Bridge	530	2012	Hejiang, China
4	New River Gorge Bridge	518	1977	Fayetteville, North Carolina
5	Bayonne Bridge	510	1931	Kill Van Kull, New York
6	Sydney Harbor Bridge	503	1932	Sydney, Australia
7	Wushan Bridge	460	2005	Chongqing, China
8	Mingzhou Bridge	450	2011	Ningbo, China
9	Zhijing River Bridge	430	2009	Dazhiping, China
10	Xinguang Bridge	428	2008	Guangzhou, China

9.1.1 Classifications of arch bridges

An arch bridge has many variations according to structural arrangements, structural behaviors, and materials. Based on the arrangements of the main arch and the deck system, arch bridges are usually classified as (1) deck arch bridge, (2) half-through arch bridge, and (3) through arch bridge (Fox, 2000). As shown in Figure 9.3, a deck arch bridge is one where the bridge deck locates completely above the crown of arch; a through arch bridge is one where the deck locates at the springing line of the arch; and half-through arch bridge is one where the deck locates at an elevation between a deck arch and a through arch. When choosing a type of arch bridge among these three arrangements, the deck elevation is the primary control factor.

Horizontal outward thrust at abutments distinguishes an arch bridge from other types of bridge. The counterbalance of such outward thrust from the abutments, which reduces the bending effects in the arch, however, requires foundations capable of resisting huge horizontal thrust. Situations where foundations are not permissive, the arch can be tied horizontally by the deck or external tendons. When tied, the horizontal outward thrust is balanced internally, instead of externally by foundations. In this regard, arch bridges can be classified as (1) thrusting arch bridge and (2) nonthrusting arch bridge. A nonthrusting arch bridge, which is often called a tied-arch bridge, is widely used as there is no additional horizontal thrust requirement in the foundation.

Traditionally, a deck-through arch bridge is tied as the *tie* at the deck level connecting two ends of the arch. It is the most effective way to balance the outward thrust. A half-through arch bridge can also be tied at the deck level, in which tying forces are transferred to the main arch from side arches in two side spans. Chaotianmen Bridge and Lupu Bridge (both in China as shown in Figures 9.3b and 9.8 later in the chapter, respectively) are the first two world record keepers for arch bridges by their main span. Both of them

(a)

(b)

Figure 9.3 Types of arch bridges. (a) Deck arch bridge. (New River Gorge Bridge, http://en.wikipedia.org/wiki/New_River_Gorge_Bridge.) (b) Half-through arch bridge. (Chaotianmen Bridge, China, Courtesy of China Communications Construction Company Ltd.). (Continued)

are tied half-through arch bridges. Although a deck arch bridge can be tied at the deck level in a similar pattern, a tied deck arch is not commonly used.

When an arch bridge is tied, externally, the whole structure will behave as a single span of a simply supported girder bridge. The moment distributed to the arch and tie is related to the stiffness ratio of the arch to tie. A tied-arch bridge can further be classified as (1) stiffened arch with flexible tie,

(c)

(d)

Figure 9.3 (Continued) Types of arch bridges. (c) Through arch bridge. (Pentele Bridge, Hungary, Courtesy of SkyscraperCity.com.) (d) Multispan deck arch bridge. (Paso de los Toros Bridge, Uruguay, Courtesy of Taringa.net.)

(2) stiffened arch with stiffened tie, and (3) flexible arch with stiffened tie. As local moments due to live loads are inevitable, a flexible tie girder will distribute more live loads to arch and the arch requires a higher bending stiffness to resist moments; a stiffened tie girder will distribute less live loads to arch and the arch does not need a higher bending stiffness. Stiffnesses of the arch and the tie girder are dependent on each other; it is possible to optimize the size of each according to the goal established for aesthetics and/or cost.

Multispan arch bridges are also commonly used. Compared to a single-span arch bridge, a multispan arch bridge balances horizontal thrusts due to dead loads at interior piers. Figure 9.3d shows an example of multispan deck arch bridge.

Because compression is predominated in the arch, the arch can be built by stone, concrete, or steel. A composite system of concrete-filled steel tube has also been widely used since the last two decades. High-strength pre-stressing tendons are commonly used as the tie in tied-arch bridges. In long-span arch bridges, the main arch can be made of steel trusses, as the New River Gorge Bridge shown in Figure 9.3a. More details of this type are discussed in Chapter 10—Steel Truss Bridges.

An arch bridge can be so designed and built to release live load moments at crown and/or springing. As shown in Figure 9.4, an arch bridge can be classified as (1) nonhinge arch, (2) one-hinged arch, (3) two-hinged arch, and (4) three-hinged arch.

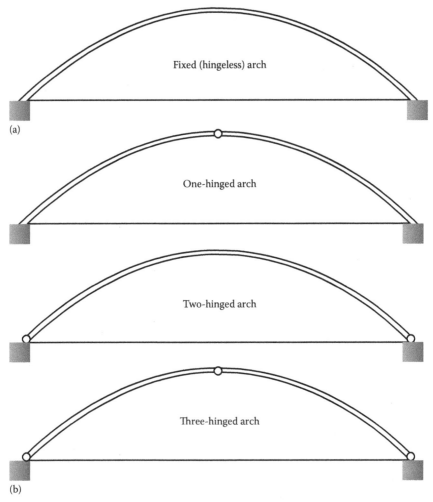

Figure 9.4 Illustration of (a) fixed and (b) hinged arch bridges.

9.2 CONSTRUCTION OF ARCH BRIDGES

Because of the curve of the arch and the restrictions of site access such as deep valleys and confined construction space, construction methods of the main arch include conventional segmental assembling, on-site rotation, and other sophisticated erecting methods. The segmental assembling construction of an arch bridge is similar to the segmental erecting of a cable-stayed bridge. The erected arch is usually hoisted by temporary cables from a temporary pylon. This conventional method is commonly used worldwide. The Hoover Dam Bypass Bridge (USA, 2009) and Lupu Bridge (China, 2003), as shown in Figures 9.5 and 9.9a later in the chapter, respectively, are examples of segmental erecting construction.

The rotation method is unique to arch bridge construction, in which the two halves of the arch are assembled in the lower position along the bridge axis or on shores perpendicular to the bridge axis. When the two half arches are assembled along the bridge axis, the arch will be lifted up and rotated vertically to closure position. Alternately, when they are assembled on shores, the arch will be rotated horizontally to closure position. The rotation method is widely used in mountain areas in China and had been greatly developed in the last two decades. For example, the Zhenzhu Bridge, as shown in Figure 9.6, was built by vertically rotating the arch from up to down, in which the two halves of the concrete arch were casted on-site

Figure 9.5 Cable-stayed segmental erecting of arch bridge. (Hoover Dam Bridge, Nevada, USA, Courtesy of galleryhip.com/hoover-dam-bridge-construction.html.)

Figure 9.6 Vertical casting and rotating of arch bridge. (Zhenzhu Bridge, China, 2008, Courtesy of Guizhou Bridge Construction Group Ltd.)

vertically and then vertically rotated to closure position. The construction of the Yajisha Bridge, which will be introduced in detail in Section 9.2.2, combines rotations in both vertical and horizontal methods.

In addition to these two predominant construction methods, a concrete arch can be designed to use CLCA method (Kawamura 1990) to cast concrete on-site. CLCA stands for concrete lapping with pre-erected composite arch. When using CLCA method to erect an arch, the concrete arch contains steel tubes as the core of the composite section, which will be acting as the falsework to form the whole composite arch. The steel tubular arch will be erected first by using either segmental erecting or rotation method and then concrete will be filled into the tube. After the core of the composite section is formed, the concrete-filled steel tube arch will be used as the falsework to support the form works to cast the outer concrete on-site.

In Sections 9.2.1 and 9.2.2, Lupu Bridge and Yajisha Bridge will be used as two examples to introduce the common construction methods of an arch bridge.

9.2.1 Lupu Bridge, People's Republic of China

The Lupu Bridge (Figures 9.7 and 9.8) is crossing over Huangpu River, Shanghai, China. Once the world record keeper, this bridge is a steel

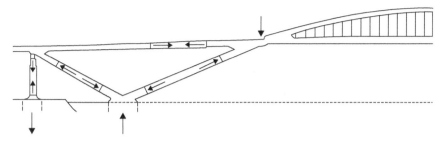

Figure 9.7 Lupu Bridge side span.

Figure 9.8 Lupu Bridge, once world record keeper, China, 2003.

half-through tied box girder arch bridge and it is also the only steel arch bridge in the world that is completely welded. High-strength strands are used as the tie connecting two ends of the deck at side spans. Horizontal thrusts due to dead loads are balanced through two side arches (Figure 9.7). The tie system is separated from the deck system. Two major obstacles, which made the construction of Lupu Bridge unique, are the assembling of large steel box arch segments in skew and the connecting of the steel box arch to concrete springing. Highlights of the construction are briefed in Subsections 9.2.1.1 through 9.2.1.3.

9.2.1.1 Foundations

Huge vertical loads plus horizontal thrust under live loads demand foundations with higher bearing capacity. As shown in Figure 9.7, foundations at side spans also require resistance to uplift. Because of the very thick soft clay of the river, the on-site geotechnical condition does not favor a mass foundation and thus leaves piled foundation the best option for Lupu Bridge. The foundations consist of long steel tube piles with large diameter. The larger surface area of the piles implies that they are frictional rather than bearing piles. The foundations are also strengthened by the use of large-diameter soil–cement stirring piles to resist the horizontal force and limit the displacement due to this force. These stirring piles are connected to each other to improve the integrity of the system. The large number of foundations and addition of stirring piles are partly due to the great working and construction loads.

Due to the horizontal force generated by live loads, the abutment and foundations are also strengthened in the horizontal direction by the use of prestressing concrete. In addition to vertical piles, inclined piles are also used to the abutments.

9.2.1.2 Arch ribs

The arches are segmentally constructed using a cable-stayed cantilever method shown in Figure 9.9a. Each section of the arch was stayed back to the temporary towers at either side of the arch after being welded to the previous section. This significantly reduces the bending stresses in the arch during construction and instead puts the constructed arch section into compression, as it would be upon completion. Cables from the temporary towers to the ground are connected at the location of the foundations that will be resisting uplift on completion of the bridge. Using the same foundations reduces the cost as extra supports were not needed during construction.

As shown in Figure 9.9b, each erecting segment contains two segments of the main arch ribs laterally, where these two segments are connected by a horizontal wind brace box section. A mobile carriage is used to lift braced arch sections up from barges. A computer-controlled system was used to synchronize the strand jacks during deck lifting. The carriage then holds the section in place while it is welded to the previous section. This secure system is favored as it reduces differential movement between the existing and new segments, allowing smoother application of the welding process.

9.2.1.3 Deck girders

For the midspan girders, a conventional suspension bridge construction method is used. After the closure of the main arch, high-strength tie cables are installed to connect the two ends of the deck at side spans. The horizontal forces from the deck ends are transferred to the main arch at springing through the side-span arches, as shown in Figure 9.9.

After the tie cables are installed and tensioned, the construction towers and stayed cables are removed. The segmental erection of the deck girder starts. The transporting and lifting of deck girder segments are similar to those of arch segments. The girders are installed from the center of the arch outward, to ensure that the sag in the horizontal cables is uniform and no distortion of the deck occurred. Another reason is that the load being put on the arch can be carried in compression whereas if spans are introduced at other points, large bending moments will be induced in the arch.

Figure 9.9 The construction of Lupu Bridge. (a) Elevation view. (b) View of lifting an arch segment. (c) Deck construction.

9.2.2 Yajisha Bridge, People's Republic of China

The Yajisha Bridge, crossing over Pearl River, Guangzhou, China, is a half-through arch bridge with a main span of 360 m and two side spans of 76 m. The main arch is fixed at springing. As shown in Figure 9.10, the main arch contains several concrete-filled steel tubes, and steel tubes are connected with other steel tubes as a truss. The bridge was started to be built in July 1998

Figure 9.10 Yajisha Bridge, China, 2000, a half-through steel tubular concrete arch bridge with a main span of 360 m.

and finished in June 2000. What distinguishes the Yajisha Bridge is not its truss-like steel tubular concrete main arch as shown in Figure 9.11, but its combined construction method of vertical rotation and horizontal rotation. Each of the two half arches is first assembled on shore at a lower vertical position. After assembled on falsework, the half arch is vertically lifted to the design elevation and then rotated horizontally to meet the bridge axis.

9.2.2.1 Cross section of the main arch

The main arch comprises six steel tubes, each with a diameter of 750 mm, as shown in Figure 9.11. Three steel tubes are connected by steel plates, horizontally forming an arch rib. Two ribs on the top and bottom are connected by steel tubes with a diameter of 450 mm as vertical and diagonal truss members, forming a composite arch cross section (Figures 9.11 and 9.12d). After closure, top and bottom ribs are filled with concrete. The composited arch section varies from 4000 mm at crown to 8039 mm at springing, while maintaining a constant width of 3450 mm.

9.2.2.2 Vertical rotation

Main arch is split into several segments, and each segment is fabricated off-site. Each of the half arch is first assembled segment by segment with the support of falsework on shore. The axis of the arch at this stage is almost perpendicular to the designed bridge axis. As shown in Figures 9.12a and 9.13,

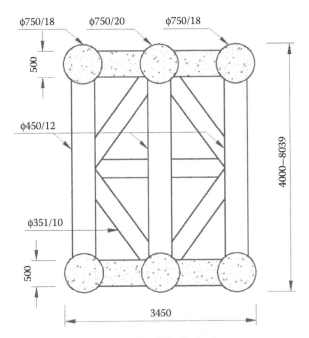

Figure 9.11 Cross section of the main arch of Yajisha Bridge.

half of the main arch is assembled at a lower position vertically. During the assembling, side-span arch and counterweight are built. A temporary construction tower is built on top of the abutment, which is connected to the foundation with a circular track of a 33-m diameter, allowing later horizontal rotation. Construction cables and temporary bracings are used to lift the arch. A temporary joint at the springing is designed and built to allow the vertical rotation of the arch.

Concrete in the top and bottom ribs is not filled at this stage. Total vertical lift weight is 2058 tons. The vertical rotation angle is 24.7014°. Figure 9.12b shows the vertical lifting. After the arch is lifted to the design position, the temporary joints at springing are fixed by reinstalling top and bottom cutout ribs. Fixing the vertical joints at the end of this stage is required to ensure stability in the next stage. After the falsework in the side span is removed, the half structure is ready to rotate horizontally.

9.2.2.3 Horizontal rotation

The horizontal rotation mechanism comprises of a fixed platform in the bottom and a lateral girder connecting two abutments on upstream and downstream sides on the top as the moving part. The rotation axis is located at the center of the lateral girder. The two half arches in upstream and downstream sides are connected laterally and rotated horizontally as a whole.

(a)

(b)

Figure 9.12 Vertical and horizontal rotation of Yajisha Bridge. (Courtesy of Guizhou Bridge Construction Group Ltd.) (a) Half of the main arch is assembled on false work at lower position along the river course. (b) Half of the main arch is rotating vertically to its design position. (Continued)

(c)

(d)

Figure 9.12 (Continued) Vertical and horizontal rotation of Yajisha Bridge. (Courtesy of Guizhou Bridge Construction Group Ltd.) (c) Half of the main arch is rotating horizontally to meet the bridge axis. (d) Before closure.

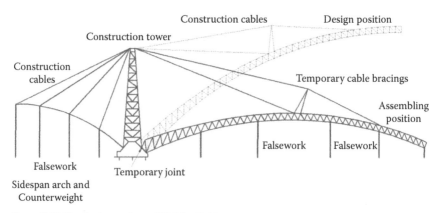

Figure 9.13 Vertical rotation of Yajisha Bridge.

The connection between the upper moving part and the lower circular track consists of 14 bearing feet, which are concrete-filled steel tubes and casted into the bottom of the upper moving part. A stainless steel plate of 3 mm thick is built on top of the track as the contact surface with the bottom of the bearing feet.

The rotating force is applied to the bottom of the upper moving part by jacking cables, which are anchored on top of the lower fixed part. The total horizontal rotated structure weighs 13,685 tons. The two half structures are rotated by 117.1117° and 92.2333°, respectively, to reach the design axis. Figure 9.12c and d shows the horizontal rotation of the arch bridge.

9.3 PRINCIPLE AND ANALYSIS OF ARCH BRIDGES

9.3.1 Perfect arch axis of an arch bridge

The shape of an arch affects the internal force distributions, and it is important to choose the best shape when designing an arch bridge. Comparing with the shape of a suspending cable, which is in tension only under a uniform dead load, it can be understood that it will be in compression if the uniform load direction is reversed. The perfect arch axis, in which the arch is in compression only under a designated uniform load, is often referred in the arch bridge design. To use the perfect arch axis under uniforme dead loads is preferable in most arch bridge designs, especially in masonry or concrete arch bridges.

As shown in Figure 9.14, an arch is under a uniform load of q. The perfect arch axis can be derived from the assumption that moment at any point p on the arch is zero:

$$qx\frac{x}{2} + Hy = \frac{ql}{2}x \tag{9.1}$$

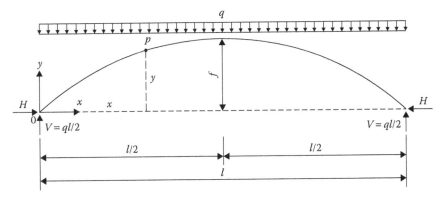

Figure 9.14 Perfect arch shape under a uniform symmetric load.

Substituting x and y with $l/2$ and f, respectively, at the crown, the horizontal thrust H can be obtained as

$$H = \frac{ql^2}{8f} \tag{9.2}$$

Substituting Equation 9.2 into Equation 9.1, the perfect arch shape can be derived as

$$y = \frac{4f}{l} x \left(1 - \frac{x}{l}\right) \tag{9.3}$$

From Equation 9.3, it can be seen that the perfect axis for an arch under uniform dead load is a parabola. In most arch bridges, the dead loads along the bridge axis do not vary much and can be assumed as uniformed, which is the reason that the parabola arch axis is commonly used. In addition to parabola, catenary and circularity can also be used as an arch axis. It should be noted that when perfect arch axis is referred it implies that the arch is under a uniform dead load.

From Equation 9.2, a common fact of arch bridges is proved that the horizontal thrust is inversely proportional to the arch rise.

9.3.2 Fatigue analysis and affecting factors

Long-span arch bridge provides a favorable driving condition for the vehicles. However, the repeated action of traffic to the bridge will lead fatigue damage to the members of the arch structure, especially the hangers in the tied-arch bridge. Most bridges were damaged not because of the load beyond capacity but because certain hangers lost the strength due to fatigue damage.

The tied-arch bridge is an internally indeterminate structure so all the parameters will affect the hangers' stress state. The factors that cause the hangers' fatigue include the size and material properties of the hangers themselves or the loads applying on the hangers. This section discusses how the arrangement of hangers affects the fatigue (Pellegrino et al. 2010).

An example of a concrete-filled steel tube tied-arch bridge was studied by Yao (2007) and is used here to explain the fatigue effect. Span is 61 m (200′), carriageway width is 15.2 m (50′), and the ratio of rise to span is 1:5. The original spacing of hangers is 5.1 m (16.7′). Hot-extruding PE high-tensile cable PES7-55 is used. The load will use normal vehicle design load in this example.

9.3.2.1 Positions of hangers

First, the difference among the stress state of different hangers will be discussed. The tied-arch bridge and the numbers of the hangers are shown in Figure 9.15. After analysis, the result is shown in Table 9.2. The stress range (SR) ratio of a hanger is its stress range over that of the middle hanger. The result in Table 9.2 shows that the middle hanger has a higher stress range, thus more prone to fatigue damage than the side hangers without considering the flexural rigidity.

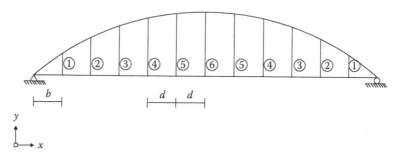

Figure 9.15 Study of a tied-arch bridge with different middle and side spaces.

Table 9.2 Study of baseline tied-arch bridge with side spaces $b = 5.1$ m (16.7′)

Hanger numbers	1	2	3	4	5	6
Maximum stress (ksi)	60.83	63.57	64.38	65.00	65.41	65.41
Minimum stress (ksi)	51.78	53.42	53.69	53.77	53.77	53.77
Average stress (ksi)	49.05	58.49	59.04	59.38	59.60	59.60
Stress range (ksi)	9.05	10.15	10.69	11.24	11.64	11.64
SR ratio	0.776	0.871	0.918	0.965	1.000	1.000

Table 9.3 Study of a tied-arch bridge with different middle spaces *d*

Hanger numbers	Side hanger (1)				Middle hanger (6)			
Spacing (ft)	13.4	16.7	25.1	33.4	13.4	16.7	25.1	33.4
Maximum stress (ksi)	46.65	60.83	90.41	121.92	51.78	65.41	96.57	128.08
Minimum stress (ksi)	39.79	51.78	77.40	104.11	42.05	53.77	78.76	104.79
Average stress (ksi)	43.22	56.30	83.90	113.01	46.91	59.59	87.67	116.44
Stress range (ksi)	6.86	9.05	13.01	17.81	9.73	11.64	17.81	23.29
SR ratio	0.758	1.000	1.438	1.970	0.836	1.000	1.530	2.000

9.3.2.2 Space of hangers

The space of hangers is an important parameter that influences the fatigue. In this example (Figure 9.15), the space d is chosen as 4.1 (13.4′), 5.1 (16.7′), 7.7 (25.1′), and 10.2 m (33.4′), respectively. The result is shown in Table 9.3, and the SR ratio of a spacing is its stress range over that of 5.1 m (16.7′).

Based on Table 9.3, a conclusion could be drawn that a closer spacing of hangers will benefit hanger fatigues more than a sparse spacing. This can be understood from the simple fact that the stress ranges due to live loads are greater in a sparse layout than a finer layout.

9.3.2.3 Distance between side hanger and arch springing

The distance between side hanger and arch springing (b shown in Figure 9.15) will be selected as 2.5 (8.3′), 5.1 (16.7′), 7.7 (25.1′), and 10.2 m (33.4′), respectively, which means the ratio of this distance to the distance between other hangers is 0.5:1, 1:1, 1.5:1, and 2:1, respectively. The result is shown in Table 9.4, and the SR ratio is based on the ratio of b to d, taking 1:1 as one.

The result in Table 9.4 draws the conclusion that the change of the distance between the side hanger and springing makes little influence on the middle hanger but makes great influence on the side hanger.

About the side hanger, either increase or decrease of the distance between the side hanger and springing will lead to higher stress range. When the distance is extended, tension in the side hanger as a point to bear the load will increase. When the distance is too small, however, the side hanger will be short so that the vertical tensile stiffness will increase and that will also lead to higher stress range. Hence, to enable the hangers have the most effective work, the distance between the side hanger and springing should be in a proper range.

The fatigue problem in hangers of a tied-arch bridge is a complicated one that many factors such as the section area of hangers, flexural rigidity of hangers, and the impact force of the vehicles will have considerable influences on the fatigue. Besides, the dimensions of the bridge such as span and

Table 9.4 Study of a tied-arch bridge with different side spaces b

Hanger numbers	1				2				6			
b (ft)	8.3	16.7	25.1	33.4	8.3	16.7	25.1	33.4	8.3	16.7	25.1	33.4
b/d	0.5:1	1:1	1.5:1	2:1	0.5:1	1:1	1.5:1	2:1	0.5:1	1:1	1.5:1	2:1
Maximum stress (ksi)	36.92	44.79	66.92	78.76	51.71	56.78	62.74	69.86	60.48	60.41	60.28	59.60
Minimum stress (ksi)	26.43	38.50	59.12	68.50	43.36	44.66	51.10	58.97	45.49	45.41	45.34	44.94
Average stress (ksi)	31.68	41.64	63.02	73.63	47.53	50.72	56.91	64.42	52.98	52.91	52.81	52.26
Stress range (ksi)	10.48	6.29	7.80	10.27	8.35	12.12	11.64	10.89	14.99	14.99	14.94	14.66
SR ratio	1.666	1.000	1.240	1.633	0.689	1.000	0.960	0.899	1.000	1.000	0.997	0.978

the ratio of rise to span as well as the performance of steel and concrete could not be ignored.

9.3.3 Measuring of hanger-cable force

For arch bridge (or any other cable-related bridges), hanger-cable force can be measured to provide an indication on the damage degree of bridge. The measurement methods of hanger force were studied by many researchers. Among them, the frequency method was the most commonly used one in practical projects. But in some measurement formula, the bending stiffness of cable must be recognized. While the true bending stiffness was difficult to measure because the hanger cables are made up of several steel wires and the stiffness of wires will be changed constantly. The measurement formula of hanger-cable force, which was based on the function of deflection curve shape, is introduced here (Li et al. 2014).

According to the conservation principle of energy, the total energy (the kinetic energy and the strain energy) of a free-damped vibrating elastic body should be unchanged at any time. Transverse vibration curve of a uniform cross-sectional cable is assumed to be a deflection curve under uniform load. The deflection curves satisfy the boundary conditions, and the maximum kinetic energy $E_{k\max}$ of the cable is

$$E_{k\max} = \frac{1}{2}\omega^2 \int_0^l m\left[Y(x)\right]^2 dx \qquad (9.4)$$

The maximum strain energy $V_{\max}$ of the cable is

$$V_{\max} = \frac{1}{2}\int_0^l EI\left[Y''(x)\right]^2 dx + \frac{1}{2}T_0\int_0^l \left[Y'(x)\right]^2 dx$$

$$= \frac{1}{2}\int_0^l q(x)Y(x)dx + \frac{1}{2}T_0\int_0^l \left[Y'(x)\right]^2 dx \qquad (9.5)$$

According to the conservation principle of energy

$$E_{k\max} = V_{\max} \qquad (9.6)$$

Natural vibration frequency ω can be obtained as follows:

$$\omega^2 = \frac{\int_0^l q(x)Y(x)dx + T_0\int_0^l \left[Y'(x)\right]^2 dx}{\int_0^l m\left[Y(x)\right]^2 dx} \qquad (9.7)$$

Approximated displacement function of the first-order vibration mode is built by using the deflection curve of a fixed-end rebar under the uniform load q.

$$Y_1(x) = \frac{ql^4}{24EI}\left[\left(\frac{x}{l}\right)^2 - 2\left(\frac{x}{l}\right)^3 + \left(\frac{x}{l}\right)^4\right](0 \le x \le 1) \tag{9.8}$$

Approximated displacement function of the second-order vibration mode is built by using deflection curve of a fixed-end rebar under the antisymmetric uniform load.

$$Y_2(x) = \frac{ql^4}{288EI}\left[\begin{array}{c} 3\left(\dfrac{x}{l}\right)^2 - 14\left(\dfrac{x}{l}\right)^3 + 12\left(\dfrac{x}{l}\right)^4 \\ + 95.7\left(\sinh\dfrac{x}{2l} - \dfrac{x}{2l}\right) \end{array}\right]\left(0 \le x \le \frac{1}{2}\right) \tag{9.9}$$

$$Y_2(x) = -\frac{ql^4}{288EI}\left\{\begin{array}{c} 3\left(\dfrac{l-x}{l}\right)^2 - 14\left(\dfrac{l-x}{l}\right)^3 + 12\left(\dfrac{l-x}{l}\right)^4 \\ + 95.7\left[\sinh\left(\dfrac{l-x}{2l}\right) - \left(\dfrac{l-x}{2l}\right)\right] \end{array}\right\}\left(\frac{1}{2} \le x \le 1\right) \tag{9.10}$$

So, inherent frequencies ω can be calculated:

$$\omega_1^2 = \frac{(4/5)\cdot(EI/l^2) + (2/105)\cdot T_0}{(1/630)\cdot ml^2} \tag{9.11}$$

$$\omega_2^2 = \frac{3.59059(EI/l^2) + 2.13849\times10^{-2}T_0}{4.44568\times10^{-4}ml^2} \tag{9.12}$$

where:
ω is the inherent frequency
T_0 is the cable tension
EI is the bending stiffness
l is the length of the rebar
m is the mass per unit length of the rebar

In Equations 9.11 and 9.12, cable tension T_0 has an explicit relationship with the inherent frequency f, so cable tension can be easily calculated from a measured frequency. When using a natural frequency f_1, cable tension T_0 is

$$T_0 = \frac{\pi^2}{3}\cdot ml^2f_1^2 - 42\frac{EI}{l^2} \tag{9.13}$$

Figure 9.16 Demonstration of dynamic testing instrument on a hanger cable.

When using the second-order natural frequency f_2, cable tension T_0 is

$$T_0 = 0.8207ml^2f_2^2 - 167.9\frac{EI}{l^2} \tag{9.14}$$

Because the second-order vibration function is an approximated function, the result of Equation 9.14 is also an approximation.

Eliminating bending stiffness in Equations 9.13 and 9.14, cable force can be calculated from the first and second natural frequencies as given in Equation 9.15, in which the effect of bending stiffness EI is considered but not needed to measure directly.

$$T_0 = ml^2(4.3865f_1^2 - 0.2742f_2^2) \tag{9.15}$$

Instead of measuring the bending stiffness of a cable, which is not practical in testing on-site, the second order of natural frequency can be obtained at the same time when the first-order frequency is analyzed from a frequency spectrum analyzer as shown in Figure 9.16. Therefore, Equation 9.15 has a great advantage of when to include the bending stiffness effect on cable forces.

9.4 MODELING OF ARCH BRIDGES

As high-strength hangers and/or tied cables are part of the structure, an arch bridge is usually considered as a cable structure with the same consideration as a cable-stayed bridge. The principle and modeling of an arch bridge is similar to a cable-stayed bridge in many aspects. For example, the analyses of an arch bridge contain complicated construction stage analyses, and cables or hangers need to be tuned to reach an ideal design state. These analyses are all specific topics in cable-stayed bridge analyses. Therefore, a special-purpose FEA package that can perform multistage construction analysis and cable-tuning is required. As the arch is under compression, nonlinear effects such as initial stress problem, stability, and even large displacement are often needed in arch bridge analyses. Basic principles in modeling an arch bridge, such as whether a three-dimensional (3D) model is necessary or not and how fine the mesh is adequate, are the same as those in modeling a cable-stayed bridge. Detailed discussion in Chapter 11 for a cable-stayed bridge can be applied to an arch bridge.

9.4.1 Arches

The cross section of an arch varies from solid-reinforced concrete to steel box, from steel truss to concrete-filled steel tubes. Compression is predominated in the arch under dead loads; however, live loads will also cause bending moment. 2D/3D frame elements are used to model an arch. The curvature of arch geometry can be simulated by straight elements, and the curve element is not quite necessary. Like girders in a cable-stayed bridge, initial stress effect may be considered in arch elements.

9.4.2 Deck

The deck of an arch bridge usually contains floor beams and stringers as in most half-through thrust arch bridges or tied cables/girders and floor beams as in most tied-arch bridges. 3D model is always encouraged so as to better simulate the stiffness of each deck component. Taking an example shown in Figure 14.15, floor beams and tied girders are modeled as 3D frame elements. When tied cables are separated, truss elements are used to model tied cables.

9.4.3 Hangers

Like cables in a cable-stayed bridge, hangers are usually modeled as truss elements. No sag effect exists in a hanger, and one hanger can be modeled as one truss element. Initial stress effect should be considered in analyses for lateral load cases and stability analysis.

9.4.4 Stability

Due to high compression in the arch under dead loads and the height of the crown from the deck, global stability, either in the arch plane or in the horizontal plane, is more important in an arch bridge than other types of bridges. Stability analysis is inevitable when designing an arch bridge. For a tied-arch bridge without lateral bracings on arches or a long-span arch bridge, lateral stability usually has a lower critical load than in-plane

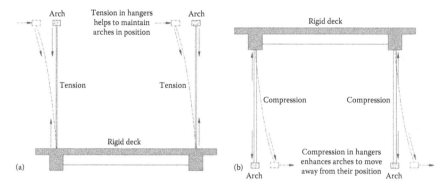

Figure 9.17 Lateral stability of arches in (a) deck-through and (b) deck arch bridges.

stability. As illustrated in Figure 9.17, the initial stress of hangers has different effects on the lateral stability of the arch from half-through arch bridges to deck arch bridges, assuming that the deck is relatively rigid in the lateral direction. In addition to initial stresses accumulated along the arch, stresses in hangers due to dead loads and tuning prestress should always be included in stability analyses. Principles of bridge stability discussed in Chapter 14 in general are applicable to arch bridge stability analysis.

9.5 3D ILLUSTRATED EXAMPLE OF CONSTRUCTION ANALYSES—YAJISHA BRIDGE, GUANGZHOU, PEOPLE'S REPUBLIC OF CHINA

As introduced in Section 9.2.2, the construction of the Yajisha Bridge combines vertical rotation and horizontal rotation. During vertical rotation, the geometry of the main arch gradually changes from a lower position to the designed position. As geometry of structural components is different from that when the bridge is complete, additional analyses are required to ensure that each component is under control. The analysis and modeling of the Yajisha Bridge are similar to a segmental erected bridge. However, the rotation of the structure requires the analysis tool to be able to process the geometry change of a component from stage to stage. In situations where such a multistage bridge analysis tool is not available, a regular FEA package can be used instead. Additional manual postprocessing of the FEA results is required to plot stress envelopes during the entire vertical rotation process.

As shown in Figure 9.13, the half arch is assembled on falsework at lower position and then is lifted to design position by jacking construction cables. Visual Bridge Design System (VBDS) Wang and Fu 2005, is used to model and analyze the vertical rotation of a half arch of the Yajisha Bridge. The truss members of the steel tubular arch, the side arch, and the construction and bracing towers are simulated by 3D frame elements, the construction cables are simulated by truss elements, and the horizontal rotation platform is simulated by rigid elements. Four stages are modeled to simulate the vertical rotation as shown in Figure 9.18. In stage 1 the arch is ready to lift, in stage 2 the arch is lifted to half way, in stage 3 the arch is rotated to the design position, and in stage 4 simulates the release of the construction cables at the design position. Figure 9.19 shows axial forces of these four stages.

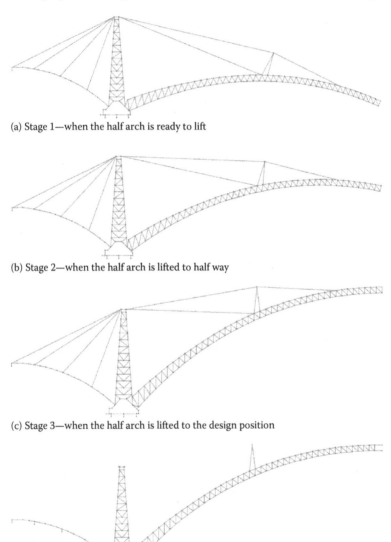

(a) Stage 1—when the half arch is ready to lift

(b) Stage 2—when the half arch is lifted to half way

(c) Stage 3—when the half arch is lifted to the design position

(d) Stage 4—when construction cables are released

Figure 9.18 (a–d) Stages simulating vertical rotation of Yajisha Bridge.

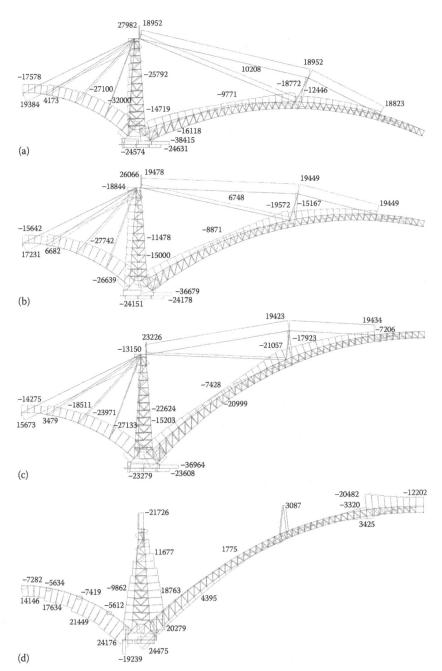

Figure 9.19 Axial forces during vertical rotation of Yajisha Bridge. (a) Axial forces in stage 1. (b) Axial forces in stage 2. (c) Axial forces in stage 3. (d) Incremental axial forces when construction cables are released in stage 4.

9.6 3D ILLUSTRATED EXAMPLE OF A PROPOSED TIED-ARCH BRIDGE ANALYSES—LINYI, PEOPLE'S REPUBLIC OF CHINA

As an arch bridge stability analysis example, later in Section 14.6 stability analyses of a proposed concrete-filled steel tube tied-arch bridge will be demonstrated. In that example, three stages are modeled as shown in Figures 14.16 and 14.17. Stage 1 is the casting of tied girder concrete on falsework; stage 2 is the installation of arches and lateral bracings, filling concrete into steel tube, and first-time jacking of hangers; and stage 3 is the installation of the deck and final jacking of hangers.

In addition to construction analyses of these stages and the stability analysis, live load analysis is also performed by using influence surface loading method. Figure 14.16 shows the axial force distribution of the live loads that cause the compression on top of one arch maximal. 3D modeling and influence surface loading clearly illustrate the lateral distribution of live loads. As shown in Figure 14.16, the maximum compression on top of one arch is 822 kN, whereas the corresponding compression on the other side is only 149 kN. Figure 14.17 similarly shows the uneven displacements on both tied girders due to live loads.

9.7 3D ILLUSTRATED EXAMPLE OF AN ARCH BRIDGE—LIUJIANG YELLOW RIVER BRIDGE, ZHENGZHOU, PEOPLE'S REPUBLIC OF CHINA

Liujiang Yellow River Bridge, crossing over Yellow River at Zhengzhou, China, was built in 2006. The bridge has a length of 9848 m in total and eight lanes carrying two bounds traffic. Most spans are simply supported prestressed concrete T girder and void slab spans. The main bridge contains four spans of concrete-filled steel tubular tied-arch spans, as shown in Figure 9.20. Traffic lanes of two bounds are separated. Each of the arch spans carries four traffic lanes with a net width of 19 m. The total width is 24.377 m, and the span length is 100 m.

In this example, the dead load and live load analyses of one arch span are introduced. The theoretical span length is 95.5 m, and a catenary arch with a factor of 1.347 is used. The ratio of arch rise to span is 1:4.5. The arch contains two vertically placed steel tubes connected by steel plates. After closure, the tubes and connecting rib are filled with concrete. The tied girders and end-floor beams are prestressed concrete box girders; the interior floor beams are prestressed concrete T girders. The deck comprises precast concrete Π modular slabs, placed on top of floor beams and connected to each other by cast-in-site segments. The hangers are high-strength steel wires.

Figure 9.20 Liujiang Yellow River Bridge, Zhengzhou, China. (Courtesy of mudedi.59706.com.)

As shown in Figure 9.21, the arch bridge is modeled three-dimensionally by using frame and truss elements. The arch, lateral bracings, tied girders, floor beams, and deck slabs are modeled by frame elements; hangers are modeled by truss elements. As there is no lateral load included in the analyses, the stringers simulating deck slabs are located at the centroids of floor beams. There are 1018 elements in total. The analyses include dead load analysis with an automated hanger tuning for a preferable moment distribution on tied girders and live load analysis.

Figure 9.22 shows moment and axial force distributions without tuning of hangers. It can be seen that the moment distribution on tied girder is similar to a simply supported girder, and tensions in hangers are low when hangers are not prestressed. Figure 9.23 shows a preferred distribution achieved by automated hanger tuning analysis, in which tied girder works like a multiple-supported continuous girders. Tensions in hangers are higher. Figure 9.24 shows the moment and axial force envelopes due to live loads. Figure 9.25 shows the moment and axial force distributions under live loads that cause extreme compression at the crown on one arch. The uneven distribution of live loads is clearly displayed.

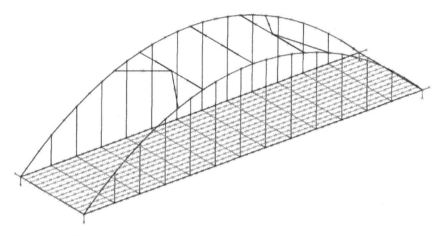

Figure 9.21 3D analysis model of Liujiang Yellow River Bridge.

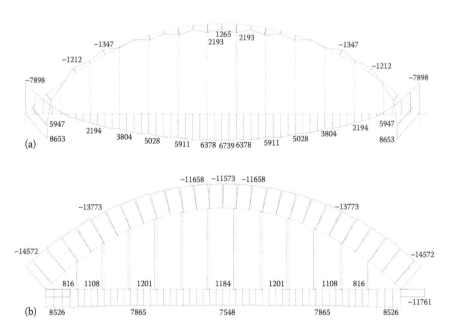

Figure 9.22 (a, b) Moment and axial force distribution due to dead loads without hanger tension tuning.

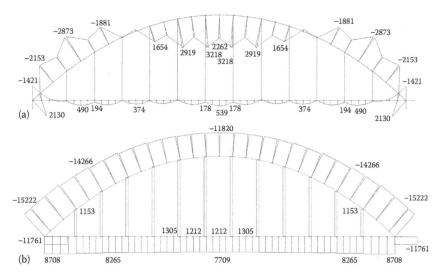

Figure 9.23 One preferred distribution after automatic hanger tension tuning. (a) Moment. (b) Axial force.

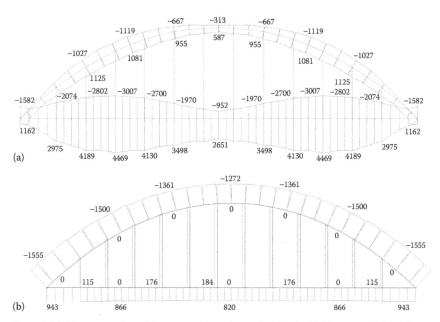

Figure 9.24 Moment and axial force envelopes due to live loads. (a) Moment. (b) Axial force.

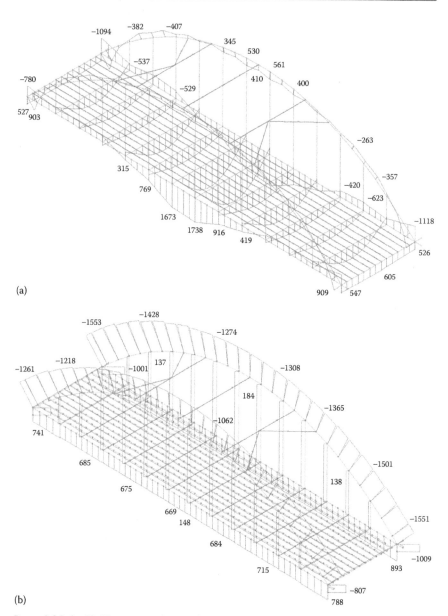

Figure 9.25 (a, b) Moment and axial force distribution due to live loads that cause extreme compression at crown on one arch.

Chapter 10

Steel truss bridges

10.1 INTRODUCTION

Trusses, in principle, behave as large beams to carry loads but are comprised of discrete members that are subjected primarily to axial loads. Joints, or nodal points, are the locations where truss members intersect and are referred to as panel points. A truss bridge is a bridge constructed using triangular units connected at joints, suspending loads through tension and compression. Traffic loads are applied to the bridge deck, which is supported by longitudinal stringers, generally placed parallel to traffic, that carry deck loads to the floor beams. Floor beams are usually set normal to the direction of traffic and are designed to transfer loads from the bridge deck to the trusses, the main load-carrying members to supports. Figure 10.1 depicts a truss bridge and terminology used.

In early years, truss bridges were built with wood. Then, metal gradually replaced wood as the primary truss bridge-building material, leading to extensive building of wrought iron bridges after 1870. The Bollman Truss, patented in 1852, used cast iron for the compression members and wrought iron for the tension, for which the 100' bridge in Savage, Maryland, is the only surviving example (Figure 10.2). The truss bridge is one of the oldest types of modern bridges, and it became popular because of its economical design and relatively affordable construction. There are a large variety of truss bridge types, with most having been built between the 1870s and the 1930s. Truss bridges have been widely used to carry automobile and railroad traffic.

Many steel truss bridges built in early years are now either renovated or replaced. One example is the George P. Coleman Bridge that carries Route 17 over the York River in Yorktown, Virginia. In 1993, the state of Virginia widened the existing 1143-m (3750') two-lane bridge to four lanes using the existing substructure. The original bridge was 9.5 m (31') wide with no shoulders, and the new structure would be 23.6 m (77') with full shoulders (Bergeron 2004). Figure 10.3 shows barges floating a truss segment into place. Some of the truss bridges in Europe are built

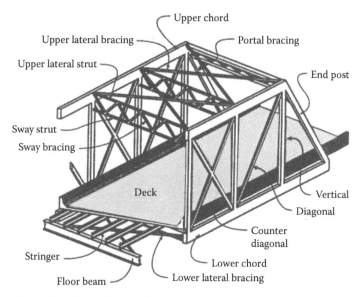

Figure 10.1 Truss bridge and terminology.

Figure 10.2 Bollman truss bridge.

of concrete or composite, such as Mangfallbrücke in Austria by the cast-in-place segmental techniques, Viaduc de Sylans of precast segmental concrete combined with external and internal tendons, and bridge over the Roize in France of composite truss with concrete for the top and bottom chords and steel sections for the open webs. Concrete or composite

Figure 10.3 Construction of George P. Coleman Bridge in segment. (Courtesy of VDOT, Virginia.)

segmental trusses are unique for long spans and offer very efficient use of materials. Due to relatively few live examples, they are not discussed in this chapter.

Lateral cross bracing in the plane of both the top and bottom chords of the trusses is essential. Its main purpose is to provide shear stiffness on these planes so that sufficient torsional stiffness of the truss bridge as a whole will be ensured. Also, it will enhance lateral stability and help to distribute lateral loads applied on truss members. Sway bracing is provided between the trusses in the plane of either verticals or diagonals, and its primary purpose is minimizing the relative vertical deflections between the trusses. Portal bracing is a sway bracing placed in the plane of the end posts.

Based on the deck location, there are three basic truss types: (1) deck, (2) through, and (3) half-through trusses (Figure 10.4). For deck trusses, the entire truss is below the bridge deck. Deck trusses are generally desirable in cases where vertical clearance below the bridge is not restricted. Through trusses are detailed so that the bridge deck is located as close to the bottom chord as possible and are generally used when there is a restricted vertical clearance under the bridge. Half-through trusses carry the deck high enough that sway bracing cannot be used above the deck. It is very difficult to design a half-through truss if the chosen truss type does not have verticals. Many of the recent trusses designed in the United States have been designed without verticals to achieve a cleaner

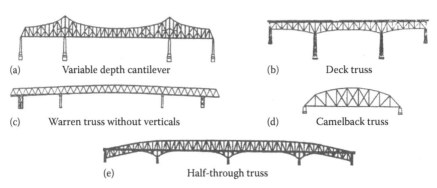

(a)   Variable depth cantilever   (b)   Deck truss

(c)   Warren truss without verticals   (d)   Camelback truss

(e)   Half-through truss

Figure 10.4 (a–e) Various truss bridge types.

and more contemporary appearance, thus minimizing the use of half-through trusses.

There are several geometric guidelines for determining truss configurations. AASHTO requires minimum truss depths of one-tenth (1/10) of the span length for simple spans. For continuous trusses, the distance between inflection points can be used as the equivalent simple-span length to determine the minimum truss depth. It is generally desirable to proportion the truss panel lengths so that the diagonals are oriented between 40° and 60° from horizontal. This keeps the members steep enough to be efficient in carrying shear between the chords. This angular range also allows the designer to maintain a joint geometry that is relatively compact and efficient.

Floor systems can use a series of simple-span stringers framing into the floor beam webs (Figure 10.5a) or continuous stringers sitting on the top of the floor beams (Figure 10.5b). Whenever there is depth restriction requirement, framed systems serve to reduce the overall depth of the floor system by the depth of the stringers.

Traditionally, when modeling and analyzing, truss can be idealized assuming that the members are pinned at the joints (free to rotate independent of other members at the joint) so that secondary stresses ordinarily need not be considered in the design, except certain cases like the half-through bridges built in China (as illustrated and discussed more in Section 10.7), which look like a truss type, but more like a frame-type bridge where joints are taking bending moments. To exclude the bending effect, joints are typically detailed so that the working lines for the diagonals, verticals, and chords intersect at a single point. However, bending stresses resulting from the self-weight of the members should be considered in the design. This idealization of a truss bridge simplifies the modeling and lowers the analyzing effort.

(a)

(b)

Figure 10.5 Truss bridge stringer-floor system. (a) Framed stringers (Courtesy of Department of Public Works, Hunterdon County, New Jersey.); (b) Nonframed stringer (Courtesy of Geiger Brothers.).

As computer technologies advanced, it is no longer difficult or costly to model and analyze a truss bridge truly reflecting its actual assembling and behavior. No modern truss bridges are true trusses. As members of a truss bridge will bend, no matter how small, at connecting nodes, truss bridge members can be more truly modeled as two- (2D) or three-dimensional

(3D) frame elements. Compared with axial force, moments at connecting nodes and along truss members are negligible for a perfect truss bridge, in which all connecting members are so assembled that all centroid lines intersect at a working point (WP) with deck loads transferred through nodal points only. When any member is assembled off its theoretical position by purpose or due to construction error, its secondary bending effect can be obtained from this modeling. Another reason that the frame model should be encouraged whenever possible is the inaccuracy in counting for the secondary bending effect in an idealized truss model. When an offset exists for a member and if the idealized model is adopted, the bending moment for calculating its extra bending stress in the design phase can only be obtained by multiplying its axial force with the offset. In reality, this bending moment may be redistributed because all other connecting members do have bending stiffness. With a frame model, all will be considered internally and automatically.

Steel truss bridges are generally considered to be fracture-critical structures. The simplified approach during design has been to designate all truss members in tension and members subjected to stress reversals as fracture-critical members (FCMs). Fracture-critical studies can be performed based on analyses that model the entire framing system, including the bracing systems and member end fixities, to determine whether certain lightly loaded tension or reversal members are truly fracture critical. In many cases the number of FCMs can be reduced through this process, which reduces fabrication costs. More details of FCM and structural redundancy are covered in Chapter 15—Redundancy Analysis.

10.2 BEHAVIOR OF STEEL TRUSS BRIDGES

10.2.1 Simple and continuous truss bridges

Simple-span truss bridge, like simply supported beam bridge, is made up of trusses spanning between only two supports. A continuous truss bridge is a truss bridge that extends without hinges or joints across three or more supports. A continuous truss bridge, which behaves the same as a continuous girder bridge as a whole, may use less material than a series of simple trusses. It is possible to convert a series of simple truss spans into a continuous truss. For example, the northern approach to the Golden Gate Bridge was originally constructed as a series of five simple truss spans. In 2001, a seismic retrofit project connected the Marin (north) approach viaduct five spans into a single continuous truss bridge (Figure 10.6).

Figure 10.6 The Golden Gate Marin (north) approach viaduct under construction.

10.2.2 Cantilevered truss bridges

A cantilever truss bridge is a structure in which at least one portion acts as an anchorage for sustaining another portion that extends beyond the supporting pier. The use of cantilevers allows for the construction of much longer bridge spans. A cantilevered bridge uses two horizontally projected beams that are supported on piers. Counterbalancing spans called anchor arms provide tension and suspension through the truss. Cantilevered truss bridges remained popular through most of the twentieth century until cable-stayed bridges became more common. The most famous early cantilever bridge is the Forth Rail Bridge (Figure 10.7). This bridge held the record for the longest span in the world for 27 years only to be surpassed by the Quebec Bridge in 1917, which is still the current record holder. The Tydings Bridge of Maryland is an illustrated example of this type of cantilever bridge and will be shown later.

Steel truss cantilevers, as shown in Figure 10.8, support loads by the tension of the upper members and compression of the lower ones. Commonly, the structure distributes the tension via the anchor arms to the outermost supports, whereas the compression is carried to the

Figure 10.7 Forth Rail Bridge, Queensferry, Scotland.

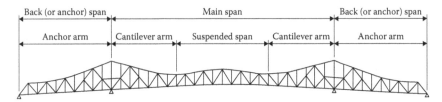

Figure 10.8 Cantilever truss bridge and its terminology.

foundations beneath the central towers. Many truss cantilever bridges use pinned joints and are therefore statically determinate with no members carrying bending moment.

Although some continuous truss bridges resemble cantilever bridges and may be constructed using cantilever techniques, there are important differences between the two forms. Cantilever bridges need not connect rigidly midspan, as the cantilever arms are self-supporting. Even though some cantilever bridges appear continuous due to decorative trusswork at the joints, these bridges will remain standing if the connections between the cantilevers are broken or if the suspended span (if any) is removed. Conversely, continuous truss bridges rely on rigid truss connections throughout the structure for stability. Removal or deterioration of any truss member in midspan of a continuous truss will endanger the whole structure. However, continuous truss bridges do not experience the tipping forces that a cantilever bridge must resist, because the main span of a continuous truss bridge is supported at both ends (Kulicke 2000).

10.2.3 Truss arch bridges

A truss arch bridge, such as Francis Scott Key Bridge (I-695) in Baltimore, Maryland, as shown in Figure 10.9, combines the elements of the truss bridge and the arch bridge. The actual resolution of forces will depend on the design. As long as the horizontal movement of the top chord is restrained, like a regular arch bridge, horizontal thrusting force will be generated and therefore the top chord of a truss arch bridge will be under compression. When the top chords are free to move horizontally, no horizontal thrusting forces will be generated and this arch-shaped truss bridge works essentially as a bent beam. If horizontal thrust is generated but the apex of the arch is a pin joint, it is termed a *three-hinged arch*. If no hinge exists at the apex, it will normally be a two-hinged arch. A tied-arch bridge is an arch bridge in which the outward-directed horizontal forces of the arch, or the top chord, are borne as tension by the bottom chord (either tie rods or the deck itself) rather than by the ground or the bridge foundations. Deck loads including live loads are transferred, as tension, by vertical ties of the deck to the curved top chord, tending to flatten it and thereby to push its tips outward into the abutments, like other arch bridges. However in a tied-arch or bowstring bridge, these movements are restrained not by the abutments but by the

Figure 10.9 Francis Scott Key Bridge (I-695) in Baltimore, Maryland.

bottom chord, which ties these tips together, taking the thrusts as tension, rather like the string of a bow. Therefore, a tied-arch bridge is often called a bowstring-arch bridge. The structure as a whole was described as nonredundant; failure of either of the two tie girders would result in the failure of the entire structure.

10.3 PRINCIPLE AND MODELING OF STEEL TRUSS BRIDGES

For truss bridges, a 2D truss model with planar truss only or a 3D finite element model of the whole superstructure can be defined. For the 2D truss model, truss on only one side is modeled and the vertical load coming from the deck is considered linearly distributed between two parallel trusses and loaded at the connection points between truss and floor beams. For the 3D truss model, two trusses plus floor beams and stringers are modeled as their actual position in space.

When modeling a truss member, as introduced in Section 10.2, 1D-truss/2D-frame or 1D-truss/3D-frame elements can be used in 2D and 3D truss models, respectively. The deck is represented by a combination of transverse beam elements and plate elements. The beam elements provide the load transfer characteristics of the concrete deck, whereas quadrilateral plate or steel elements are used only to receive the wheel loads and distribute the wheel loads to the beams. To provide the ability to represent the actual boundary conditions, hinges, rollers, or linear displacement springs, depending on the bearing situation, can be placed at the truss support locations.

It is regarded that pin-connected analysis model is applicable and accurate as long as the truss bridge is properly cambered (Kulicke 2000). Further, most long truss bridges are already on a vertical curve. Thus, in many practical truss bridges, a parabolic curve exists over at least part of the length of the bridge. When a truss is analyzed as a three-dimensional (3D) assemblage with moment-resisting joints, the inclusion of camber, usually to a *no-load* position, becomes even more important. If the truss is indeterminate in a plane, just like any other type of indeterminate structure, it will be necessary to use realistically close cross section areas for the truss members and may be important to include the camber of the members to get realistic results in some cases. A sample calculation of the cross section is shown in Figure 10.10.

An influence line is a graphical presentation of the force in a truss member as the load moves along the structure. If the truss is statically indeterminate, then the influence lines will be a series of chords to a curve, not a straight line like the statically determinate case. It is often found efficient to calculate the influence lines for truss members using the

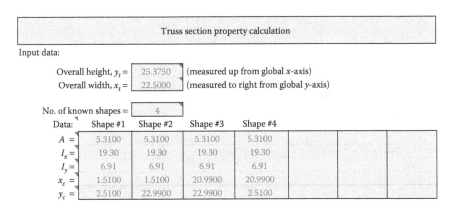

Truss section property calculation

Input data:

Overall height, $y_t =$ | 25.3750 | (measured up from global x-axis)
Overall width, $x_t =$ | 22.5000 | (measured to right from global y-axis)

No. of known shapes = | 4 |

Data:	Shape #1	Shape #2	Shape #3	Shape #4			
$A =$	5.3100	5.3100	5.3100	5.3100			
$I_x =$	19.30	19.30	19.30	19.30			
$I_y =$	6.91	6.91	6.91	6.91			
$x_c =$	1.5100	1.5100	20.9900	20.9900			
$y_c =$	2.5100	22.9900	22.9900	2.5100			

No. of rectangles = | 5 |

Data:	Rect. #1	Rect. #2	Rect. #3	Rect. #4	Rect. #5		
Length, $l_x =$	0.5000	20.0000	0.5000	20.0000	12.0000		
Length, $l_y =$	24.0000	0.3750	24.0000	0.5000	0.5000		
$x_c =$	0.2500	11.2500	22.2500	11.2500	11.2500		
$y_c =$	12.7500	25.1875	12.7500	0.2500	0.2500		

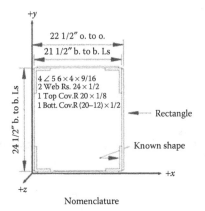

Nomenclature

Results:

Centroid location and area:

x_c(left) =	13.493
x_c(right) =	9.007
y_c(top) =	11.687
y_c(bot) =	13.688
$A =$	46.120

Centroidal axes properties:

$I_x =$	4048.93
S_x(top) =	346.46
S_x(bot) =	295.79
$I_y =$	4205.15
S_y(left) =	311.66
S_y(right) =	466.87
$I_{xy} =$	−97.072
$r_x =$	9.370
$r_y =$	9.549

Principal axes properties:

$I_x' =$	4002.44
$I_y' =$	4251.64
$I_{xy}' =$	0.000
$r_x' =$	9.316
$r_y' =$	9.601
$\theta =$	−25.588

Figure 10.10 Calculation of a sampling truss member cross section.

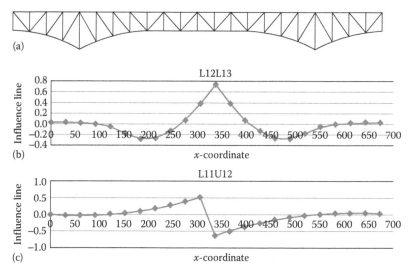

Figure 10.11 Influence lines of a truss bridge. (a) Main span with two cantilever arms. (b) Influence line sample of a top chord. (c) Influence line sample of a diagonal member.

Mueller–Bresslau principle as adopted in many customized bridge software. Two influence line samples of a top chord and a diagonal member are shown in Figure 10.11b and c, respectively.

10.4 3D ILLUSTRATED EXAMPLE—PEDESTRIAN PONY TRUSS BRIDGE

The pony truss bridge as shown in Figures 10.12 and 10.13 has been considered for the case study. It is a pedestrian steel truss bridge with 57.6 m (189′) length, 4.0 m (13′) height, and 28 panels, located in New York suburban area. The sections used in this truss bridge are shown in Table 10.1.

A 3D model is developed using STAAD.Pro as shown in Figure 10.14 where truss elements are used for truss and bracing members, beam elements are used for floor beams and stringers, and plate elements are used for the deck. The bridge is considered fixed in all three directions at one end supports and x (longitudinal) direction released at another end supports. Shadow area shows a 127-mm (5″) thick concrete deck (Figure 10.13). Loads based on AASHTO LRFD (U.S.) code are used in this study. The self-weight of every active element is calculated and applied as a uniformly

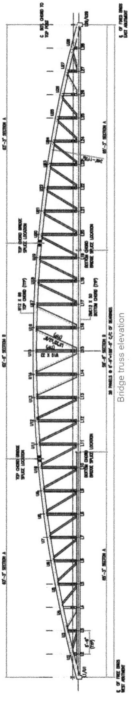

Figure 10.12 Elevation view of the case study pedestrian truss bridge.

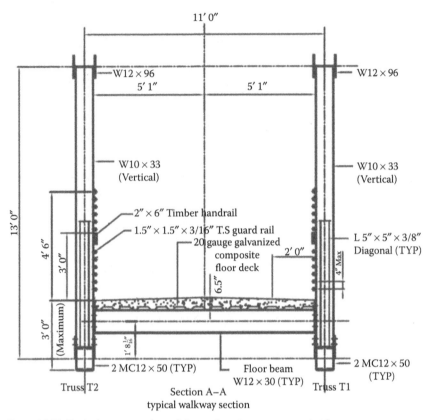

Figure 10.13 Typical cross section of the pedestrian pony truss bridge.

Table 10.1 Truss sections of pony truss bridge

Member	Size
Top chord	W12 × 96
Bottom chord	2MC12 × 50
Vertical	W10 × 33
Diagonals	L 5″ × 5″ × 3/8″ (127 × 127 × 10 mm)
Floor beam	W12 × 30

distributed member load using the information from the section properties in the structural modeling phase. In this case study, all steel members and concrete deck are considered. Bridges that are designed only for pedestrian/bicycle use should design the live load as 4.1 kN/m² (85 psf) and typical panel width is 2 m (6′–9″).

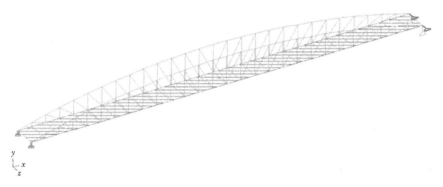

Figure 10.14 The case study pedestrian bridge model.

As for horizontal wind, STAAD.Pro provides the wind-load generation utility for analyzing wind loadings. The utility takes wind pressure at various heights as the input and converts it to joint loads in specific load cases. Meanwhile, an upward vertical linear load of 9.6×10^{-4} MPa (0.020 ksf) times the width of the deck should be applied to windward quarter-point of the deck.

Various methods can be used for performing earthquake analysis. Response spectrum analysis is used in this example. STAAD.Pro provides a utility to specify and apply the response spectrum loads for dynamic analysis. The graph of frequency–acceleration pairs are calculated based on the input requirements of the command and as defined in the code. As mass is processed in a form of directional load in STAAD.Pro, self-weight that represents the structure mass has to be applied to all x, y, and z directions so that accelerations in all these directions will be considered in the 3D dynamic analysis.

Alternatively, time history analysis can be adopted for earthquake analysis. This case study uses the explicit definition with the time versus acceleration data of "IMPERIAL VALLEY 10/16/79 0658, WESTMORELAND FIRE" from USGS database to generate the time history analysis table.

STAAD.Pro covered the information of total applied load and structural reaction for each load case and the response spectrum analysis results including modal base actions, participation factors, and the eigenvalue solution for each mode. The first six eigenvalue solutions are 1.759, 1.957, 2.951, 4.146, 4.733, and 6.156 cycles per second. Figure 10.15 shows the first two mode shapes where the first mode is mainly in lateral (z) direction and the second mode is in vertical (y) direction.

Maximum reactions for all degrees of freedom are presented in Figure 10.16. Reactions output from the analysis are checked first

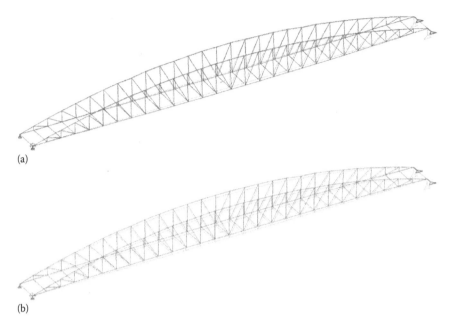

(a)

(b)

Figure 10.15 First two mode shapes of the case study pedestrian bridge. (a) First mode shape (lateral). (b) Second mode shape (vertical).

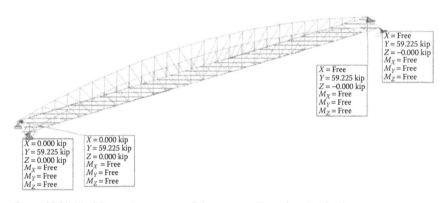

Figure 10.16 Nodal reaction report of the case study pedestrian bridge.

against applied loads so as to essentially eliminate simple errors in modeling.

A similar pony truss bridge of a 43.5-m (142′–8 1/2″) span with the full strength developed at joints used to demonstrate the structure redundancy is shown in Chapter 15.

10.5 2D ILLUSTRATED EXAMPLE—TYDINGS BRIDGE, MARYLAND

The Millard E. Tydings Memorial Bridge (Figure 10.17) is a steel deck truss structure that spans the Susquehanna River on I-95 about 40 miles (64.4 km) north of Baltimore, Maryland, since 1961. The design of this bridge was for HS-20 truck loading. The design temperature range is from

Figure 10.17 Perspective view of Tydings Bridge, Maryland.

Figure 10.18 Configuration of Tydings Bridge, Maryland.

−10°F (−23.3°C) to 120°F (48.9°C), and also it employs a combination of riveted as well as bolted connections.

The 13-span, 1540-m (5051′) structure possesses a deck width of 26.6 m (87′–4″), with 11.9 m (39′–0″) of roadway width in each direction. The bridge uses two parallel truss structures, spaced at 13.7 m (45′–0″), center to center and each consisting of three unique truss panel arrangements as shown in Figure 10.18 with the combination of a suspended span and an anchor span with two cantilever arms. Each of the six suspended spans consist of eight truss panels spaced at 9.3 m (30′–7 1/2″), providing a total length of 74.7 m (245′). Five anchored spans of seven panels for a total length of 65.3 m (214′–4 1/2″) for the center span and 10 cantilevered arms with four panels make up the 37.3-m (122′–6″) span. The entire bridge consists of these two types in a repetitive fashion, essentially permitting one to analyze the entire structure with a simplified approach. The bridge also uses 13 piers, with piers 2 and 13 supporting the end of the truss suspension spans and thus carrying identical loads, whereas piers 3 through 12 support equal load. All members are built-up plate sections constructed with one of two possible materials: high-strength low alloy structural steel or the typical structural carbon steel. For analysis purposes, specific section properties, such as member area, yield stress, and radius of gyration, are of vital importance to ensure accuracy throughout the analysis and were obtained from the original construction documents.

Supported by the trusses, the bridge is comprised of just over 300 floor beams of three unique types. The beams are designated as F1, F2, and F3. Beams F1 and F3 are plate girders with (1) 1524 mm × 8 mm (60″ × 5/16″) web, (2) 356 mm × 19 mm × 17 m (14″ × 3/4″ × 56′) cover plates, and (3) 200 mm × 150 mm × 17 mm (8″ × 6″ × 3/4″) angles. F2 beams are composed of (1) 1524 mm × 8 mm (60″ × 5/16″) web, (2) 325 mm × 13 mm × 17.3 m (13″ × 1/2″ × 56′–10½″) cover plates, and (3) 150 mm × 150 mm × 14 mm (6″ × 6″ × 9/16″) angles. F3 beams can be seen at all floorbreaks, whereas F2 beams are located at each expansion joint, and F1 beams are at all the remaining panel point locations. Carried by the floor beams, seven different stringers were used. Designated A through G, each stringer spans between the floor beams, stiffening the

Table 10.2 Stringer sections for Tydings Bridge

Stringer	End spans	Intermediate spans
A, B, C, D, F	W24 × 76	W24 × 76
E, G	W24 × 84	W24 × 76

Table 10.3 Stringer spacing for Tydings Bridge

Spacing	Distance
A–B, B–C, C–D, F–G	2.06 m (6'–9")
D–E	1.89 m (6'–3/16")
E–F	1.32 m (4'–4")

entire structure. The sections used in the design are shown in Table 10.2, while the spacing is shown in Table 10.3.

10.5.1 Thermal analysis

This study finds the main cause of the premature cracks of the expansion plates as shown in the insertion of Figure 10.17 (Fu and Zhang 2010). A 2D truss model is built by TRAP to study the bridge behavior under thermal loads. In long-span truss bridges, spandrel-braced arch bridge, or called cantilever truss bridge, is a very popular type. Rigid arms extend from both sides of two piers. Diagonal steel trusses, projecting from the top and bottom of each pier, hold the arms in place. The arms that project toward the middle are supported only on one side, like strong cantilever arms, and support a third, central span. Changes of temperature cause material to contract or expand due to the effect of thermal contraction or expansion.

Originally, the bridge sliding plate system was designed assuming that plates would slide on horizontal surfaces when the bridge contracts or expands. However, a closer scrutiny of the behavior indicates that the sliding plate action was affected by the complex movement between anchor spans and suspension spans as well as the force-release systems. A thermal model as shown in Figure 10.19 was generated for the thermal analysis.

The temperature change is assumed to the extreme of 130°F (54.4°C), the difference between the highest and the lowest temperature for the sliding plate design. x- and y-movements are plotted along the panel point. Figure 10.20 shows the expansion of the x-movement of the

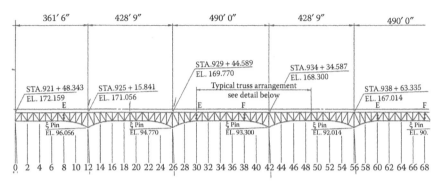

Figure 10.19 Five-span thermal analysis model.

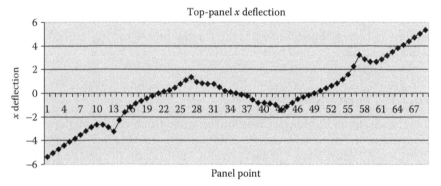

Figure 10.20 Top-panel horizontal movement due to temperature rise.

top-panel points where the left-panel points move toward the negative direction, whereas the right-panel points move toward the positive direction. Figure 10.21 shows the y-movement (i.e., vertical movement) of the top-panel points where the panel points near supports move upward, whereas the panel points away from supports move downward. Noticeably, discontinuity is formed at the expansion joints, which means there is an angular movement at sliding plate locations. For comparison, x- and y-movements of the bottom-panel points are also plotted on Figures 10.22 and 10.23, respectively. It is clearly seen that the x-movement is much less at the bottom-panel points on the anchor span. This displacement pattern reveals that the archlike anchor span will bend up when temperature arises. Because the vertical movements at expansion joints are not even, it is numerically proved that sliding plates as noticed in the field do not fully bear the stiffened plates on the bottom as a designed sliding plate system. Gaps are formed between plates, and sizes of the gaps depend on the temperature. The formation of the gap is

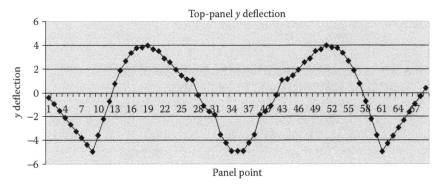

Figure 10.21 Top-panel vertical movement due to temperature rise.

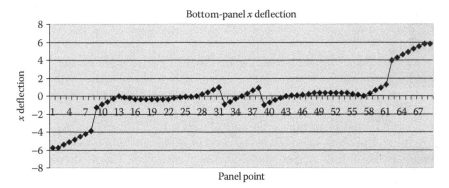

Figure 10.22 Bottom-panel horizontal movement due to temperature rise.

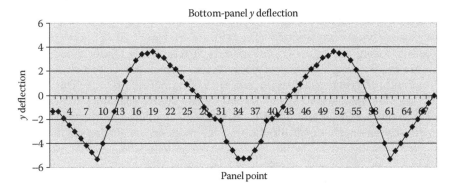

Figure 10.23 Bottom-panel vertical movement due to temperature rise.

evident gaps between sliding top plate and fixed bottom plate and is the main cause of plate crack due to bending.

10.6 3D ILLUSTRATED EXAMPLE—FRANCIS SCOTT KEY BRIDGE, MARYLAND

There are 22 bridges in various types that account for 4.1 miles (6.6 km) of the 10.5-mile (16.8-km) facility, which includes the Francis Scott Key Bridge over Patapsco River and its I-695 approaches in the state of Maryland. This bridge structure combines the behaviors of an arch, truss, as well as cantilever. The main section of the Key Bridge (Figure 10.9) is a three-span, 219–366–219 m (720′–1200′–720′) for a total of 805 m (2640 ft) through truss bridge. The as-is bridge is first modeled and analyzed in 2D by using Win-TRAP program. In this simple 2D model, only one main truss is considered and each truss member is modeled as the truss element—one-dimensional axial element. One main truss is modeled by 416 truss elements. Figure 10.24 shows the elements of the 2D truss model.

As introduced in the beginning of this chapter, a truss bridge can be modeled as frame elements in 3D if analysis tools permit. To illustrate this more sophisticated 3D approach and make a comparison to the simple 2D truss model, the Francis Scott Key Bridge is also modeled with frame and truss elements in 3D and illustrated in this section. As shown in Figure 10.25, all components of a truss bridge including main trusses, bracings, sway frames, floor beams/trusses, stringers, and its diaphragms and hangers are modeled.

In this model, the connections between floor beams/trusses and main trusses are framed together at their centroid positions. The stringers are placed at their centroid positions, and rigid bodies are used to connect them to floor beams/trusses. As bottom chord bracings are aligned on the bottom chord plane, rigid bodies are also adopted to connect them to the floor beams/trusses in the middle. Stringers are modeled as four- or eight-span continuous beams with joints inserted at as-is locations. Figure 10.26 shows the end portal portion, and Figure 10.27 shows its corresponding photo. Figure 10.28 shows the detailed modeling of floor systems. There are 11,618 elements in total with 62 truss elements for hangers and 11,556 3D frame elements for all other components. The total 3D finite element analysis nodes and degrees of freedom are 9,124 and 54,744, respectively. One hundred thirty different cross sections are used in the model. The modeling and analyses are conducted by using VBDS (Visual Bridge Design System, Wang and Fu 2005).

The axial force and moment distributions due to structural weight are shown in Figures 10.29 and 10.30, respectively. Figure 10.31 shows moment distributions in part of stringers.

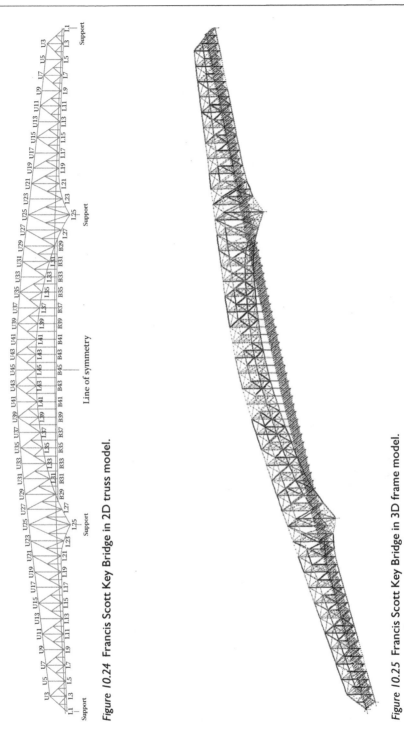

Figure 10.24 Francis Scott Key Bridge in 2D truss model.

Figure 10.25 Francis Scott Key Bridge in 3D frame model.

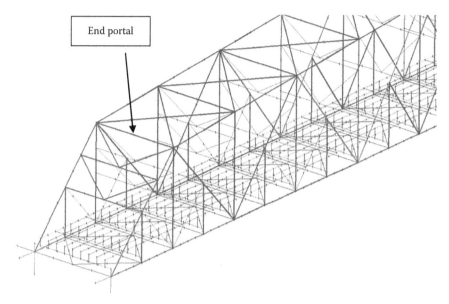

Figure 10.26 3D Francis Scott Key Bridge model in detail—end portal.

Figure 10.27 End portal of Francis Scott Key Bridge.

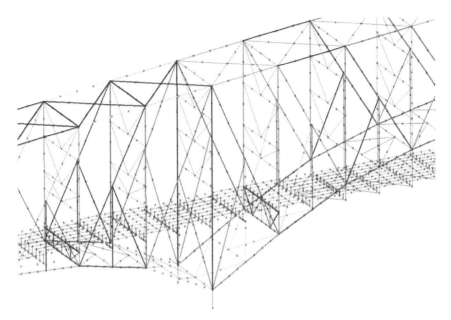

Figure 10.28 3D Francis Scott Key Bridge model in detail—floor system.

Live load analysis for four-lane HS-20-44 is conducted in this example. Figures 10.32 and 10.33 show the combined dead and live load results on half of a main truss.

10.7 3D ILLUSTRATED EXAMPLES—SHANG XIN BRIDGE, ZHEJIANG, PEOPLE'S REPUBLIC OF CHINA

This recently constructed (2010) Shang Xin Bridge in Zhejiang, China (shown in Figure 10.34), is considered as a half-through three-span continuous steel semitruss bridge with minimum top bracing. The span lengths are 62–100–62 m, and the total width is 30 m. The unique design of this bridge is that the hinge connections are replaced by semirigid connections on both truss panels. So, it behaves more like a frame than a truss even though its members maintain the triangular shape in geometry as a truss bridge does. However, to meet safety requirement, nodal and middle segments of members are designed as a hinge-connected structure with only axial forces, whereas end segments of truss members are designed

Figure 10.29 Axial forces of half of one main truss due to structural weight (kip).

Figure 10.30 Vertical bending moment of half of one main truss due to structural weight (kip-ft).

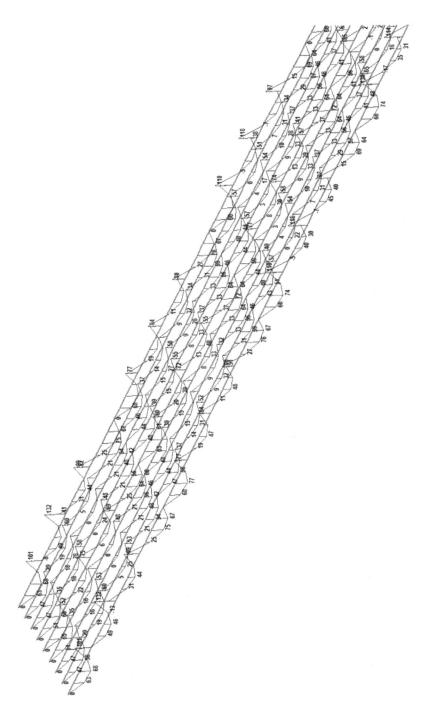

Figure 10.31 Vertical bending moment distributions in part of stringers due to structural weight (kip-ft).

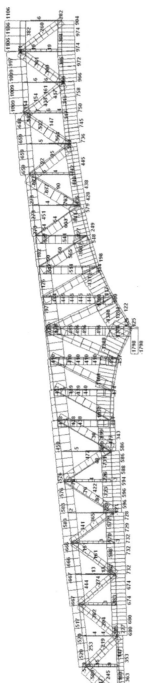

Figure 10.32 Extreme axial forces in main truss due to four-lane HS-20-44 (kip).

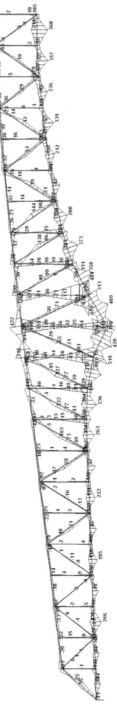

Figure 10.33 Extreme bending moment in main truss due to four-lane HS-20-44 (kip-ft).

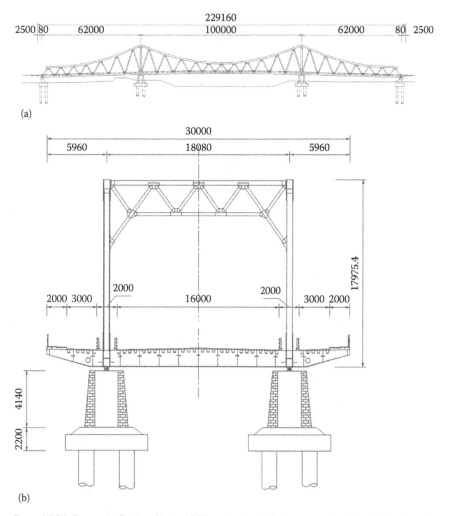

Figure 10.34 Shang Xin Bridge, China. (a) Elevation (mm); (b) cross section (mm). (Continued)

as a frame structure with combined action of axial force and bending moments.

As shown in Figures 10.34b and 10.35, the deck system comprises of steel deck plates and floor beams. Steel plates are stiffened longitudinally by U-shape stiffeners. The top of floor beams are cut out to allow stiffeners to pass through. Thus, the deck clearance is minimized. There is no stringer used in the deck system. The bottom lateral bracings are crossed and placed in the lower part of the floor beam (Figure 10.35). The top

(c)

Figure 10.34 (Continued) Shang Xin Bridge, China. (c) Perspective view of the completed trusses.

Figure 10.35 Floor beams and bottom lateral bracings of Shang Xin Bridge.

lateral sway bracings for wind (as shown in Figure 10.34a and c) are provided only at the two highest posts at pier locations. Panel members are all made of tube sections, and they are semirigidly connected as shown in Figure 10.36. The trusses are supported by temporary supports underneath and launched forward as shown in Figure 10.36. There are 25 triangular panels with each of them as one erection unit.

Finite element models were generated with MIDAS Civil (Figure 10.37a) and ANSYS (Figure 10.37b) for cross-checking the simulation of all

Figure 10.36 Rigid connections of Shang Xin Bridge.

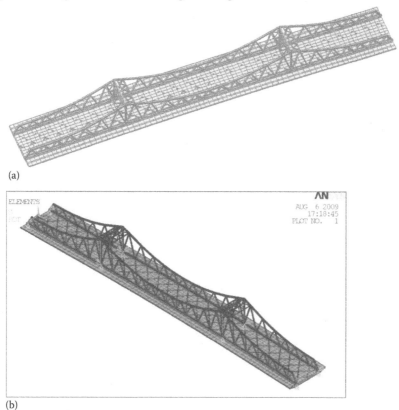

(a)

(b)

Figure 10.37 Finite element models of Shang Xin Bridge. (a) MIDAS model; (b) ANSYS model.

construction stages. Both MIDAS and ANAYS models are using a combination of beam and shell elements, with 2798 nodes and 6763 elements in total (1694 shell elements for deck and U-shape gusset plates and 5069 beam elements for the rest of the members) for 52 construction stages.

Chapter 11

Cable-stayed bridges

11.1 BASICS OF CABLE-STAYED BRIDGES

In the case of a continuous girder, as shown in Figure 11.1, the zone close to the middle support area is in compression at the bottom and in tension at the top. The major tensile principal stress in the middle support area is about 45° downward from the supports. For longer-span bridges, girder height in the middle support areas can be designed taller than that in the middle span areas. In comparison, a cable-stayed bridge has a similar load distribution path to a prestressed concrete (PC)/reinforced concrete (RC) continuous bridge by replacing bent-up prestress tendons/rebars with external cables.

A typical cable-stayed bridge, as shown in Figure 11.2, consists of a continuous girder, stay cables, two pylons, and two end piers. The span between two pylons is called the main span. The main span length is a key design parameter of cable-stayed bridges. From the perspective of engineering efficiency and cost-effectiveness, a cable-stayed bridge is very competitive with other bridge types that have a main span range of 200–500 m (656' to 1640'). For an overall satisfaction in both structural performance and economy, cable-stayed bridges can be built in a span of up to 1000 m (3280') (Chen and Duan 1999). Several cable-stayed bridges with a main span over 1000 m have been built since 2005 (i.e., Russian Russky Bridge, 1104 m [3622'] in 2012, and China's Sutong Bridge, 1088 m [3570'] in 2008].

Classified by bridge elevation, variations of cable-stayed bridge include (1) typical cable-stayed bridge with one main span and two side spans (Figure 11.2), (2) typical cable-stayed bridge with one main span and two side spans with auxiliary side-span piers (Figure 11.3), (3) single-pylon cable-stayed bridge (Figure 11.4), and (4) multiple main span cable-stayed bridge (Figure 11.5). The cable-stay layout can be a fan, modified fan, or parallel (harp) (Figure 11.6). In cable-stayed bridges with a very long main span, main stay cables are crosstied by groups of transverse secondary cables (crossties) to counter cable oscillations due to wind and rain (Figure 11.7). In the transverse direction, the cable stays can be in two planes or in only one plane (Figure 11.8).

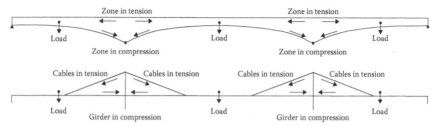

Figure 11.1 From reinforced to cable-stayed bridge.

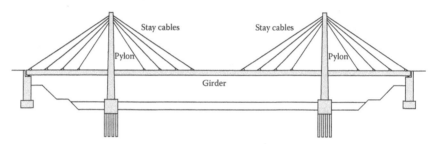

Figure 11.2 A typical cable-stayed bridge.

Figure 11.3 A cable-stayed bridge with auxiliary piers.

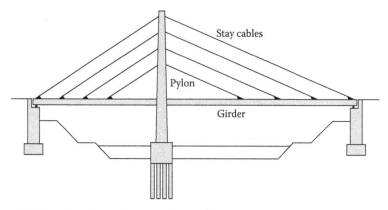

Figure 11.4 A single-pylon cable-stayed bridge.

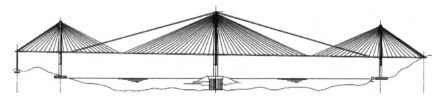

Figure 11.5 A three-pylon cable-stayed bridge.

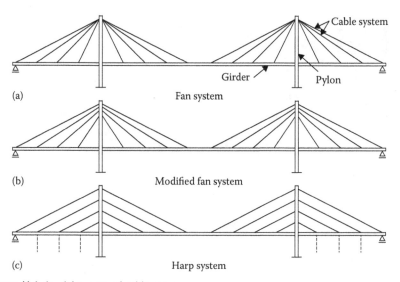

Figure 11.6 (a–c) Layouts of cable stays.

Figure 11.7 Crosstie of cables.

Figure 11.8 A cable-stayed bridge with only one stay plane.

Connections of stay cables to a pylon can be anchored to the pylon or cradled through the pylon. When a stay cable is cradled through the pylon, the cable is continuous from the deck on one side of the pylon to the deck on another (Figure 11.9).

In a cable-stayed bridge, the cross section of the main girder can be a multiple-cell box, two I-section girders, or trusses. Concrete box girders

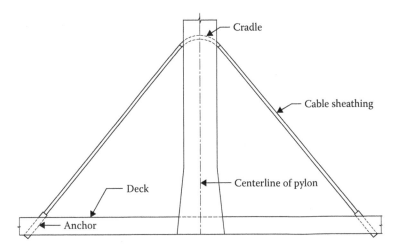

Figure 11.9 Cradle stay system.

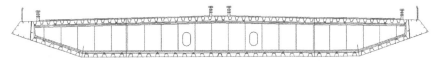

Figure 11.10 Cross section of a typical steel box girder.

are widely used in short-span ranges, whereas steel box girders dominate the long-span range (Figure 11.10).

Based on NCHRP Report (National Cooperative Highway Research Program; Tabatabai 2005), several types of cables are available for use as *stays** in cable-stayed bridges. The form or configuration of the cable depends on its make-up; it can be composed of parallel wires (no longer commercially available in the United States), parallel strands, parallel solid bars (larger diameter and lower allowable stress and fatigue resistance than comparable parallel strand stays), and single or multiple arrangements of structural strands or locked-coil strands (no longer used in the United States). Nowadays, high-strength steel strands, same as those used in PC bridges, are commonly used for stay cables. Each strand usually contains seven high-strength steel wires. Each stay cable contains a number of strands inside and a polyethylene (PE) jacket on the outside. The number of strands in one cable varies from as few as 10 to more than 100. The corrosion protection of cables is a common concern in the design of cable-stayed bridges. Usually the strand itself can be epoxy-coated, galvanized, or greased. The stay cable is sheathed by PE jacket, which can effectively protect the

---

* *Stay* is defined as a large strong rope usually made of wires used to support a mast.

cable from ultraviolet radiation, atmospheric moisture, precipitations, and temperature fluctuations. In most situations, the PE jacket will be filled with cement grout.

Because cables can also provide the initial support by transferring the loads to the pylons, girders can be erected segment by segment. For steel girders, girder segments can be fabricated in the factory and installed on-site (Figure 11.11). For concrete girders, a traveling carriage can be built to support the casting of concrete on-site (Figure 11.12). Once a segment is in place, a pair of cables will be jacked to support the erected segment.

Figure 11.11 Lifting and erecting of one segment.

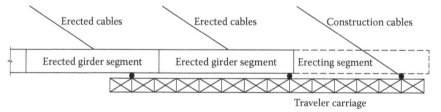

Figure 11.12 Traveling carriages for casting in-site segment.

Table 11.1 Top 10 longest cable-stayed bridges in the world

No.	Name	Main span (m)	Year of built	Location
1	Russky Bridge	1104	2012	Vladivostok, Russia
2	Sutong Bridge	1088	2008	Jiangsu, China
3	Stonecutters Bridge	1018	2009	Hong Kong, China
4	Edong Bridge	926	2010	Huangshi, China
5	Tatara Bridge	890	1999	Seto Island, Japan
6	Pont de Normandie	856	1995	Le Havre, France
7	Jingyue Bridge	816	2010	Jingzhou, China
8	Incheon Bridge	800	2009	Incheon, Korea
9	Zolotoy Rog Bridge	737	2012	Vladivostok, Russia
10	Shanghai Yangtze River Bridge	730	2009	Shanghai, China

The erection of girders on both sides of the pylon usually proceeds simultaneously. This method is referred to as the *balancing erection method*. This cantilever erection method is a valuable and practical advantage that is unique to cable-stayed bridges. This benefit has resulted in designers commonly selecting cable-stayed bridges over other bridge types. For a given bridge site, this construction method could make the cable-stayed bridge the only option.

Long-span cable-stayed bridges have rapidly developed since the turn of the century, with some having main spans that exceed 1000 m (3280′). This was previously considered the extreme limit of cable-stayed bridges. Achieving longer spans, cable-stayed bridges have demonstrated that they are structurally competitive to suspension bridges. Table 11.1 lists the recent top 10 longest cable-stayed bridges in the world.

11.2 BEHAVIOR OF CABLE-STAYED BRIDGES

The idea of a cable-stayed bridge is simple: to provide intermediate support for the girder by using cables that are anchored to the pylon at the other end. This extends the length to which the girder can span. The mechanical behavior of such structural components like the continuous girder, cables, and a pylon is clearly shown in Figure 11.13. Loads are mainly vertical loads on the girder due to its structural weight and live loads, cables are under tension so as to pass loads on the girder to the pylon, the pylon is under compression due to the downward forces from cables and its own structural weight, and the girder encounters axial compression due to the horizontal load components from cables and bending moments due to vertical loads.

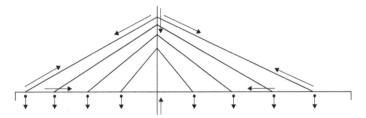

Figure 11.13 Mechanical behavior of a cable-stayed bridge.

11.2.1 Weakness of cable supports

However, the distribution of vertical loads from the girder to the cables is far less than a continuous girder with intermediate rigid supports; in another words, the spring supports' stiffness of cables are far less than that of real supports. To illustrate the weakness of cable supports, a five-span continuous girder and a girder stayed by four cables, as shown in Figure 11.14, are taken as an example. Comparison between figures demonstrates that a girder stayed by long vertical cables has less support stiffness than that supported by bearings. This is due to the fact that cables are more flexible than regular bearings, and further, a girder stayed by slanted cables has less support stiffness than that stayed by vertical cables. In a fan cable system, the smaller the cable angle to the girder, the less vertical support stiffness can cable provide. The stiffness that anchor cables in side spans and end cables in the middle span can provide is much less than that cables close to pylons can provide, which is one factor that limits the main span capacity when the height of the pylons is limited. Compared with the continuous girder, distributions of moment and axial forces, as well as bridge displacements (under a uniform load on the girder such as the girder's structural weight), reflect the weakness of the cable's support capacity. Figure 11.15 shows this phenomenon. This holds true in response to the live loads as well.

From these distributions shown in Figure 11.15, it is clear that the axial forces in cables are low, bending moments along the girder are high, and the vertical displacements of the girder are large. If the structural weight of the girder and superimposed deck loads are not redistributed, as behaved in an RC bridge that is built by casting concrete directly in its setting location all at once, the span capacity of a cable-stayed bridge would be similar to a continuous girder bridge. By using high-strength steel wires or strands as cables and prestressing them at a much higher level than what it would be distributed due to structural deformation under dead loads, as shown in Figure 11.15, the dead loads will be transferred to the pylon so bending moments on the girder will be reduced. Therefore, the weakness effect under dead loads is improved and the span capacity is increased. The structural advantages of a cable-stayed bridge,

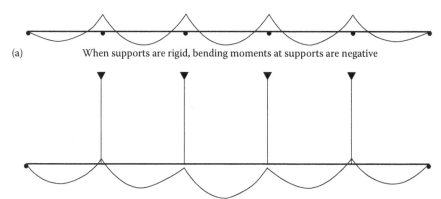

(a) When supports are rigid, bending moments at supports are negative

(b) When supports are vertical cables, bending moments at supports become less or even positive

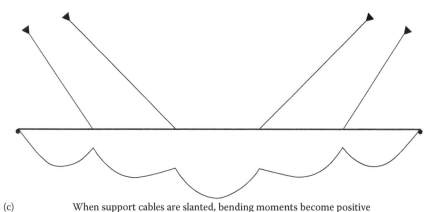

(c) When support cables are slanted, bending moments become positive

Figure 11.14 (a) Continuous girder; (b) vertical cable supports; (c) slanted cable supports.

as well as demonstrating the general engineering concept and the practical/economic benefits of building such a bridge, are covered in this section.

11.2.2 Ideal state

What would be the best jacking stress of each cable in terms of increasing the girder span capacity? This is a unique question to cable-stayed bridges during structural analyses and design. From the girder capacities' perspective, the answer is found when the maximum bending moments due to dead loads on the girder are the same as those of a continuous girder as shown in Figure 11.16. Although the live load distribution does not depend on the cable stress level and does not change once a cable-stayed bridge is structurally determined, dead loads dominate in cable-stayed bridges. Therefore, as long as the dead load distribution reaches a desired state, the span capacity can be increased. This simple idea is based on the fact that

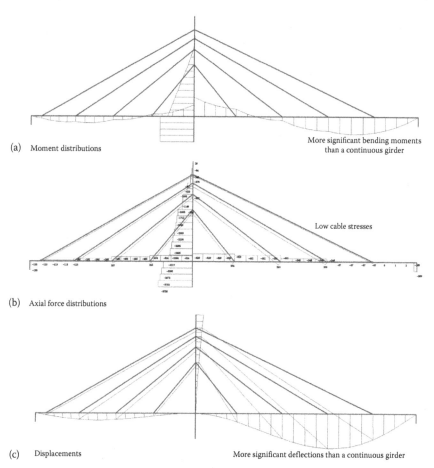

(a)   Moment distributions

More significant bending moments
than a continuous girder

Low cable stresses

(b)   Axial force distributions

(c)   Displacements

More significant deflections than a continuous girder

Figure 11.15 (a–c) Weak supports from cables.

both the girder and the pylons are more efficient under axial compression than under tension and compression due to large bending moments.

For any particular cable-stayed bridge, such an ideal state of dead load distribution will also depend on the pylons. Sometimes, when girder spans are not symmetrical around a pylon and most of girder dead loads are transferred to the pylon as well, certain moment will be created at the bottom of the pylon as shown in Figure 11.16. The girder moments, the horizontal displacements at the top of the pylons, and the longitudinal bending moments at the bottom of the pylons are usually the primary control points to determine an ideal state. It should be noted that the so-called ideal state, as shown in Figure 11.16, is only to demonstrate that dead loads on the girder should be redistributed to pylons by adjusting the jacking forces of cables. In more real situations, such an ideal state is unsuitable,

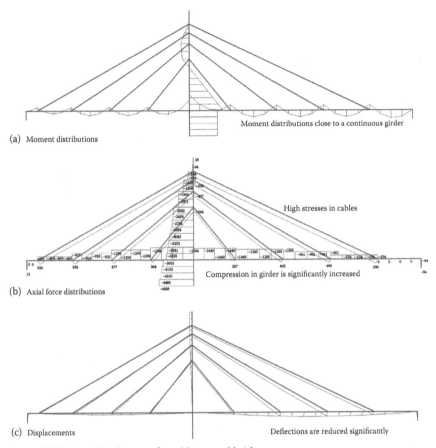

(a) Moment distributions

Moment distributions close to a continuous girder

High stresses in cables

Compression in girder is significantly increased

(b) Axial force distributions

(c) Displacements

Deflections are reduced significantly

Figure 11.16 (a–c) Ideal state of a cable-stayed bridge.

unattainable, or simply not economical. What bridge engineers define is only one *preferred state*.

During the schematic design of a cable-stayed bridge, the concept of continuous girder behavior can be used to estimate cable quantities by hand. Figure 11.17, for example, shows one erection segment of the girder. If 100% of the dead loads on one girder segment are redistributed to the pylon by a pair of cables, the cable forces F would be

$$F = \frac{W}{\sin a} \qquad (11.1)$$

where:
   W is the dead loads on a girder segment
   α is the cable angle

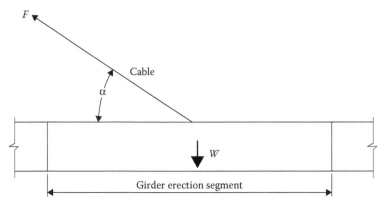

Figure 11.17 Determine cable quantities during scheme design.

Having cable forces due to the dead loads, plus an estimated percentage of the live loads, the strength of cable strands, and a guided safety factor of cables, quantities of each can be determined quickly. In most cases, it is preferred that all cables, except anchor cables and end cables, be the same size. The quantities determined earlier can be used in initial analyses.

11.2.3 Desired state

As the dead loads on the girder can be redistributed to pylons by jacking cables, tuning cables will reach a desired moment distribution on the girder (Wang and Fu 2005). However, determining the jacking stress of each cable so as to reach a desired state is a unique question in the design of a cable-stayed bridge. If only the girder is considered, it is easy to conclude that the ideal state of a cable-stayed bridge is the state in which the total bending energy accumulated along the girder is minimal. In practice, it is equivalent to adjusting the girder moment at anchor to zero (or even negative) or vertical displacements to zero. If the pylons have to be considered together with the girder, having no longitudinal displacement or no bending moment would be perfect. Because most bridges are not symmetrical about pylons, bearing a minor moment is unavoidable.

Moment and displacement distribution along the girder and towers can reach the ideal state by adjusting cable stresses. The moment or the displacement of an ideal state Z can be written as (Wang and Fu 2012)

$$Z = \left\{ z_1 \quad z_2 \quad \dots \quad z_n \right\}^T \tag{11.2}$$

where:
n is the total number of targets that need to be satisfied
T stands for the transformation of a matrix or a vector

The approach to an ideal state is achieved when the variables in Equation 11.2 are as close to the desired values as possible. The minimum square error method is one of the most effective ways to obtain the optimal Z, in which the resulting cable stresses S can be written as

$$S = \{s_1 \quad s_2 \quad \cdots \quad s_m\}^T \tag{11.3}$$

where m is total number of cables to be tuned.

By analyzing the response of a unit stress applied at each pair of tuning cables, the influence values of all targets can be obtained, and the influence matrix A can be written as

$$A = \begin{bmatrix} a_{11} & a_{12} & \cdots & a_{1m} \\ a_{21} & a_{22} & \cdots & a_{2m} \\ \cdots & \cdots & \cdots & \cdots \\ a_{n1} & a_{n2} & \cdots & a_{nm} \end{bmatrix} \tag{11.4}$$

where a_{ij} is the response at target i due to a unit stress at cable j. Thus, their relationship can be written as

$$A \times S = Z \tag{11.5}$$

If the number of tuning cables is the same as the number of targets, cable stresses can be obtained by solving the linear equation (11.5). In this case, engineering experience is required in selecting cables and the targets. A bad or contradictory tuning of cables and targets may cause matrix A not to be a diagonal dominant matrix or Equation 11.5 in ill condition. If, as in most cases, m is less than n, cable stresses can be optimized by minimizing the error between the desired state and the state that can be reached. D, which has the same form as Z, is the desired target value. The error E can be written as

$$E = D - Z \tag{11.6}$$

The optimization goal is to minimize Ω, which is the square of E, and can be written as

$$\Omega = (D - Z)^2 \tag{11.7}$$

From the variation principle, it is known that the condition to have Ω minimized is

$$\frac{\partial \Omega}{\partial S_i} = 0, i = 1, 2, 3, \ldots m \tag{11.8}$$

By using the matrix differential and considering Equations 11.7 and 11.5, the following equation can be obtained:

$$A^T A \times S = A^T D \qquad\qquad (11.9)$$

After solving S from the linear equation group in Equation 11.9, the optimized target values will be obtained from Equation 11.5 (Wang and Fu 2005).

The following procedures are commonly used in approaching the ideal state in the complete stage* after the deck is superimposed:

1. Select all the cables to be tuned.
2. Perform static analysis under structural weight and superimposed dead loads.
3. Select negative girder displacements at each anchor in step (2) as D. This step varies in different situations.
4. Evaluate S as above.
5. Similar to jacking loads, reapply S on the structure and perform a round of full analysis.
6. The sum of steps (2) and (5) is the ideal state at the complete stage.

11.2.4 Anchor of pylons

Due to the weakness of cable supports as stated in Section 11.2.1, the bending moments and vertical displacements under live loads can be significant on the girder. This may become a control factor for the maximum length a cable-stayed bridge can span. Figure 11.18 shows how the horizontal stiffness of pylons influences the vertical stiffness of the girder. In very long-span cable-stayed bridges, the anchor cables in the end spans and end cables in the main span have smaller angles to the girder; the vertical stiffness they provide to the girder becomes smaller. The upper part of a middle pylon in multiple-span cable-stayed bridges lacks enough horizontal anchor stiffness; hence, the vertical stiffness of the girder is not sufficient.

Such behavior will lead to excessive displacement under live loads. Anchorage of pylons is a common issue in two-pylon bridges with a very long main span or multiple-span cable-stayed bridges. As shown in Figure 11.18, it is obvious that (1) adding secondary or auxiliary piers at side spans in very long-span cable-stayed bridges, such as Pont de Normandie Bridge (shown in Figure 11.3) and Sutong (illustrated example in Section 11.5) Bridge, and (2) using cross cables to anchor a middle pylon to the deck where it has a strong vertical stiffness (adjacent pylon area) are both effective geometry

---

* The complete stage of a cable-stayed bridge is defined as the stage when the erecting girder is closed and the deck is superimposed.

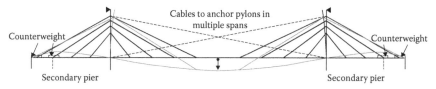

Figure 11.18 Anchor of pylons.

configurations in improving the live load stiffness. In long-span cable-stayed bridges, it is common that the anchor cables contain more strands than other nonanchor cables or they are simply composed of two or three cable stays. Increasing the working area of a cable will increase its axial stiffness so that the horizontal stiffness in the upper part of the pylon will be improved.

In cases that the secondary piers in side spans are used to improve the main span vertical stiffness, sand boxes may be used as counterweight measures on the top of secondary pier areas. As the main span length increases, the extreme live load reactions of secondary piers may exceed their reactions due to structural weight and superimposed dead loads. Uplift may happen without counterweight.

11.2.5 Backward and forward analyses

The *ideal state* is defined in the complete stage when a bridge is ready for traffic. Although rejacking some particular cables after closure is possible, retuning all cable stresses so as to reach a desired state is impractical. To simplify the construction procedure and reduce each erection cycle, it is best to jack a cable to the correct level at that stage when it is erected, which guarantees its final stress level in the ideal state after the deck is superimposed. How much is the correct jacking stress of each pair of cables to reach the expected final ideal state? The answer to this question leads to a unique analysis method or technique in cable-stayed bridge analyses—backward and forward analyses.

Backward analysis simulates the reverse process of erection, and forward analysis simulates the normal construction process of erection. Given the state after being superimposed, backward analysis will show the state before each girder segment and cables are erected. Given the erection parameters such as girder segment properties, structural and other construction weighs, and jacking stresses, forward analysis will show the state after the erection cables are jacked. Theoretically, a full forward analysis using jacking stresses obtained from backward analysis should meet the ideal state predefined at the stage after being superimposed.

By removing superimposed dead loads and disassembling girder segments and cables stage by stage, the bridge state* of each erection can be obtained, which will be used to control girder displacements and cables stressing during forward erection. It should be noted that it is impossible for an actual forward stage to reach precisely the state obtained by backward analysis. This can be understood from the fact that the closure segment stress will not reach zero after the superimposed dead loads are removed in backward analysis, whereas in reality it is zero after closure.

Unlike backward analysis, forward analysis based on the actual state of any construction stage can predict the state when the bridge closes in the middle span. This prediction is very important for cable tunings at any stage. Because retuning every pair of cables will increase on-site labor dramatically and hence slow the construction pace, usually only the newly installed pairs of cables are jacked according to the analysis results backward to that stage. If, however, tuning one pair of cables cannot keep the state of the bridge in control, retuning of multiple cables will be required. The retuning is required at least in the complete stage.

To exactly simulate removal and installation of some components, a dedicated analysis program is required. The backward analysis can be performed as follows:

1. Apply negative nodal forces of the removed components in the previous stage
2. The sum of step (1) and the state before removal equals the state after removal

The forward analysis can be performed as follows:

1. Analyze the new stage with the application of the installed components' weight.
2. Analyze jacking loads, if applicable.
3. The sum of steps (1) and (2) and the state before the installation equals the state after erection.

11.2.6 Geometric nonlinearity—P-Delta effect

The girder of a cable-stayed bridge works as a continuous girder with a spring support at each anchorage of the cable. However, as the girder is under compression, its bending stiffness will be reduced due to the P-Delta effect. Similar to the girder, the pylons are under compression. Its bending behavior would also be affected by the P-Delta effect.

---

* A bridge state is defined as the bridge's geometry configuration, internal forces, and structural displacements.

The P-Delta effect can be categorized as the *initial stress* problem in mechanics, which is that the existing stress condition of a component will affect its behavior when new loads are acting on it. Hence, the superposition principle will be no longer valid. The change of component stress under new loads will further affect its behavior. The iteration process is inevitable. The P-Delta phenomenon is common in bridge structures. It can be ignored in many situations during preliminary analyses. However, in cable-stayed bridges, it has to be considered as the initial stresses accumulated in the girders and pylons are significant.

By using nonlinear iteration process, the P-Delta effect is fairly easy to be accounted for in dead load analysis. As the positions and magnitudes of dead loads are known and do not vary, loads can be scheduled into several different steps. For each step, the analysis can be linear and the stresses obtained will be considered when evaluating the stiffness of the next step.

However, it could be extremely complicated to reach the theoretical solution for live load analyses as the positions, magnitudes, and/or load patterns vary. For most live load standards, seeking theoretical solutions is impractical. In general, the following steps are used as a practical way to consider P-Delta effects in live load analyses:

1. Include the effects of axial forces in the girder and pylons after secondary dead loads are imposed when evaluating influence values.
2. Use regular methods to obtain extreme live load positions and magnitudes.
3. Apply the earlier extreme live loads as a dead load case on the bridge and conduct a nonlinear analysis so as to adjust the extreme results.

Some researchers even suggest moving all axle positions obtained in step (2) equally from left to right and step by step to further search for the true extreme positions. This method assumes the linear results are very close to reality. However, as a general rule to bridge modeling and structural analyses, this method lacks theoretical support and should be studied case by case.

As there will be many points of interest in a cable-stayed bridge that need to perform the earlier tedious live load analyses, a finite element analysis (FEA) and live load analysis package specifically developed for cable-stayed bridges is essential.

11.2.7 Geometric nonlinearity—Cable sag effect

Due to its own weight, a cable between two anchors will sag downward and will not remain straight. Taking a horizontal cable as an example,

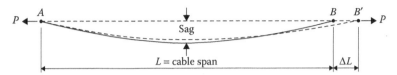

Figure 11.19 Sag of a horizontal cable (to show axial deformation under force P).

Figure 11.19 illustrates the cable geometry in reality compared to what it is modeled mathematically.

The axial stiffness of a cable is simply defined as the axial force required for causing a unit axial deformation or the elongation along the axis of two anchors. When a cable is straight, the total elongation is the deformation of the cable so that the axial force that causes such an elongation is higher. For example, ΔL as shown in Figure 11.19 is the deformation of cable AB if sag does not exist. When cable sags away from its axis, not all of the elongation is due to deformation, yet it is due to the geometry change, so the axial force required to cause the same amount of elongation is lower. As shown in Figure 11.19, ΔL, the elongation of the sagged cable AB is the sum of the cable deformation and shortage of cable geometry. That is how a sagged cable behaves as if the material has a lower Young's modulus.

One fact about the cable sag is that the higher the existing axial force is, the smaller the sag and thus the stiffness of the cable is closer to a straight cable.

In preliminary analyses, one cable is usually meshed into one element by its two anchors. In a regular FEA package, the stiffness of such a cable is calculated based on a straight line between two anchor points. Its stiffness is, therefore, overcalculated. The Ernst formula (Ernst 1965), as shown in Equation 11.10, is usually adopted to calculate the cable's equivalent stiffness or Young's modulus based on a given cable stress.

$$E_{eq} = \frac{E}{1 + [(\omega H)^2 AE/(12P^3)]}$$

(11.10)

where:
E is the Young's modulus of a cable, in kN/m^2
ω is the unit weight of the cable, in kN/m
H is the cable span in horizontal direction, in m
A is the cable area, in m^2
P is the cable force, in kN

Cable forces will be redistributed in the next phases after they are initially jacked, or, in another words, cable forces are never constant from stage to

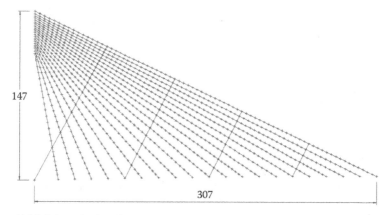

147

307

Figure 11.20 Submeshed cables with crossties.

stage. Again, a specialized FEA package considering the effective cable stiffness calculation under multiple construction stages is preferred.

For long-span cable-stayed bridges, the geometric nonlinear effects, including the P-Delta effect, cable sags, and large displacements, become significant. Full geometric nonlinear analysis is required in certain detailed design and study situations. It should be noted that the effective stiffness approach for sag effects is suitable only for cases where each long cable is modeled as only one element by its two anchor points. When a cable is submeshed into small segments to investigate the large displacements in detail or when cable crossties are considered (Figure 11.20), effective stiffness calculation is no longer needed during the iteration processes. During iteration in large displacement analysis, the stiffness of an element will be evaluated at current geometric locations. The axial stiffness of such a cable segment is very close to the actual stiffness as the sag between two end points of a cable segment becomes negligible.

11.2.8 Geometric nonlinearity—Large displacements

As the main span of the bridge increases, the global stiffness decreases and the displacements (not the deformation) become significant. There are two features that can be used to help understand what will impact a regular linear analysis when displacement becomes large. The first one is that the difference of stiffness at current geometry configurations and at its original positions is no longer negligible. It will cause major errors to evaluate the responses of an incremental load at the current configuration when using the stiffness obtained from the original geometry configuration. The second aspect is the coupling between displacements and forces, or the stiffness of one degree of freedom (DOF), depends on the existing stress of another degree. For example, the so-called P-Delta effect, or initial stress problem, is one such phenomenon. As introduced in Chapter 3, all nonlinear effects of initial stresses, large displacements, and cable sag are due to geometric nonlinearities. When a large displacement problem is considered in cable-stayed bridge analysis, a full geometric nonlinear analysis should be employed.

Large displacement behavior in long-span cable-stayed bridges should be investigated case by case. In general, when large displacements are considered, the lateral stiffness of a bridge will be enhanced due to the cables' geometric stiffness under significant tensions, that is, the tendency to maintain its lateral positions. When the girder is cambered as a shallow arch, as most long-span bridges are, the girder will behave as with stronger stiffness than the girder not considering large displacement. This characteristic comes from the geometric stiffness along the shallow-arch girder, similar to the behavior of a shell under pressure.

11.2.9 Stability

Stability is one of the factors governing long-span bridge design and analysis. It will be discussed in more detail in Chapter 14. Stability includes static stability and aerodynamic stability. Static stability can be further categorized as elastic stability and ultimate plastic stability. Elastic stability deals with scenarios where material is assumed linear but geometric deformations and stresses are coupled. The P-Delta effect in columns and beams is one of these types of problems. Plastic stability focuses on scenarios where material enters a plastic stage so that local components yield. In general, the geometric nonlinearity in long-span bridges is more significant than material nonlinearity and the elastic stability should be investigated first. On the other hand, the material nonlinearity in middle- and short-span bridges is more significant than geometric nonlinearity, and the plastic stability becomes more important. Elastic stability can further be grouped as bifurcated stability (Class I), which considers the coupling at only the current geometric configurations, and full geometric stability, which traces the changes of geometric configurations under each increment of load. Both elastic stabilities in cable-stayed bridges should be analyzed. When plastic stability is considered, geometric nonlinearity will be considered at the same time, or the so-called dual-nonlinear analysis will be performed.

Aerodynamic stability, which includes structural and cable oscillations under wind and rain, is a critical issue for long-span cable-stayed bridges. The design of a cable-stayed bridge should follow special guidance for aerodynamic issues. Wind tunnel testing may be unavoidable for the design of long-span cable-stayed bridges. The aerodynamic stability issue is not covered in this chapter. However, a simple formula to estimate the critical wind speed will be introduced in the dynamic analysis section.

As the span increases, the compression in the girder increases; additionally, the stability of the girder becomes critical to the design and building of a long-span cable-stayed bridge. Stability analysis in both lateral and vertical directions is required, especially before the closure.

For the initial stresses, the axial forces in the girder under dead loads can be used. The critical load can then be obtained from solving the eigenvalue problem. Although the Class I stability result gives only the upper limit of the critical loads due to the fact that a perfect stability problem rarely happens in actual engineering situations, it can serve as an initial guidance for the stability analysis.

The process of static stability analysis with consideration of large displacements can be the same as a regular static nonlinear analysis, except that the loads should be selected to reflect the nature of the structure. Also, the FEA system should allow for the increase of some of the loads step by step in search of the ultimate loads. For instance, if the issue of temporary construction loads is a concern for lateral stability before closure, minor lateral wind loads, structural loads, and cable prestressing loads should be applied as constant loads. The construction loads, as the main loads, should be increased step by step. The level of the major loads at which the structure fails is the critical load of the stability analysis.

A long-span cable-stayed bridge rarely fails in static geometric nonlinear analyses. Even for lateral stability, it is easy to understand that the transverse components of the high-stressed cable tensile in a changed geometry configuration will help to prevent large lateral displacements. In terms of static stability, a full analysis by counting both geometric and material nonlinearities is inevitable. More details will be discussed in Chapter 14.

11.2.10 Dynamic behavior

Compared with other girder-type bridges, cable-stayed bridges are relatively slender and more flexible. In seismic design, a cable-stayed bridge is preferable because of its low natural frequency. On the contrary, when aerodynamic stability is of concern, a stiffer bridge is preferred. Certain special measurements will have to be taken into account for a long-span cable-stayed bridge for both seismic and aerodynamic requirements, for example, installing damping devices in girders so as to improve responses to dynamic loads from vehicles, adopting a wind-resisting girder cross-sectional shape so as to improve aerodynamic response, and cross-tying long cables to reduce wind and rain oscillations of cables. In both aspects, the natural modes of a cable-stayed bridge should be investigated carefully.

The connections between girders and pylons influence the dynamic behaviors. If girders and pylons are rigidly connected, as they do in single cable-stayed plane bridges, the girder is stiffer and the first mode will be the bending mode with higher frequency. If the girder is designed to move free longitudinally and independently from the pylons, as commonly adopted in long-span cable-stayed bridges, the girder will behave like a suspending pedal and the first mode will be the horizontal swinging mode. Many cable-stayed bridges use special connection mechanisms such as thin concrete blocks to limit the girder horizontal movements. These are carefully designed so that when there are severe movements, such as earthquakes, they will break and lose their function, and the girder therefore behaves more effectively to absorb dynamic energy. For long-span cable-stayed bridges, as the longitudinal displacements due to temperature and wind loads are both significant, more sophisticated connection systems are needed to resolve these contradictory requirements. Figure 11.21 shows the horizontal damping systems used in the Sutong Bridge, in which gradual displacements due to temperature are released, dynamic displacements due to wind loads are reduced by the viscous damper systems, and excessive displacements are blocked by their movement-stopping mechanism.

In regard to the aerodynamic behavior of a slender structure, the shape of the girder cross section plays an important role. Because the height is much less than its other two dimensions, the girder can be treated as a flat slab in most cases. The side of the cross section is usually cosmetically modified sharply toward the outside, as shown in Figure 11.10, or wind fairings are installed, which can be the first technique used to improve the aerodynamic

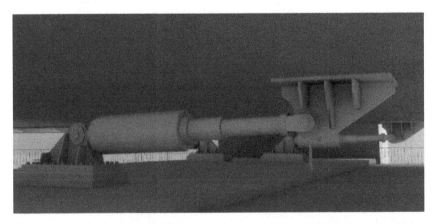

Figure 11.21 Horizontal damping systems between girder and pylon in Sutong Bridge. (Data from You, Q. et al., "Sutong Bridge—A Cable-Stayed Bridge with Main Span of 1088 Meters, ABSE Congress Report, 17th Congress of IABSE, Chicago, 2008, pp. 142–149.)

behavior. Damping devices in the main span are also helpful to improve the aerodynamic response. From a structural design point of view, separating the girder's torsional mode away from its first bending mode as far as possible can be considered the next method in terms of increasing the critical flutter wind speed. When wind attacks from the sides of a bridge, the coupling of bending and torsional forces of the girder is a key factor that leads to the collapse of the bridge. During the preliminary design, several different formulas can be used to estimate the critical flutter wind speed for girders that have slablike cross sections as suggested by the Wind-Resistant Design Specification for Highway Bridges Ministry of Transport of China, 2004, which became mandatory to comply with in China since 2005. For example, the Van der Put formula (Equation 11.11) considers the ratio of the first torsion frequency to the first bending frequency.

$$V_{cr} = \left[1 + (\varepsilon - 0.5) \bullet \sqrt{0.72 \bullet \left(\frac{r}{b}\right) \bullet \mu}\right] \bullet \omega_b \bullet b \qquad (11.11)$$

where:

V_{cr} is the critical flutter wind speed, in m/s

$\varepsilon = f_t/f_b$ is the ratio of the first torsion frequency to the first bending frequency

f_t is the first torsional frequency, in Hz

f_b is the first bending frequency, in Hz

r is the $\sqrt{(I_m/m)}$, the mass radius of gyration

I_m is the mass inertia per unit length of the girder, in kg $\bullet$ m^2/m

m is the mass per unit length of the girder, in kg/m

b is the half width of the deck, in m

μ is $(m/\pi\rho b^2)$, the ratio of mass to air density

ρ is the air density, in kg/m^3

ω_b is $2 \pi f_b$, the angular frequency of the first bending

Equation 11.11 is further simplified by Tongji University as

$$V_{cr} = 2.5\sqrt{\mu \bullet \frac{r}{b}} \bullet 2b \bullet f_t \qquad (11.12)$$

For very long-span cable-stayed bridges, girders and pylons are usually tied down by anchor cables. This is to increase the wind stability during construction.

Oscillation of cable stays due to wind and rain in long-span cable-stayed bridges could be significant and should be considered. Crossties of cable stays as shown in Figure 11.7 and a stay damping system as shown in Figure 11.22 are commonly used measures to counter cable oscillations.

Figure 11.22 Cable-stay damping systems in Arthur Ravenel Jr. Bridge.

11.3 CONSTRUCTION CONTROL

The unique construction method of cable-stayed bridges brings up a distinctive topic to cable-stayed bridges—construction control. The girder is erected segment by segment, and the cables are jacked pair by pair during erection. Engineering errors commonly exist in any step of this long process, and as a result, what engineers expect may not be achieved at the end. Among many structural measurements of a bridge state, girder geometry and cable forces are the two most critical ones. Too many errors in girder geometry may cause the closure segment hard to fit and adjust, and too many errors in cable forces may cause cable forces to exceed their allowable range. The importance of reducing these errors is obvious. Engineering error does exist in any bridge construction. The control of engineering errors is important especially in cable-stayed bridges.

11.3.1 Observation errors

There are two types of errors: (1) observation errors and (2) construction errors. The observation errors are due to the measurement systems, which occur in measuring the following characteristics:

1. Girder elevations
2. Cable stresses or forces
3. Horizontal displacement at the top of pylons
4. Stresses on the bottom of pylons
5. Stresses on the top and bottom of the girder at any point of interest
6. Environmental temperature, and so on

Minimizing errors in the measurement systems should be taken in the first place whenever and wherever a structural behavior is measured. Reliable field measurement methods or technologies are critical in control analyses. In cases where a strain gage is used to measure stresses, nonglobal stress-related strain such as temperature changes and creep and shrinkage strains should be carefully investigated. Multiple strain gages usually are needed to measure stress at one point, and its configuration should be studied based on location. In the control of Yamen Bridge (a PC box girder cable-stayed bridge with a main span of 338 m [1109'] and a single stay plane in Guangdong, People's Republic of China), the girder stress at the neutral axis can be simply derived from cable forces to calibrate the stress measurements and to identify the strains due to creep and shrinkage.

11.3.2 Measurement of cable forces

Among the structural responses that determine the state after erection, cable forces are the most important measurement, and obtaining them is a relatively simple and reliable process. The fundamental frequency method, which is fast and accurate, is widely used to measure the tension force of cables. By collecting random vibration signals of cables under ambient excitation, the fundamental frequency f can be obtained by time- and frequency-domain analyzers. The string vibration equation 11.13 can be used to calculate the cable force:

$$T = 4mLf^2 \tag{11.13}$$

where:
    f is the fundamental frequency of cable, in Hz
    L is the length of cable, in m
    m is the mass of cable, in kg
    T is the tension force, in N

11.3.3 Construction errors

Construction errors, which may cause incorrect assumptions in structural analyses, are due to the quality control of construction and may include the following features:

1. Material properties such as errors in Young's modulus, temperature expansion factor, and material densities
2. Sectional properties such as errors in girder dimensions due to installation or formwork deformations
3. Temporary construction loads
4. Creep and shrinkage properties for a concrete cable-stayed bridge, and so on

In the control of Sutong Bridge (a steel box girder with the main span of 1088 m in Jiangsu, China), which was once the world record holder for main span length, sensitive analyses revealed that deviations of creep and shrinkage of pylons, girder segment weight, length of the girder segments and length and Young's modulus of cable stays were the primary control parameters that would significantly influence the girder elevation. The deviations of the height of the steel anchor boxes, cable weights, and girder stiffness were the secondary control parameters that had moderate influences. Effects of other factors such as Young's modulus of pylons, the height of anchors at the deck end, the verticality of steel anchor boxes, and the shrinkage of welding between girder sections were considered negligible. Wind and temperature effects are also sensitive to girder elevation.

It is clear that all errors can be minimized only by improving measurement systems and quality control processes, and it is impossible to eliminate these errors completely. However, knowing the errors and incorporating them into engineering assumptions so as to better predict a countermeasure, that is, cable jacking stresses in the next erection to control the girder geometry or the primary target to meet design requirements, is achievable. That is the whole purpose of the construction control.

11.3.4 General procedures of construction control

The procedures of construction control and sensitive analysis are two important issues in the construction of a cable-stayed bridge. The steps, which include considering property errors and modifying FEA models, may be complicated and tedious. In general, these steps can be simplified as follows:

1. Use the theoretical model in initial stage and jack the first pair of cables at theoretical stress level.
2. Compare the observations with forward analysis results and analyze the errors.
3. Adjust the model when the errors in step (2) exceed a preset tolerance.
4. Forward-analyze the next erection based on the modified model to obtain the jacking stress of the next erection cables.

During these procedures, sensitive analyses may be required to rank these numerous construction errors. Once the differences between observations and expectations are known and a few variables that are ranked as the most sensitive are identified, certain error analysis methods or algorithms can be used to determine the variations, which will be the basis of the model adjustment. Lin (1983) first applied the Kalman filter method to the construction control of Maogang Bridge, a PC cable-stayed bridge located in Shanghai that is marked as a milestone of cable-stayed bridge construction in China. Nowadays, this method is widely used to analyze the construction errors in certain key parameters, such as girder segment weights, concrete creep, and shrinkage properties. Other prediction models based on grey prediction theory are also practiced to identify errors (Chen et al. 2011). The back propagation neural network method is also used in the construction control of cable-stayed bridges (Li et al. 2007).

Incremental jacking of erection cables and cable jacking stress adjustment are common during erection and for the purpose of construction control, for example, jacking the erection cables to a certain level initially when the traveler carriage is positioned and jacking the erection cables again when the girder segment is positioned. When needed, forces of a group of cables can be adjusted by rejacking to meet certain control goals. A special-purpose analysis tool is preferred to guide the adjustment of multiple cables. The expected cable stresses to meet these goals are usually not the same as the jacking stresses because the jacking process is usually conducted one by one. When one pair of the cable is rejacked, stresses of all the other cables will be redistributed. The jacking sequences should be carefully scheduled, and the analysis should truly reflect the sequences.

11.4 PRINCIPLE AND MODELING OF CABLE-STAYED BRIDGES

There are many considerations in modeling cable-stayed bridges. The first is to identify an analysis tool. Different FEA packages have different features regarding the special requirements for analyzing a cable-stayed bridge. The following lists a few items that need to be identified for any particular FEA package:

1. How a desired state is determined?
2. How backward and forward analyses are processed?
3. How jacking a cable is simulated?
4. How sag effects, initial stresses, and large displacement are considered?
5. How the live load envelops are obtained?

The second question is whether to build the model in 2D or 3D. This was an important question several years ago when 3D analyses, including 3D preprocessing and postprocessing, were more expensive than it is now. Because the lateral dimension is much less than the longitudinal and vertical dimensions, it is adequate to use 2D modeling to conduct analyses for preliminary design purposes. Nowadays considering that advanced 3D processing tools are widely available and the computing capacity and performance are significantly advanced, the 3D modeling should be used whenever feasible. By using 3D modeling, not only can the analysis be more realistic and accurate, but the stiffness and weight of pylons and their connection to the girder become easy to simulate. For some particular analyses, such as bridge instability in wind and spatial flutter analyses, 3D modeling must be used and the torsional stiffness of the girder from the cables has to be counted for.

Using an appropriate mesh density or element length also needs to be considered in modeling a bridge. In general, due to the advancement of computer technologies, the total number of DOFs is no longer a restriction. A model of 100,000 DOFs, or 20,000 3D nodes, is very common nowadays. Similarly, when meshing a component, computer capacity is no longer an issue. However, an appropriate density should be overviewed and controlled. Considering the common bridge dimensional scales in reality, 1 m (about 3′), in longitudinal and vertical directions, could be taken as the minimum distance of elements. It is not necessary to mesh the girder or the pylons smaller than 1 m. In the lateral direction, 1/2 m (about 1 1/2′) could be used, respectively. Wherever there is a specific point of interest, it should be meshed regardless. In most cases, cables can be simply simulated by one element described by its two anchor points. When large displacements, crosstie cables, or local natural modes of cables are of interest, cables will be submeshed into smaller segments.

11.4.1 Main girders

Box girders (steel or concrete) and composite steel I-girders are two types of girders commonly used in cable-stayed bridges. A box girder, as shown in Figure 11.23, can be modeled as a beam at the centroid of its cross section in the longitudinal direction. In the transverse direction, the rigid connection from anchor point A to the beam centroid B is adequate and should be used by default. From the perspective view, the girder looks like a fish bone, as shown in Figure 11.24. The widely used Hambly formula (1991) to simulate the vertical bending stiffness of transverse equivalent beams is not necessary as the lateral distribution is no longer a concern in the global analysis of a cable-stayed bridge. If stiffened transverse beams, instead of rigid bodies, are used to simulate the connection of cables and the centroid of the girder, its bending and shearing stiffness along the

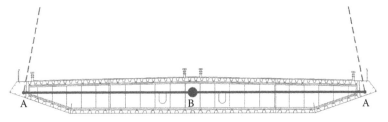

Figure 11.23 Model of a typical steel box girder.

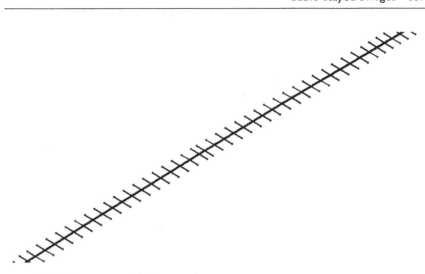

Figure 11.24 Fish bone model of box girder.

length of the bridge should be carefully calculated. As the cables are not perpendicular to the girder, the longitudinal stiffness of the connections between the anchor and the girder centroid will influence the live load distributions.

For a cable-stayed bridge, such as Alex Fraser Bridge (also known as Annacis Bridge, Greater Vancouver, BC) or Nanpu Bridge (Shanghai, China), that uses composite I-girders as shown in Figure 11.25, the girder can be modeled as a grid. In addition to the transverse direction, stringers can be modeled as beam elements. When using an advanced graphical

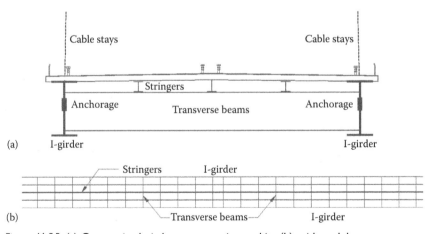

Figure 11.25 (a) Composite I-girder cross section and its (b) grid model.

preprocessing tool, modeling a girder in such detail will be simple and fairly easy. Load distributions, on the other hand, will be more accurate and structural weight calculations will be simplified. When influence surfaces[*] are used in live load analyses, such a grid model will also help to improve the interpolation of influence values as the interpolation triangles are getting smaller and more regular.

11.4.2 Pylons

In the transverse direction, shape of a pylon can be in a single solid/hollow column H, invert Y, diamond, or other shapes. Figure 11.26 shows two alternative pylon plans of Sutong Bridge (China). 3D beam elements are usually used to model pylons. When 2D modeling is used for preliminary analyses, cross-sectional properties should be calculated carefully

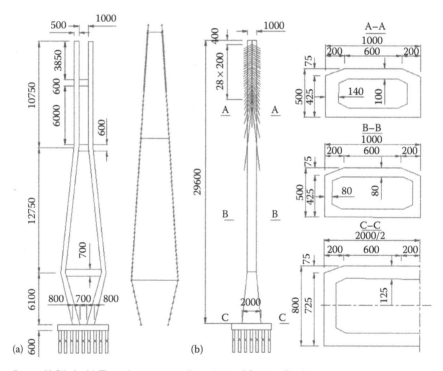

Figure 11.26 (a, b) Two alternative pylon plans of Sutong Bridge.

---

[*] Influence surfaces of all deck nodes are built from the results of analyses for a series of unit vertical loads applied to the deck and stored for later usage for finding the extreme load effects.

to truly reflect the pylon stiffness in the elevation plane. The geometric complexity of pylons is one of the considerations when deciding to use a 3D model.

11.4.3 Connections between girder and pylon

Connections between girders and pylons could be in full separation, rigid connection, or vertical support only, which must be modeled correctly. Figure 11.27 shows a perspective view of elements of a fully separated system (solid lines). A rigid connection can be simulated simply by connecting elements of the girder and pylon. When the pylon provides only vertical supports to the girder, a transverse rigid body simulating the transverse beam or diaphragm and vertical truss elements connecting the rigid body and transverse beam of pylon can be used (dash lines as shown in Figure 11.27).

When a 2D model is used, a vertical support-only connection should be carefully modeled. Usually two FEA nodes are inserted in the same position. One node is used to represent pylon elements and the other one for the girder elements. A master–slave relationship technique, which can link two separated DOFs by a linear relation, will be applied to these two nodes so that both have the same vertical displacements. In that case, the two nodes in the 2D model have only five DOFs in total, rather than six. The disadvantage of using the master–slave relationship is that the

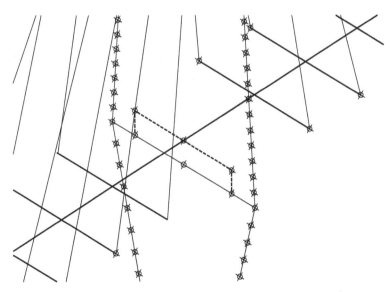

Figure 11.27 Full separation and vertical support of the connection between the girder and the pylon.

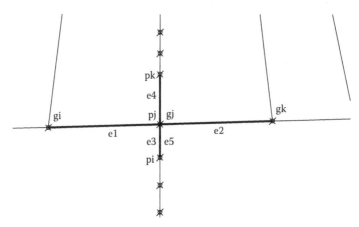

Figure 11.28 Vertical supports in 2D model.

reaction between these two nodes cannot be obtained directly. If the bearing reactions are of interest, a similar connection truss element has to be built in a 2D model. As shown in Figure 11.28, a separated node gj is added. Girder elements e1 and e2 connect nodes gi and gj and gj and gk, respectively. Pylon elements e3 and e4 connect nodes pi and pj and pj and pk, respectively. Truss rigid element e5, which simulates the bearings, connects nodes pi and gj. The bearing reactions can be obtained from the internal force of element e5.

For cable-stayed bridges in which longitudinal semifloating systems or damping systems are used as shown in Figure 11.21, connections between girder and pylons should be simulated carefully when the displacement is beyond its allowed movements. In general, a horizontal truss element can be added between the girder and pylon. Its stiffness has to be modified to a very small value when the displacement is within its limit and to the correct stiffness of the links when the displacement is beyond its limit.

11.4.4 Cables

In most analysis scenarios, each cable can be simply modeled as one truss element by its two anchor points. When sag effects have to be considered, the equivalent Young's modulus has to be calculated according to Equation 11.10 either manually or automatically by the analysis package. As mentioned in Section 11.2.7, the equivalent Young's modulus changes when the erection phase changes; a special FEA package with the capability to handle this issue is always preferable.

When any of the followings is of interest, each cable needs to be modeled as many truss elements with a density of 5–10 m, as shown in Figure 11.20.

1. Sag itself or a more accurate effect of it
2. Crossties have to be simulated
3. Local natural modes of cables

If this detailed model is elected, the initial stress in cable and/or large displacements have to be considered. Otherwise, the analysis will fail due to no stiffness attached to each internal nodes of a cable in its perpendicular direction. When large displacements are considered, loads will be loaded incrementally and the element's full stiffness will be automatically computed at each iteration step.

Figure 11.20 is an example that shows local natural modes are required for wind-raining oscillation study. In such a detailed dynamic mode analysis, initial stresses of all cables are obtained separately from an ideal state analysis and are entered as known parameters to the dynamic mode analysis model.

When cable stays are cradled through pylons, which are rarely used nowadays, the cradling point can be treated as an anchor point, thus eliminating the need to simulate the possible relative movements between the stays and the saddles. The fraction between them is large enough to balance the difference of cable forces between the two sides of the pylon. However, extreme cable forces due to live loads should be investigated case by case.

11.5 ILLUSTRATED EXAMPLE OF SUTONG BRIDGE, JIANGSU, PEOPLE'S REPUBLIC OF CHINA

Sutong Bridge crosses Yangtze River about 100 km upstream from Shanghai. It connects Suzhou and Nantong, two major cities in Yangtze River Delta area. The bridge name, Sutong, comes from the combination of these two cites' names and was built in 2008. Its once-world-record-breaking main span length, 1088 m, made it one of the most famous long-span cable-stayed bridges. Technically, the motivation of building such a long-span cable-stayed bridge comes from a feasibility study of building a cable-stayed bridge with a main span over 1200 m (3937′), which was conducted in the early 1990s. This example is based on the feasibility study of Sutong Bridge started in the late 1990s. All analyses in this example were conducted by Visual Bridge Design System (Wang and Fu 2003, 2005).

Figures 11.26 and 11.29 show pylon dimensions and the elevation of an alternative plan of Sutong Bridge, respectively. The main girder is a steel box girder as shown in Figure 11.10, with a total width of 37 m (121′) or eight traffic lanes. The design live load is super Qi-20, which allows one heavy vehicle and many normal vehicles in each traffic lane. The total axle weight of the heavy vehicle is 55 tons whereas the normal vehicle is 20 tons. The minimum distance between normal and heavy vehicles is 10 m (33′) and 15 m (49′) between normal vehicles. The analyses focus on dead loads and live loads and ideal state and static wind stability. Three typical stages (1) before reaching the first auxiliary pier in the side span, (2) before closure, and (3) in-service are selected to investigate the nonlinear effects such as sag, initial stress, and large displacements.

The bridge in the service stage is modeled as 1032 elements and 1035 nodes in total. Figure 11.30 shows the perspective view of half of the bridge. Figure 11.31 shows one of the preferred ideal states obtained by automatic cable tuning. Figure 11.32 shows the live load stress envelope. Table 11.2 compares the extreme live load displacements with and without geometric

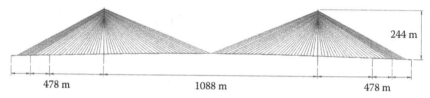

244 m

478 m   1088 m   478 m

Figure 11.29 The elevation of an alternative plan of Sutong Bridge.

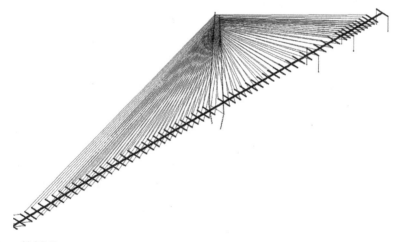

Figure 11.30 Perspective view of the 3D FEA model of Sutong Bridge.

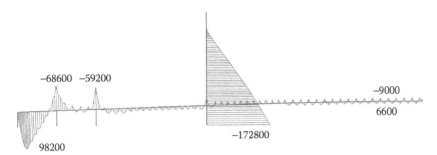

Figure 11.31 Moment distribution (kN-m) of one preferred ideal state after superimposed.

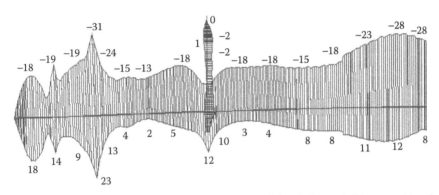

Figure 11.32 Live load stress envelope (on the top of the girder and side-span side of tower, MPa).

Table 11.2 Live load extreme displacements (mm)

Position	Linear[a]	Nonlinear[b]
Girder (in the middle of main span)	1081	935
Top of pylon	263	242

[a] Computed by direct influence line loading, which is obtained by the application of unit forces while initial stress and sag is considered.
[b] Obtained by reanalyzing extreme live loads of load case a with consideration of initial stress, cable sag, and large displacements.

nonlinear effects. The nonlinear analysis of live loads shows that initial stresses accumulated along the flat arch-like girder will increase the girder stiffness if large displacement is considered.

The wind pressure is designated by the bridge site and varies at different altitudes along the towers. With regard to longitudinal connections between the girder and towers, three alternatives are studied. The first, the recommended one, is that the girder is restrained with one tower only. The second is not to restrain at all; the third is to restrain with two towers.

The girder is restrained in both vertical and lateral directions with towers in all three cases. Under longitudinal wind loads, displacement at the top of the tower is 1052 mm (41.4″), and girder displacement is 988 mm (38.9″) if there is no longitudinal restraint at all. In the lateral direction, the maximum tower displacement is 276 mm (10.9″ in the maximum dual-cantilever stage).

Six load patterns are studied to search for load safety factors in the stability analyses:

1. In the complete stage, maintain dead loads and cable jacking loads and increase vehicle loads. At 40 times, displacements abruptly reached 42 (138′) and 13 m (42.7′) in the middle of the main span and the top of the pylons, respectively.
2. In the complete stage, maintain cable jacking loads and increase dead loads. At three times, displacements abruptly increased.
3. In the maximum dual-cantilever stage, maintain dead loads and cable jacking loads and increase construction loads. At 240 times, displacements abruptly increased.
4. In the maximum dual-cantilever stage, maintain dead loads and cable jacking loads and increase lateral wind loads. The bridge still remains stable when the lateral loads are increased by 50 times and the lateral displacement at the end of the girder reaches to 7 m (23′) accordingly.
5. In the maximum single-cantilever stage, maintain dead loads, cable jacking loads, and lateral wind loads and increase construction loads. At 46 times, vertical displacement at the end of the girder increased to over 100 m (328′), accompanied with 42 m (138′) of lateral displacement.
6. In the maximum single-cantilever stage, maintain dead loads and cable jacking loads and increase lateral wind loads. At 48 times, the lateral displacement at the end of the girder increased over 100 m (328′).

In all six patterns, only the construction load, which includes a 1000 kN (225 kip) crane at the end of the girder and a uniform load of 10 kN/m (0.685 kip/ft) in the maximum single-cantilever stage (5), shows the coupling of bending in vertical and lateral directions. Figure 11.33 shows the displacements in load pattern 5. The stability analysis also shows that the bridge is more vulnerable before closure in the main span than before reaching the second auxiliary pier in the side span. Although results of these six loading patterns show that the bridge has sufficient stability against live, wind, construction, and dead loads, full nonlinear ultimate analysis, in which material nonlinearity is also considered, and aerodynamic stability analysis are required.

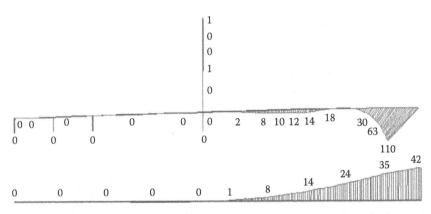

Figure 11.33 Girder displacements (m) in vertical (top) and lateral (bottom) directions when construction loads increase to 46 times at the maximum single-cantilever stage.

11.6 ILLUSTRATED EXAMPLE WITH DYNAMIC MODE ANALYSIS OF PANYU BRIDGE, GUANGDONG, PEOPLE'S REPUBLIC OF CHINA

In this example, a concrete cable-stayed bridge with a main span of 380 m (1247′) is briefly introduced for the purpose of dynamic mode analysis. This bridge is located in Panyu, China, and is a typical concrete cable-stayed bridge. It is modeled as 714 elements and 475 nodes in total by using Visual Bridge Design System (Wang and Fu 2005). Figure 11.34 shows its elevation.

A full floating system is used in this bridge; the first natural mode is in a horizontal pendulum movement. Figure 11.35 shows the first bending mode with a radian frequency of 3.78 rad/sec ($f = 1.06$ Hz). Figure 11.36 shows the first torsional mode with a radian frequency of 4.23 rad/sec ($f = 1.18$ Hz).

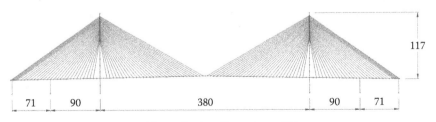

Figure 11.34 The elevation of Panyu Bridge, Guangdong, China.

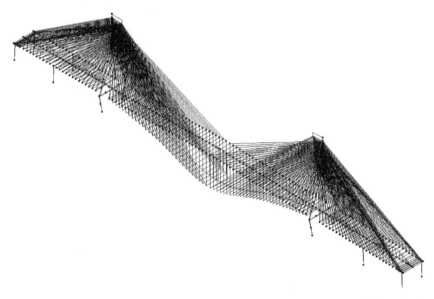

Figure 11.35 The first bending mode of Panyu Bridge (radian frequency = 3.78 rad/sec).

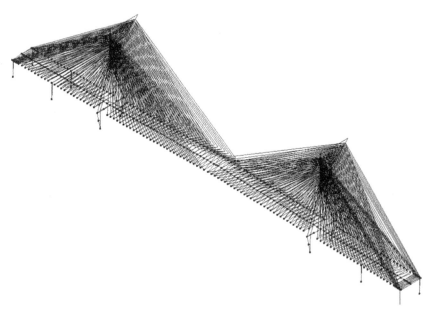

Figure 11.36 The first torsional mode of Panyu Bridge (radian frequency = 4.23 rad/sec).

11.7 ILLUSTRATED EXAMPLE WITH DYNAMIC MODE ANALYSIS OF LONG CABLES WITH CROSSTIES

In this example, cable stays in the main span side of a cable-stayed bridge are modeled separately, as shown in Figure 11.20. It is modeled as 992 truss elements and 963 nodes in total. The analysis was conducted by VBDS (Wang and Fu 2005) and verified by ANSYS. The initial tensile forces of crossties have the same value of 50 kN, which only serves the purpose of tying down the main stays. The initial tensile forces of main cable stays are obtained from a separated ideal state analysis, which range from 1532 to 4807 kN (344.4 to 1080.7 kip).

Figure 11.37 shows the first mode with a radian frequency of 5.39 rad/sec (f = 1.51 Hz). Figure 11.38 shows the second mode with a radian

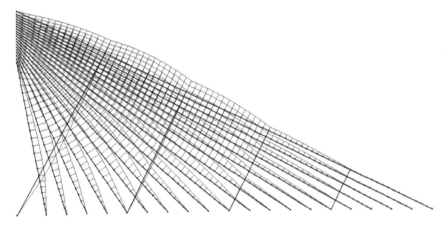

Figure 11.37 The first mode of long cables with crossties (radian frequency = 5.39 rad/sec).

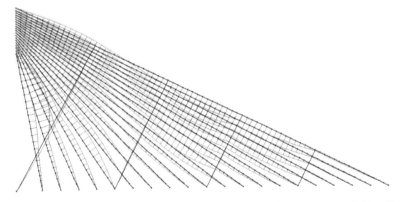

Figure 11.38 The second mode of long cables with crossties (radian frequency = 7.38 rad/sec).

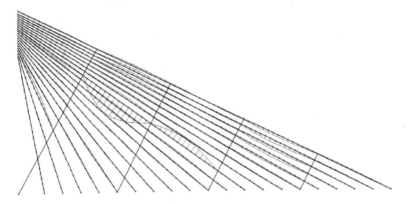

Figure 11.39 The eighth mode of long cables with crossties (radian frequency = 9.31 rad/sec).

frequency of 7.38 rad/sec (f = 2.07 Hz). These two low-frequency modes show the global movements of the cable net. At a higher mode, the eighth mode has a radian frequency of 9.31 rad/sec (f = 2.61 Hz); for example, as shown in Figure 11.39, local movements of single cables will appear.

Chapter 12

Suspension bridges

12.1 BASICS OF SUSPENSION BRIDGES

A typical three-span suspension bridge, as shown in Figure 12.1, consists of main cables, two pylons, stiffened girder, and hangers. The weight and vehicular loads from the deck are transferred to the main cables by vertical hangers or suspenders. Unlike cables in a cable-stayed bridge that are anchored on the deck on both sides of a pylon at an angle, hangers of a suspension bridge are perpendicular to the deck and will not create any horizontal force on the deck. Except in the self-anchored suspension bridges, the main cables carry and transfer loads to anchorages that are separated from the bridge. For this respect, cable-stayed bridges are self-anchored systems, whereas most of the suspension bridges are externally anchored systems. As horizontal forces in cables are transferred to ground rather than to the girder, the stiffened girder will not have the P-Delta effects as in cable-stayed bridges and therefore the spanning capacity is much increased.

In terms of span layout, as shown in Figures 12.2 and 12.3, variations of suspension bridges include single-span and multispan suspension bridges. The stiffened girders can be two hinged or continuous at the locations of pylons. The two-hinged stiffened girder, which is commonly used, is discontinued from the side span to the main span and simply supported by the pylons. When deck continuity is required, the continuous stiffened girder can be used (Chen and Duan 1999).

Vertical hangers are commonly used. Diagonal hangers, as shown in Figure 12.4, are also used to enhance structural damping so as to improve aerodynamic behaviors.

Most suspension bridges are externally anchored, in which the main cables are anchored into anchor blocks that are built on ground. This type of anchorage relies on the gravity of the massive anchor blocks. Where such massive anchor blocks are not feasible, main cables can be anchored

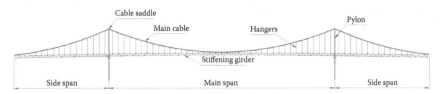

Figure 12.1 A typical suspension bridge.

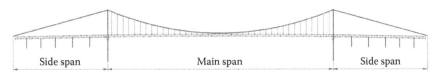

Figure 12.2 A single-span suspension bridge.

into the stiffened girders at the end of side spans. This is the so-called self-anchored suspension bridge. As the stiffened girders will resist axial compression from main cables, the span capacity of self-anchored suspension bridges is limited.

Pylons in most long-span suspension bridges are usually designed not to resist longitudinal bending moment due to the structural weight of stiffened girders. This type of pylons is flexible in the longitudinal direction. For short- or multispan suspension bridges, pylons may be designed as rigid to resist longitudinal bending due to dead or live loads. The high-strength parallel wires are widely used for the main cables in modern suspension bridges.

The suspension bridge has a long history (Kawada 2010). Its original forms that two suspending ropes carrying walking boards directly were constructed in ancient China. The development of modern suspension bridges started in the early nineteenth century. Jacob's Creek Bridge, which had a center span of 21.3 m, was built in the United States in 1801. Its main cables were made of iron chains. Niagara Falls Bridge, in which parallel wire cables were used for the main cables, was built in 1855. It had a main span of 251 m and was the world's first working railway suspension bridge. Due to the great increase of railway loads, it was later replaced by Whirlpool Rapids Bridge in 1897. The Golden Gate Bridge, with a main span of 1280 m, was built in the San Francisco Bay area in 1937. It had the longest main span till 1964. The Severn Bridge, with a main span of 988 m, was built in England in 1966. Box girder and diagonal hangers were used in the Severn Bridge. In 1981, the Humber Bridge, with a main span of 1410 m, was built in England. It had the longest main span till 1997. The Akashi Kaikyo Bridge, with the world's longest main span of 1991 m, was built in 1998. Table 12.1 lists the top 10 world's longest suspension bridges so far.

Figure 12.3 A multispan suspension bridge.

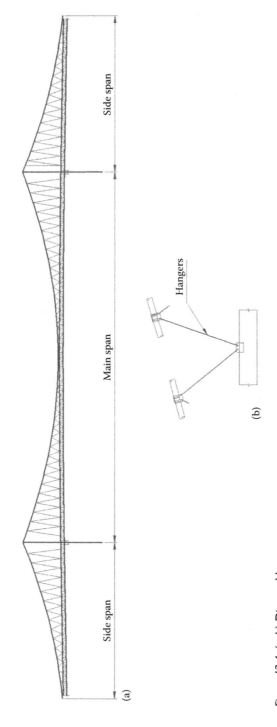

Figure 12.4 (a, b) Diagonal hangers.

Table 12.1 Top 10 longest suspension bridges in the world

No.	Name	Main span (m)	Year of built	Location
1	Akashi Kaikyo Bridge	1991	1998	Kobe-Awaji, Japan
2	Xihoumen Bridge	1650	2009	Zhoushan, People's Republic of China
3	Great Belt Bridge	1624	1998	Halsskov–Sprogo, Denmark
4	Yi Sun-Sin Bridge	1535	2012	Gwangyang–Yeosu, South Korea
5	Runyang Bridge	1490	2005	Jiangsu Province, People's Republic of China
6	Humber Bridge	1410	1981	Hessle–Kingston, England
7	Jiangyin Bridge	1385	1999	Jiangsu Province, People's Republic of China
8	Tsing Ma Bridge	1377	1997	Hong Kong
9	Verrazano–Narrows Bridge	1298	1964	New York City
10	Golden Gate Bridge	1280	1937	San Francisco, CA

12.2 CONSTRUCTION OF SUSPENSION BRIDGES

The construction method and procedures are critical and play a very important role in the design and analyses of a suspension bridge. The design, analyses, and construction procedures of a suspension bridge are completely corelated to each other. Figure 12.5 shows a typical construction process of a suspension bridge.

12.2.1 Construction of pylons and anchorages and install catwalk system

Pylons and anchorages are critical components of a suspension bridge. In the longitudinal direction, pylons of a suspension bridge can be designed as rigid or flexible in terms of resisting horizontal forces from cables on the top of pylons. Most pylons of long-span suspension bridges are designed as flexible. This type of pylons is mainly under compression due to dead loads and minor bending deflection due to live loads. Both steel and concrete are commonly used on pylons. Concrete pylons may have advantages over steel pylons in terms of the cost of construction and maintenance. Examples of steel pylons include the Golden Gate Bridge and Akashi Kaikyo Bridge, whereas most suspension bridges, such as Xihoumen Bridge and Jiangyin Bridge built in People's Republic of China, have concrete pylons.

During the erection of stiffened girders and after applying superimposed dead loads on the deck, the cable forces in the main span and side spans are

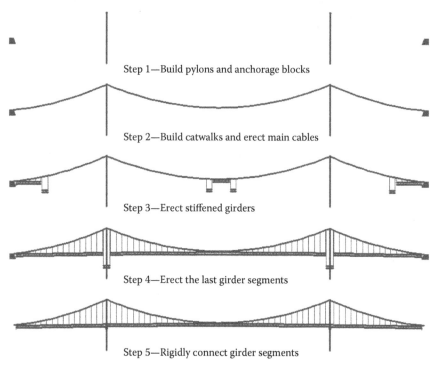

Step 1—Build pylons and anchorage blocks

Step 2—Build catwalks and erect main cables

Step 3—Erect stiffened girders

Step 4—Erect the last girder segments

Step 5—Rigidly connect girder segments

Figure 12.5 Construction process of a suspension bridge (showing from tower to installing stiffened girder).

not balanced, and the pylons will be deflected toward the main span. For these flexible pylons, saddles on the top of pylons have to be adjusted so as to release most deflection of pylons due to dead loads.

Anchors are the components that distribute the cable forces to ground so that the main cables are sustained. The main cables in most suspension bridges are anchored externally. The anchors of these bridges can be classified as gravity anchor and rock tunnel anchor. The gravity anchor is built by massive concrete to balance the cable forces. Where the geology is permitting, an inclined tunnel can be excavated down to the bedrocks and the anchor beams or bars can be built into ground with backfilled concrete.

Steel wires of the main cable will have to be sprayed out by going through a splay saddle in front of the anchorage so as to be anchored wire by wire or to group several wires together. One main function of the splay saddle is to change the cable tangent more downward so as to make the anchorage easy. The change of the cable tangent at the end will also reduce the anchor forces directly to the anchorage as part of the cable forces will be distributed to splay saddle. In terms of the bending in the main cable plane, the splay saddle can be rigid or flexible. Figure 12.6 shows a typical rigid splay saddle that is

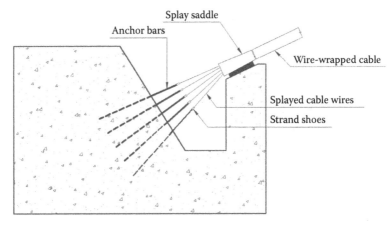

Figure 12.6 Gravity anchor and rigid splay saddle.

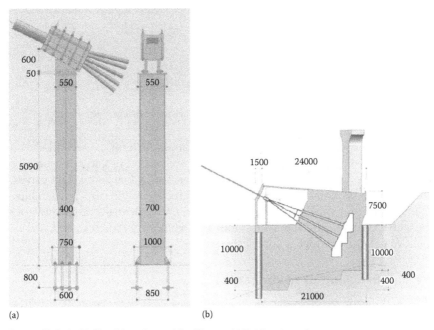

Figure 12.7 (a, b) Flexible splay saddle (Kanazaki Bridge, Japan).

built together into the entire anchor block. Figure 12.7 shows a flexible splay saddle that is supported by a steel column, which provides less bending stiffness in the longitudinal than in the transverse direction. Figure 12.8 shows a flexible splay saddle built as a rigid body. The type of splay saddle has to be considered into the design and analysis of a suspension bridge.

Figure 12.8 Flexible splay saddle as a rigid body (Hardanger Bridge, Norway).

A catwalk system, which will be the main platform of cable works, will be constructed after the pylons and anchors are built. The catwalk system is more like an ordinary suspension bridge supported by few lightweight cables, in which the deck or the platform is built directly on the suspended cables. The catwalk system will have the same geometry as the main cables and goes close to the main cables. It serves only as a temporary structure and will be removed after the bridge is completed.

12.2.2 Erection of main cables

The making of main cables from steel wires is an important step in the construction of a suspension bridge. There are two methods to erect main cables from individual steel wires: aerial spinning (AS) method and prefabricated parallel wire strands (PWS or PPWS) method.

The AS method was first developed by John A. Roebling in the mid-nineteenth century. In the AS method, a looped moving cable will be built first over the main cable and along the main cable from one anchor to another. The looped moving cable carries a spinning wheel that runs like modern suspended cable cars. The individual steel wires are stored in and fed from wire reelers located close to one anchor. Initially, the spinning wheel is docked at the end where the wire reelers are located. The live end of the wire unreeled from the reeler is passed through the spinning wheel in a way of top-in and bottom-out. The live end is then tied to anchor shoes. When the spinning wheel moves away from the wire reeler side (the approach route), it brings two wires to the other side. The wire laid from the bottom of the wheel is called dead wire, which will be placed into the cable former while the wheel is moving. The wire laid from the top of the wheel is called live wire, which will be stored on top of the cable former while the wheel is moving toward the other end. The feeding speed of the wire reeler is double compared to the wheel moving speed. When the wheel reaches the other anchor, the looped end of the two wires will be prepared to connect to an anchor shoe. Then, the wheel will return back to the wire reeler side (the return route). The top wire (the live end) will be laid into the cable former while the wheel moves back. The wire reeler will stop feeding while the wheel returns. Once it reaches the anchor at the wire reeler side, the wire will be taken off the wheel to form a looped end so as to connect to an anchor shoe and then is taken on the wheel again to start another round.

After each wire is laid out, it has to be adjusted. When all wires of a strand are erected, they will be banded into the strand shape. Apparently, the AS method is simple and needs less equipment on site. However, erecting cable wire by wire is time consuming and the process relies on weather conditions.

Instead of erecting wire by wire, in the PPWS method, all wires of a strand are shop-fabricated and socketed into the final shape and packaged on reels. When erecting, a complete strand will be pulling from one end to another by hauling cables. As the strand is prefabricated earlier, when it is ready to anchor, anchor shoes will be adjusted according to the difference in temperature so as to make the tension of each strand as even as possible. Compared with the AS method, PPWS method will save time on site significantly. Individual wire adjustment is eliminated too. As the strand is formed in shop, quality can be well controlled and cost will be lowered.

After all strands are erected, the cable will be squeezed by squeezing machine and lashed at a certain pitch to form a round shape. Cable clamps or bands will be installed at each design locations for suspender connections. As the cable will be deformed much while the stiffened girder is erected, offsets to the suspender locations should be already considered in the design locations. The final step for cable erection is to lash the cable with wrapping wires and to protect the cable with coating treatment.

12.2.3 Erection of stiffened girder

The erection of stiffened girders, as shown in Figure 12.5, can start from either center of the main span or end of the side span or pylons. Considering the change of the suspender angle to the cable clamps, the sequence of starting from center of the main span is better as these changes are smaller. Especially when the last segments close to the pylons are installed, most deformation of the cable is completed if the erection is from center of the main span to the pylons. However, to erect girders starting from the pylons is easier.

No matter what sequence is adopted, one common goal should be achieved during the erection of the stiffened girder, which is to minimize the bending moment in the girder at the hanger locations. The ideal situation is that all the girder weight is evenly distributed onto the cables. To reach this ideal situation, girder segments are usually connected only at the top part of the girder and the bottom part is left unconnected during erection. These joint connections will be changed to rigid connections after all girder segments are installed and the suspenders are adjusted. The change from joints to rigid connections is done before the deck is superimposed.

As more and more segments of the stiffened girders are erected, bending moments at the bottom of pylons and deflections on the top of pylons will be accumulated. Such a distribution of girder weights to pylons that are designed to be flexible should be released by adjusting the horizontal position of saddles on the top of pylons. Ideally, the pylons should be under pure compression after all girder segments are erected. Figure 12.9 shows the adjustment of the saddle position so as to release the deflections of pylons by jacking the saddles. This adjustment may be required several times during the erection of girders.

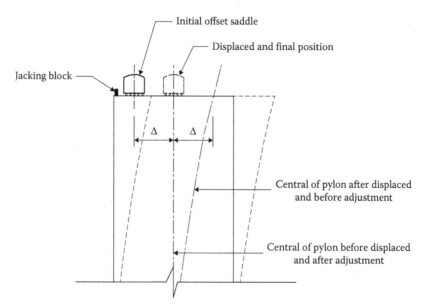

Figure 12.9 Offset of saddle and release of the deflection of pylon by the jacking of saddle.

12.3 BEHAVIOR OF SUSPENSION BRIDGES

As the flexible cables are the main structural component carrying dead loads and live loads and the span length is longer than most other types of bridge, the global vertical stiffness of a suspension bridge is very low. The principles and characteristics of suspension bridges are distinguished from others. Due to the flexible cables and its huge deflections from the very beginning of construction to the stage of operations, the geometry configuration of a suspension bridge can no longer be treated as constant. The changes of geometry configuration, hence the construction procedures, are deeply involved and crucial in the design and analyses. Obviously, geometric nonlinear analysis is the basis of the structural analysis of suspension bridges.

12.3.1 Basis of cable structures—Initial stress and large displacements

Taking a simple symmetric truss structure as shown in Figure 12.10 as an example, tension in truss elements under external load can simply be obtained from the force balance equation at node B as

$$T = \frac{PL}{2H} \tag{12.1}$$

The strain of the truss element is obtained as

$$\varepsilon = \frac{PL}{2EAH} \tag{12.2}$$

where E and A are the Young's modulus and cross area of the truss element, respectively.

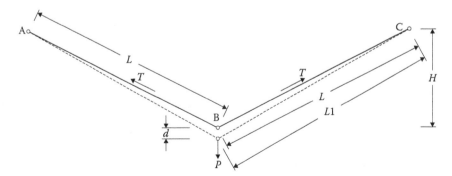

Figure 12.10 A simple symmetric truss structure.

The vertical displacement d at node B under external load P can be derived from geometric consistency.

$$d = \frac{L^3}{2EAH^2} P \tag{12.3}$$

The solution of displacement by geometric consistency can be improved by the principle of minimum total potential energy, which can be simply described as that the change of total strain energy equals the work done by external loads or the total energy does not change. For the case shown in Figure 12.10, it can be described as

$$\delta\Pi = \delta(U + V) = 0 \tag{12.4}$$

where:
  Π is the total energy
  $U = 2AL\cdot(1/2)\sigma\varepsilon$ is the total strain energy (internal energy)
  $V = -dP$ is the total potential energy (external work)
  σ and ε are stress and strain, respectively, and linear elastic is assumed

Figure 12.11 shows the stress and strain relationship and the density of strain energy.

$$\sigma = E\varepsilon \tag{12.5}$$

From geometric consistency at node B as shown in Figure 12.10, the strain of truss elements is

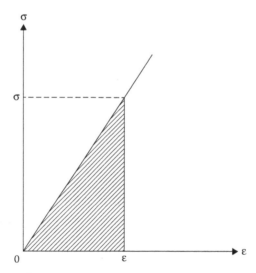

Figure 12.11 Density of strain energy.

$$\varepsilon = \frac{H}{L^2}d \tag{12.6}$$

Equation 12.4 can then be rewritten as

$$\delta\Pi = \delta\left(\frac{EAH^2}{L^3}d^2 - Pd\right) = 0 \tag{12.7}$$

From Equation 12.7, the same result of the displacement as Equation 12.3 can be obtained.

However, in the earlier approach to the displacement under external load, it is assumed that the two truss elements have no stress at all when the external load is applied. What it would be if there was an initial stress and strain when the external load is applied? The principle of minimum potential energy is still valid. However, the stress in Equation 12.5 will become

$$\sigma = E\varepsilon + \sigma_0 \tag{12.8}$$

where $\sigma_0 = T_0/A$ is the initial stress in the truss elements. By substituting Equation 12.8 into the equation of total strain energy, Equation 12.4 can then be rewritten as

$$\delta\Pi = \delta\left(\frac{EAH^2}{L^3}d^2 + \frac{H}{L}T_0d - Pd\right) = 0 \tag{12.9}$$

From Equation 12.9, the displacement at node B can be derived as

$$d = \frac{L^3}{2EAH^2}\left(P - \frac{H}{L}T_0\right) \tag{12.10}$$

where T_0 is the initial tension force when the external load is applied.

Equation 12.10 reveals that the displacement under external loads will be reduced due to initial stress, and the higher the initial stress, the more it is reduced. In general, the vertical stiffness is enhanced by initial stresses in cables.

Note that the strain and displacement relationship in Equation 12.6 is obtained with the assumption that d is very small compared with L or H. As illustrated in Chapter 3, large displacement can also be considered in this simple truss structure as follows:

$$L_1 = \sqrt{d^2 + 2Hd + L^2} \tag{12.11}$$

where L_1 is the truss element length after deformed. Using Maclaurin series to expand Equation 12.11 and taking only the second order, Equation 12.11 becomes

$$L_1 = L + \frac{H}{L}d + \frac{1}{2}\left(\frac{1}{L} - \frac{H^2}{L^3}\right)d^2 \tag{12.12}$$

and Equation 12.6 becomes

$$\varepsilon = \frac{H}{L^2}d + \frac{1}{2}\left(\frac{1}{L^2} - \frac{H^2}{L^4}\right)d^2 \tag{12.13}$$

Follow the same procedure to derive Equation 12.10 from Equations 12.8 and 12.9; a polynomial equation with unknown variable of d can be obtained:

$$a_3d^3 + a_2d^2 + a_1d + a_0 = 0 \tag{12.14}$$

where:

$$a_3 = \frac{EA}{L}\left(\frac{1}{L} - \frac{H^2}{L^3}\right)^2 \tag{12.15}$$

$$a_2 = \frac{3EA}{L}H\left(\frac{1}{L^2} - \frac{H^2}{L^4}\right) \tag{12.16}$$

$$a_1 = \frac{2EA}{L}\left(\frac{H}{L}\right)^2 + \left(\frac{1}{L} - \frac{H^2}{L^3}\right)T_0 \tag{12.17}$$

and

$$a_0 = \frac{H}{L}T_0 - P \tag{12.18}$$

By resolving Equation 12.14, the displacement under external load P with consideration of initial stress and large displacement can be obtained.

By comparing Equation 12.14 with Equation 12.10, which only illustrates how an existing stress in cables influences their behavior, it is clear that it is easier to consider only initial stress. If the initial stress is the predominant issue, for certain purposes of analyses, only initial stress issue may be considered to save analysis time and cost. During the preliminary design, for example, initial stress due to dead loads can be estimated first and further be considered in live load analysis so as to quickly estimate extreme deflections.

12.3.2 Basics of suspension bridge analysis

Like suspension bridges, the analytical theories of them have a long history of development. They can be classified as elastic theory starting from the

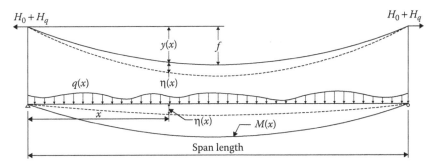

Figure 12.12 A single-span suspension bridge.

early nineteenth century, deflection theory in the late nineteenth century, and finite deformation (large displacement) theory nowadays.

Considering one span of the simple support girder as shown in the upper part of Figure 12.12, its moment distribution is

$$M(x) = M_0(x) \tag{12.19}$$

If the girder is multiple supported by hangers from the cable as shown in the lower part of Figure 12.12, when loads that cause moment distributions as shown in Equation 12.19 are applied on the girder, the cable will be tensioned and the moment distribution on the girder will be reduced as

$$M(x) = M_0(x) - H_q(x)y(x) \tag{12.20}$$

where $H_q(x)$ is the horizontal component of cable tension due to loads distribution of $q(x)$. Equation 12.20 represents the elastic theory. Its differential equation form is

$$q(x) = EI\frac{d^4\eta}{dx^4} - H_q(x)\frac{d^2y}{dx^2} \tag{12.21}$$

One example of using Equation 12.20 is to calculate the cable tension in the middle of the span. Assuming the loads distribution $q(x)$ is constant as q_0 and is completely distributed to the cable. Thus, the moment distribution in the girder in Equation 12.20 will be zero. The cable tension in the horizontal component at the middle of the span can be derived by considering the moment in the middle span of a simple support beam.

$$H_q\left(\frac{l}{2}\right) = \frac{q_0 l^2}{8f} \tag{12.22}$$

Assume the structure is balanced before $q(x)$ applies and the horizontal component of its initial cable tension is $H_0(x)$. After $q(x)$ applies, the increase of cable tension in the horizontal component is $H_q(x)$ and the structure is balanced at a deflection of $\eta(x)$. Equation 12.20 can be rewritten as Equation 12.23 if the deflection and the initial cable tension are considered.

$$M(x) = M_0(x) - H_q(x)[y(x) + \eta(x)] - H_0(x)\eta(x) \tag{12.23}$$

In cases where dead loads that cause $H_0(x)$ are predominated in comparing with $q(x)$, $H_q(x)\eta(x)$ is negligible. Therefore Equation 12.23 can also be simplified as

$$M(x) = M_0(x) - H_q(x)y(x) - H_0(x)\eta(x) \tag{12.24}$$

Equations 12.23 and 12.24 reflect the stiffness enhancement due to initial cable tension over deflection. These equations represent the deflection theory. Similar to Equation 12.21, the differential form of Equation 12.23 is

$$q(x) = EI\frac{d^4\eta}{dx^4} - H_q(x)\frac{d^2y}{dx^2} - [H_0(x) + H_q(x)]\frac{d^2\eta}{dx^2} \tag{12.25}$$

Since 1960s, the computer application and the finite element method (FEM) have been advanced greatly. Especially because of the extreme development of both computer hardware and software in the twenty-first century, the geometry nonlinearities of suspension bridges are commonly considered by using modern FEM analyses. As introduced in Chapter 3, a full geometric nonlinear analysis, in which the initial stress and large displacement are considered, will not only cover the second-order problem showing as $H_0(x)\eta(x)$ or $H_q(x)\eta(x)$ in Equations 12.23 and 12.24 but also establishes the balance on the deformed configuration. This full-scale geometric nonlinear analysis is often referred as finite deformation method.

12.3.3 Live load analyses of a suspension bridge

As illustrated in Equations 12.10, 12.14, 12.23, and 12.24, the initial stress due to dead loads in cables affects the succeeding live loads response. The live load analyses of a suspension bridge have to consider the initial stress due to dead loads, specifically the initial stress in main cables. The priority nonlinear issues to be considered in live load analyses are the following:

1. The initial stress in main cables due to dead loads
2. The initial stress in pylons due to dead loads
3. The large displacements under live loads

In preliminary design, for example, live load analysis can be performed with the consideration of the initial stress in the main cables. For simplification, the geometries of bridge defined by the design plan can be used.

The analysis can be simply considered linear. The cable stress under dead loads can be estimated by the elastic theory. Further, the vehicular loads and their locations obtained by the regular live loading process can be reapplied to the structure and reanalyzed with full scale of geometric nonlinear analysis. The extreme internal forces and displacements can be so adjusted to reflect all the nonlinear effects. This approach is the same as the live load analysis of a cable-stayed bridge in Chapter 11.

For detailed design analyses, the state to be used for influence value analysis including the geometries of the bridge and initial stresses should be the final state obtained in construction control analyses.

12.3.4 Determination of the initial configuration of a suspension bridge

Due to the flexibility of main cables and thus the large displacements under dead loads and live loads, the difference of geometric configuration from one state to another is no longer negligible as in most other types of bridge. The design plan of a suspension bridge including the main cable geometry and the stiffened girder elevations refers to the final deformed state after the deck is superimposed. Unlike other moderate and short-span bridges, such as concrete girder bridges, the design state of a suspension bridge is far different from its initial state in which there is no load acting on the structure. Only certain types of load analyses, such as live load analyses, can be based on this design state in terms of geometric models. The initial state of a suspension bridge is also referred as zero-stress state, which is the basis of any kind analysis or construction control. It is the foundation of the design and analyses of long-span bridges.

How to determine the zero-stress state based on the design plan is a well-known issue in suspension bridge design and analyses. Backward and forward iteration analyses are commonly adopted. Based on the fact that unloading a load and/or removing a component will restore the structure back to its previous state, if the current state is assumed to be accurate, backward analyses will be able to restore the structure to its initial state. Once an initial state is obtained, forward analyses can be performed as the sequence of the bridge is constructed to reach a final state. By comparing the final state with the design state, adjustments to the initial state can be estimated. Iterating this process, the initial state can then be determined. As mentioned in Section 12.3.2, both initial stress and large displacements have to be considered in both backward and forward analyses. The following description illustrates the steps to determine the initial state in general.

1. Estimate cable forces in the design state. The initial stress in cables due to dead loads plays an important role in backward analysis. To better calibrate the starting state of backward analysis, the cable forces have

to be obtained or estimated first. In this step, all dead loads including the structural weight of cables and stiffened girders plus all superimposed dead loads are assumed to be acting on the main cable. The cable geometry can be assumed the same as the design geometry. The elastic method mentioned in Section 12.3.2 can be used.

2. Establish the starting state of backward analysis by building a model using the design geometry and including the estimated cables in step (1) as the initial stress.

3. Remove superimposed dead loads at all hanger locations. This removal may be split into several stages according to the sequence of the imposed dead loads on deck. In addition to a full scale of geometric nonlinearity, including initial stress and large displacements, a special displacement restraint at cable ends should be applied to consider the tangent change over saddles during the load increments and iterations in nonlinear analysis, which will be described in Section 12.3.5.

4. Reverse the jacking of saddles on the top of pylons if saddles are adjusted after all girder segments are erected.

5. Remove dead loads due to stiffened girder weights at all hanger locations. This removal should truly reflect the girder erection process. If saddles are jacked during the erection of stiffened girders, correspondent reversal actions should be inserted. The analysis method is the same as in step (3).

6. Remove cable loads due to cable weights and other additional loads such as wrapping wires and sheathing. The analysis method is the same as in step (3).

7. Take the current geometric state as the initial state.

8. Conduct forward analyses simulating the loading of cables, erecting of stiffened girder segments, adjustment of saddles, and application of superimposed loads after joints between girder segments are changed to rigid connections. Difference will be found from the final state obtained in this step and the design state. The bending of the stiffened girder due to the superimposed dead loads should be considered in backward analysis after the first round of forward analysis. Rebuild the starting state including hangers and stiffened girder; repeat steps (3) to (7). Superimposed dead loads should be removed directly from the stiffened girder.

9. The initial state obtained in step (7) may be manually adjusted so as to match the final state obtained in forward analyses and the design state. This adjustment could be simply change of the chord height of the cable in zero-stress state. The adjusted initial state will be the basis of all other succeeding analyses. The total length of cable in zero-stress state can also be calculated from the initial state.

12.3.5 Consideration of cable tangent changes

Unlike a free node of any component or element whose movement completely depends on stiffness and forces, nodes of cable ends that connect to saddles are restrained by the shape of the saddle. When cable deforms, the displacements at cable ends fall in the cable slots on top of the saddles (Figure 12.13). In the iteration of large displacement analysis, any incremental displacements at the ends of main cables should be adjusted to reflect that the trace of the displacement is on the arc of the saddle.

Considering that the load step is small enough and the incremental displacement in each of many iterations for one load step is sufficiently small, the adjustment to the incremental displacement of the cable end at saddle can be simplified as follows:

1. Assume the deformed saddle center is S_c, which will be derived from the new position of the rigid body that is used to simulate the saddle; the saddle radius is R; the deformed cable end at saddle is N_0, the other end is N_1; and the adjusted cable end at saddle is N_0'.
2. Construct a tangent line from N_1 to an arc of radius R, centered at S_c with a tangent angle as the same as line N_0 N_1. The tangent point is the adjusted cable end N_0'. The offset from N_0' to N_0 is the adjustment to the incremental displacements of the cable end at saddle.

This process shows that a general-purpose FEA package with nonlinear analysis feature may not be sufficient in suspension bridge analyses. Additional displacement constraints to the cable ends due to the saddle slot have to be built into the analysis package.

However, it should be noted that the consideration of saddle slot constraint to the cable ends does not affect the analysis of internal forces and

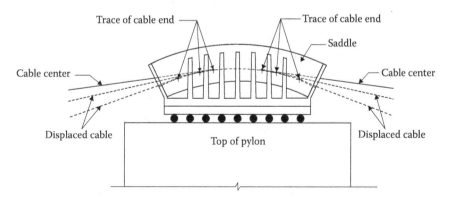

Figure 12.13 Displacements at cable ends fall in slot on top of saddles.

displacements much. In most analyses of the suspension bridge, this can be ignored. But in the backward analysis to obtain the initial state, the accurate tangent point of the cable and the saddle influences the accuracy of the zero-stress state, and the total cable length under no stress is derived from the zero-stress state.

12.3.6 Offset of saddles and release of the deflection of pylons

When unbalanced horizontal forces in the main cables on the top of a pylon exist during the erection of the stiffened girders in the main span and side spans, the saddle would not be able to move freely due to frictions between the saddle bottom and the top of the pylon, and such a move is forbidden during the erection of girders. Thus the pylon will deflect toward the main span as the dash lines shown in Figure 12.9.

For pylons that are designed as flexible, the deflection due to the erection of stiffened girders and superimposed dead loads has to be released. The release is accomplished by adjusting the horizontal position of the saddle. Depending on the design of the pylons, usually the deflection has to be released several times, during the erection and after the deck loads are superimposed.

Because the final position of the saddle has to be centered, when a saddle is installed, its initial position has to be offset from the center of the pylon toward the side span. The offset value is the total horizontal displacements of the saddle starting from the erection of cables to finishing of superimposed dead loads. When adjusting the position of the saddle, jacking force is applied between the saddle and the jacking block on the side span side of the pylon, as shown in Figure 12.9. Due to cables, the horizontal stiffness of the saddle is much higher than the horizontal stiffness of the pylon at the top. When jacking, the saddle will remain still and the pylon will move toward the side span so the deflection is released.

12.3.7 Low initial stress stiffness of the main cable close to pylon

As illustrated in Section 12.3.2, the initial stress of the main cable due to dead loads plays an important role in the enhancement of live load stiffness. The initial stress stiffness is perpendicular to the cable, and therefore the vertical stiffness enhancement to the stiffened girder reaches maximum in the middle of the main span. As the angle of the cable to the stiffened girder increases in the area close to the pylon, the initial stress stiffness in vertical projection decreases. One phenomenon relating to this behavior is that the vertical displacements on the stiffened girder under live loads are larger in the area close to the pylon than others.

For extreme long-span suspension bridge proposals, such as the Gibraltar Strait Bridge or cable-stayed-and-suspended hybrid bridge as shown in Figure 12.14, the girder is stiffened by stay cables anchored from pylons or hangers from rigid components in areas close to the pylon.

12.4 PRINCIPLE AND MODELING OF SUSPENSION BRIDGES

Similar to modeling a cable-stayed bridge, the modeling of a suspension bridge needs to identify the analysis tool. The following specific issues for suspension bridge analyses make general-purposed FEA packages not a suitable tool for many types of analysis in general.

1. Analyses to determine the initial state from the design state
2. Geometric nonlinear construction and control analyses
3. Simulation of saddle adjustment
4. Tracking the changes of cable ends in saddles
5. Live load analyses with the consideration of initial stresses and large displacements

Whether or not to use a 3D model depends on the purpose of the analyses. A 3D model is always preferable for all types of analyses, not only because the lateral distribution can be included but also because the modeling of pylons and stiffened girders can be simplified in 3D modeling. For example, the discrete truss members can accurately and easily reflect the properties of the stiffening girder than beam elements. Having a modern graphical tool aided, modeling a suspension bridge in 3D is no longer a challenge as it was many years before. Also the computer capacity and performance nowadays guarantees that a full-scale nonlinear analysis simulating multiple-stage construction in 3D is doable.

12.4.1 Main cables

How to model the main cables is an important question in modeling a suspension bridge, and it should be answered first. Due to its special characteristics such as flexibility, large displacements, and catenary behaviors, some analysis tools may have a catenary element type included. When a catenary cable element type is adopted for modeling cables, some special considerations should be taken as properties to describe that such a cable element may vary among different tools. For example, one may use stressed state to describe the cable geometry and the other may use its zero-stress state. In general practice, the simple truss element type can be used to model the main cables.

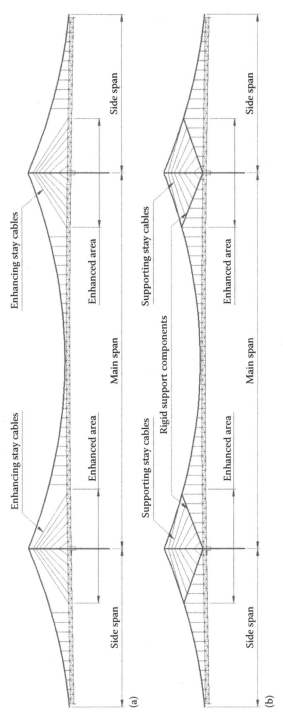

Figure 12.14 (a, b) Hybrid suspension bridge with stay cables to enhance stiffened girders.

When using the simple truss element, the main cables have at least to be meshed at the hanger locations. For longer spacing of hangers, cable has to be meshed in between hangers. Generally speaking, a meshed cable segment length of 10 m is adequate considering the long span of a suspension bridge. Along with the way the cable is modeled, how the initial stress and large displacement iteration are considered should be clearly understood. When a cable segment between two adjacent hangers is submeshed, for example, the analysis may fail as there may be no stiffness perpendicular to the cable initially if the initial stress is not addressed correctly.

As discussed in Sections 12.3.4 and 12.3.5, features regarding how the zero-stress state is obtained by iterations and how saddle curves are considered in the analysis tool should be studied too.

12.4.2 Hangers

The hangers are simple components and can be simply modeled as single truss elements. For rigid connections between main cables and the stiffened girder in the middle of the main span as shown in Figure 12.15, truss or beam elements can be used.

12.4.3 Stiffened girder

The modeling of stiffened girders is similar to the main girder of a cable-stayed bridge. For a box girder, as shown in Figure 11.23 of Chapter 11, the fish bone model as shown in Figure 11.24 of Chapter 11 can be used. The transverse

Figure 12.15 Rigid connections between main cables and stiffened girder (Runyang Bridge, China). (Data from Ji, L. and Feng, Z., *Construction of Suspension Bridges across the Yangtze River in Jiangsu, China*, IABSE Workshop—Recent Major Bridges, May 11–20, 2009, Shanghai, People's Republic of China.)

rigid bodies or beams should locate in all hanger locations. For truss stiffened girder, it is preferable to use the 3D model. Each truss member can be modeled as a beam or truss element so the girder's properties can be accurately modeled. For cases where the deck is comprised of floor beams and stringers, stringers too can be included as beam elements. The superimposed dead loads will be applied only on main beams, main trusses, and/or stringers.

12.4.4 Pylons

The modeling of pylons is similar to cable-stayed bridges. 3D beam elements are usually used to model pylons. The longitudinal bending stiffness of pylons is an important factor to influence the analysis of saddle offsets. When 2D model is used for preliminary analyses, cross-sectional properties should accurately reflect the pylon stiffness.

12.4.5 Saddles

The modeling of saddles and their connections to pylons and cables are critical in the entire bridge model, especially for construction control analyses in which large displacement iterations will be involved. The moving between saddle and pylon usually is locked during the erection of the stiffened girder and is unlocked when horizontal adjustment is needed between erections. After the deck is superimposed, as in most suspension bridges, the connection between the saddle and the top of the pylon will be changed to rigid so the pylon will work to resist unbalanced cable forces due to live loads. This change should be incorporated into modeling according to the type of analyses.

Figure 12.16a shows a general model of the saddle and its connection to cables and the pylon. A temporary horizontal rigid truss element is needed to simulate the locking between saddle and pylon during erection and adjustment. By applying an initial displacement of Δ, as shown in Figure 12.9, on the temporary right truss element, a load case of jacking saddle can be simulated. Changing its stiffness to significantly small or simply removing it, free moving between the saddle and pylon can be simulated during live load analyses. However, when conducting certain types of analysis, the saddle and its connections can be simplified as shown in Figure 12.16b. For example, when only extreme live load responses are of concern in schematic analysis, a single rigid truss or rigid body can be used. If the saddle is designed not to rigidly connect to pylon when bridge is in services, a simple truss from the top of the pylon to the intersection of cables can be used.

Figure 12.16c shows a general model of splay saddle and cable anchorage. When the splay saddle is built into anchor box or a rotational splay saddle is used as shown in Figure 12.8, the splay saddle shown in Figure 12.16 is either the arc center of the saddle surface or the saddle hinge. When a

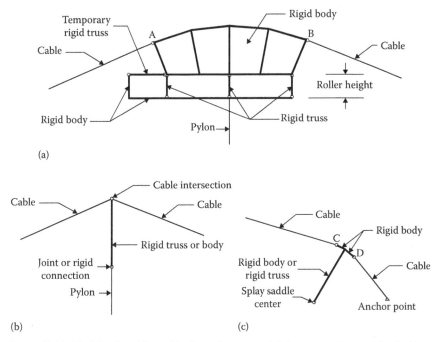

Figure 12.16 Model of saddles. (a) Complicate model incorporating saddle jacking. (b) Simple model for certain types of analysis. (c) Splay saddle model.

flexible splay saddle as shown Figure 12.7 is used, the splay saddle column will be fixed at the bottom and the true stiffness of the column will be used.

If any analysis is targeting the establishment of the initial state (zero-stress state) according to the design plan, the constraints of displacements at cable ends A, B, C, and D should be applied during large displacement iterations as discussed in Section 12.3.5.

12.5 3D ILLUSTRATED EXAMPLE OF CHESAPEAKE BAY SUSPENSION BRIDGE, MARYLAND

The main shipping-channel bridges of Chesapeake Bay Bridge, Maryland, 1952, also known as Bay Bridge, are suspension bridges where the east-bound bridge has a main span of 487.68 m and two 205.74-m long suspended side spans. The tower is 107.7 m high, and the truss stiffened girder passes through it at about its center (Figure 12.1) (Wang and Fu 2012).

The purpose of the analyses is the cost allocation study, which tries to reveal the contribution to stresses on main components by each designated live loads. The analysis type is linear with only initial stress considered, and the geometry configuration as planned is used as the initial state for dead

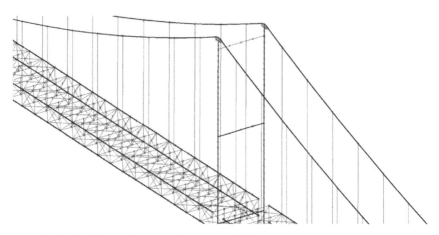

Figure 12.17 3D FEA model of Bay Bridge by VBDS.

load analysis and the final state for live load analysis. The model is built in 3D using VBDS program (Wang and Fu 2005) and cross-checked using SAP2000 (2007). The main cable and hangers are modeled as truss elements; all others are modeled as beam elements, including members of the truss stiffened girder. For simplicity, deck stringers are not modeled, but its weight plus all other superimposed dead loads are included. All members of each truss floor beam are also modeled as beam elements (Figure 12.17). The entire model contains 5264 elements and 2798 nodes.

The saddle is not indicated as rigidly connected to the pylon when the bridge is in services. So the modeling of the saddle and its connections are simplified as shown in Figure 12.16(b).

In addition to dead loads, live loads in the main span and far side span for extreme bending moment at the bottom of tower leg and loads due to temperature change are analyzed. As shown in Figures 12.18 and 12.19, the extreme bending moment due to live loads is −8456 kN-m, and its corresponding deflection in the middle of the center span is 1372 mm. Table 12.2 lists other results of the static analyses.

−8456   2395

Figure 12.18 Extreme live load bending moment (kN-m) in one pylon leg of Bay Bridge.

Figure 12.19 Extreme live load displacements (mm) of Bay Bridge.

Table 12.2 Static analyses results of Bay Bridge

| | | Tower leg reaction | |
Loads	Cable reaction (kN)	Axial (kN)	Moment (kN-m)
Dead loads	29,304	−31,169	0
Live loads[a]	7,192	−5,026	−8,464
26°C temperature drop	555	−147	1,901
11°C temperature rise	−369	98	−1,268

[a] Three lanes with each of 0.87 kN/m + 80 kN concentrated load and lanes discount of 0.9 are used.

The dead load analysis is performed at first. Several rounds of iterations are needed to consider the initial stresses in cables due to dead loads. These initial stresses will be automatically considered in succeeded analyses of live loads and temperature loads. For the purpose of cost allocation study, a truck load moving from one end to another is simulated by using different dead load cases.

Part III

Special topics of bridges

Chapter 13

Strut-and-tie modeling

13.1 PRINCIPLE OF STRUT-AND-TIE MODEL

Structural concrete members used in bridges can be subdivided into two regions, B- and D-regions (Figure 13.1). In the B-region, Bernoulli's hypothesis holds valid, where it is assumed that a normal cross-sectional plane remains plane and normal to the reference lines when the beam deforms. Bernoulli's hypothesis facilitates the flexural design of reinforced concrete structures by allowing a linear strain distribution for all loading stages, including an ultimate flexural capacity. Design of the B- (Bernoulli or beam) region is well understood, and the entire flexural behavior can be predicted by simple calculations. For torsion, the sectional shape and size in its own sectional plane are assumed to be preserved during torsion, and the cross section can warp freely out of its plane.

In the D-region (disturbed or discontinued portion), Bernoulli's hypothesis does not apply. Some examples of D-regions are corbels, dapped beams, deep beams, regions near the support or concentrated load, sudden changes of the cross section, holes, joints, and so on. All these are considered two-dimensional (2D) applications of the strut-and-tie model (STM). Three-dimensional (3D) STM are required when the structure and loading are considerably spread over all three dimensions, such as pile caps with two or more rows of piles.

According to St. Venant's principle, the localized effects caused by any load acting on the body will dissipate or smooth out within regions that are sufficiently far enough from the load location (Figure 13.2b). This is applied in the analysis of D-regions.

Design of the B-region has long been established and can be easily calculated. However, even for the most common cases of D-regions, the ability to predict capacity by traditional methods is either empirical or requires finite element analysis to reach an estimation of capacity. An STM closes this gap and offers engineers the ability to develop a conservative capacity without sophisticated modeling. D-regions can be idealized as consisting of concrete struts in compression, steel ties in tension, and nodes where more than one member are joined together.

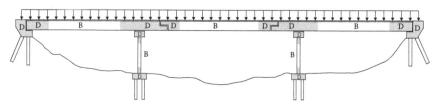

Figure 13.1 B- and D-regions in a common bridge structure. (Data from Kuchma, D., "Strut-and-Tie Website," 2005, http://dankuchma.com/stm/index.htm.)

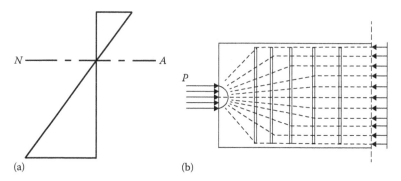

(a)   (b)

Figure 13.2 Stress diagram. (a) Linear stress distribution. (b) Load dissipation. (Data from MacGregor, J.G. et al., *Reinforced Concrete: Mechanics and Design*, 5th Edition, Prentice Hall, Englewood Cliffs, NJ, 2008.)

13.1.1 Development of STM

The steps of the STM and design are the following:

Step 1: Lay out STM. Laying out the model requires an understanding of basic member behavior and good engineering judgment. Because there could be more than one truss configuration, the design is more art than science.
Step 2: Determine the member forces.
Step 3: Decide the shapes of the struts and the nodal zone.
Step 4: Calculate the strength of the struts, ties, and the nodal zones based on the applicable code.
Step 5: Verify the anchorage of the ties.
Step 6: Apply detailing requirements.

The STM follows the lower-bound theorem of plasticity, which states that the capacity of such a system of forces is a lower bound on the strength of the structure, provided that no element is loaded beyond its capacity. A stress field that satisfies equilibrium and does not violate the yield criteria at any point provides a lower-bound estimate of capacity of elastic-perfect plastic materials. For this to be true, crushing of concrete (struts and nodes) does not occur prior to yielding of reinforcement (ties or stirrups).

Nevertheless, there are limitations to the truss analogy. The lower-bound theorem of plasticity assumes that concrete can sustain plastic deformation and is an elastic-perfect plastic material, which is not absolutely correct. To address this deviation from the theoretical concept, codes and specifications adopted the compression theory to limit the compressive stress for struts with consideration of the condition of the compressed concrete at ultimate load resistance. The prerequisites of such assumptions are the following:

- STM is a strength design method, and the serviceability should also be checked
- Equilibrium must be maintained
- Tension in concrete is neglected
- Forces in struts and ties are uniaxial
- External forces are applied at nodes
- Prestressing is treated as a load
- Detailing for adequate anchorage shall be provided

In strut-and-tie truss models, only equilibrium and yield criteria need to be fulfilled as the first two requirements. But the third requirement, the strain compatibility, is not considered. As a result of this relaxation, more than one admissible STM may be developed for each load case as long as the selected truss is in equilibrium with the boundary forces and the stresses in the struts, ties, and nodes are within acceptable limits.

With such a convenient structural analysis tool, questions in STM applications remain

- How does one construct an STM?
- If a truss can be formulated, is it adequate or is there a better one?
- If there are two or more trusses for the same structure, which one is better?

Several empirical rules that aid in generating STM are given as follows:

- Elastic stress contours generated by finite element analysis provide the general direction of the stress trajectories and are useful in laying out an STM.
- Minimum steel content is a goal to achieve. Loads are transmitted by the principle of minimum strain energy. Because the tensile ties are more deformable than the compression struts, the least and shortest ties are the best. A nonlinear finite element comparison of three possible models of a short cantilever is shown in Figure 13.3.

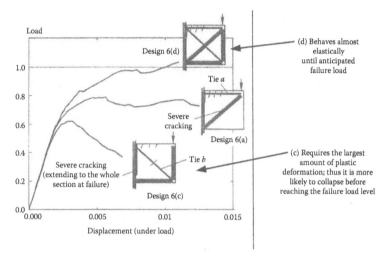

Figure 13.3 Nonlinear finite element comparison of three possible models of a short cantilever. (Data from MacGregor, J.G. et al., *Reinforced Concrete: Mechanics and Design*, 5th Edition, Prentice Hall, Englewood Cliffs, NJ, 2008.)

- The crack pattern may also assist in selecting the best STM. It is suggested by tests (MacGregor et al. 2008) that an STM developed with struts parallel to the orientation of initial cracking will behave very well (Figure 13.4a).
- The minimum angle between a strut and a tie (Figure 13.4b) that are joined at a node shall be 25° according to ACI (2002). There are several other recommendations by other codes and researchers, but they are all within close variation.
- Other than the empirical rules, the common constraints are the code requirements. ACI and AASHTO code recommendations will also be discussed.

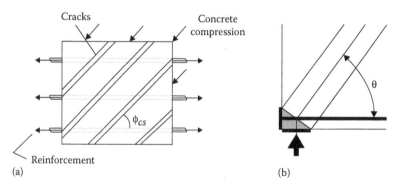

Figure 13.4 Strut. (a) Orientation of strut. (b) Angle at support. (Data from MacGregor, J.G. et al., *Reinforced Concrete: Mechanics and Design*, 5th Edition, Prentice Hall, Englewood Cliffs, NJ, 2008.)

13.1.2 Design methodology

The design of struts, ties, and nodal regions shall be based on

$$\phi F_n \geq F_u \qquad\qquad (13.1) \text{ (ACI [eq. A-1])}$$

$$\phi = 0.75 \quad \text{for struts, ties, and nodes} \qquad\qquad \text{(ACI [sect. 9.3.2.6])}$$

13.1.2.1 Struts

Compression members or struts fulfill two functions. Like the compression chord of a truss member, they resist compression due to moment. The diagonal struts transfer forces to the nodes or transfer shear to the supports. In actual function, the diagonal struts will be oriented parallel to the cracks. There are three different types of struts (Figure 13.5). The simplest one is the "Prism," with a constant cross section. The second type is the "Bottle," in which the strut expands or contracts along its length. The third type is the "Fan," where an array of struts with varying inclinations meet at or radiate from a node.

According to Appendix A of ACI-318-2002, the strength of a longitudinally reinforced strut is

$$F_{ns} = f_{cu}A_c + A_s'f_s' \qquad\qquad (13.2) \text{ (ACI [eq. A-5])}$$

The strength of an unreinforced strut is

$$F_{ns} = f_{cu}A_c \qquad\qquad (13.3) \text{ (ACI [eq. A-2])}$$

where the effective compression strength of the concrete in a strut is

$$f_{cu} = 0.85\beta_s f_c' \qquad\qquad (13.4) \text{ (ACI [eq. A-3])}$$

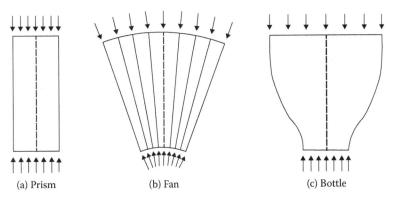

(a) Prism   (b) Fan   (c) Bottle

Figure 13.5 (a–c) Three types of struts. (Data from Schlaich et al. 1987.)

where β_s is the effectiveness factor. The factors affecting the effective concrete strength of struts are (1) load-duration effects, (2) cracking of the struts, and (3) confinement from the surrounding concrete. For (1) and (2), there is reduction of strength, but for (3) the strength is increased. For example, in pile caps, the compressive strength may be increased by the confinement resulting from the large volume of concrete all around the struts.

$\beta_s = 1.0$	for a strut of uniform cross-sectional area over its length
$\beta_s = 0.75$	for a bottle-shaped strut with reinforcement satisfying A.3.3; (ACI 318-2002)
$\beta_s = 0.6$	for a bottle-shaped strut with reinforcement not satisfying A.3.3; (ACI 318-2002)
$\beta_s = 0.4$	for a strut in the tension member or the tension flange of members
$\beta_s = 0.6$	for all other cases

Note: Crack control reinforcement requirement is

$$\sum \frac{A_{si}}{b_s s_i} \sin\gamma_i \geq 0.003 \qquad (13.5) \text{ (ACI [eq. A-4])}$$

where:

A_{si} is the area of surface reinforcement in the i-th layer crossing the strut under review

s_i is the spacing of reinforcement in the i-th layer adjacent to the surface of the member

b_s is the width of the strut

γ_i is the angle between the axis of the strut and the bars

According to *AASHTO Load Resistance Factor Design (LRFD) Bridge Design Specifications* (2013) stress limit for struts is

$$f_{cu} = \frac{f_c'}{0.8 + 170\varepsilon_1} \leq 0.85 f_c' \qquad (13.6) \text{ (AASHTO [eq. 5.6.3.3.3-1])}$$

where

$$\varepsilon_1 = \varepsilon_s + (\varepsilon_s + 0.002)\cot^2\theta_s \qquad (13.7) \text{ (AASHTO [eq. 5.6.3.3.3-2])}$$

where:

θ_s is the smallest angle between the strut under review and the adjoining ties

ε_s is the average tensile strain in the tie direction

f_c' is the specified concrete compressive strength (psi or MPa)

The stress limit assumes that a minimum distributed reinforcement ratio of 0.003 in each direction is provided.

13.1.2.2 Ties

The tension ties are stirrups, longitudinal tension chord reinforcements, and other special-detail reinforcements. All tension reinforcements should be adequately anchored. Inadequate development of tension reinforcement will lead to brittle failure at a lower load than at anticipated ultimate capacity.

According to Appendix A of ACI-318-2012, the nominal strength of a tie shall be taken as

$$F_{nt} = f_y A_{st} + A_{ps}\left(f_{sc} + \Delta f_p\right)$$ (13.8) (ACI [eq. A-6])

where $(f_{sc} + \Delta f_p)$ shall not exceed f_{py} and $A_{ps} = 0$ for nonprestressed member.

13.1.2.3 Nodes

Nodes are the locations where struts and ties converge. In other words, nodes are the locations where forces are redirected within an STM. Nodal zones are classified as CCC if all the compressive forces meet and CCT if one of the forces is in tension (Figure 13.6). Similarly, CTT and TTT are also possible (Figure 13.6). One way of laying out a nodal zone is to create equal pressure on each face of the node. By doing so, on a CCC node, the length of the sides of the nodes $a1:a2:a3$ becomes the same proportion as $C1:C2:C3$. If one of the forces is in tension, the length of that side of the node is calculated from a hypothetical bearing plate on the end of the tie, which exerts the same bearing pressure as the compression member. Because the in-plane stresses in the nodes are equal in all directions, such a node is referred to as *hydrostatic element* (Figure 13.7). For a CCC node, this can be easily applied but for other nodes it can be tedious. This can be simplified by considering a nodal zone formed by the extension of all the members meeting at that node (Figure 13.8). However, this allows unequal stress at

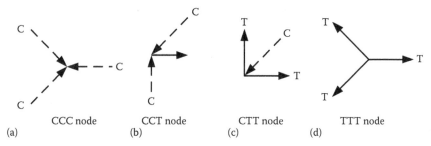

| CCC node | CCT node | CTT node | TTT node |
| (a) | (b) | (c) | (d) |

Figure 13.6 Classification of nodal zones.

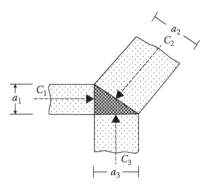

Figure 13.7 Hydrostatic element.

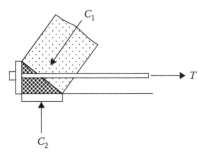

Figure 13.8 Nodal zone formed by the extension of the members.

the different faces of the node. At these nodes the following conditions need to be satisfied:

1. The resultants of the three forces coincide
2. The stresses are within the limits
3. The stress is constant on any face

Nodal zones fail by concrete crushing. Again, the anchorage of tension ties must be provided. Within a node, if a tension tie is anchored, incompatibility of tensile strain in the rebars and the compression strain in the concrete take place, which weaken the compressive strength of the concrete.

According to Appendix A of ACI-318-2012, the limiting compressive strength on a face of a node is given by

$$F_{nn} = 0.85\beta_n f_c' A_n \qquad\qquad \text{(13.9) (ACI [eqs. A-7 and A-8])}$$

where for

CCC node $\beta_n = 1.0$ (ACI [sect. A-5.2.1])
CCT node $\beta_n = 0.8$ (ACI [sects. A-5.2.2 and A-5.2.3])
CTT node $\beta_n = 0.6$

According to *AASHTO LRFD Bridge Design Specifications* 3rd Edition (sects. 5.6.3.5 and 5.5.4.2), stress limit for nodes is

$f_{cu} = 0.85\ f_c'$ when nodes are bounded by struts and/or bearing areas
$f_{cu} = 0.75\ f_c'$ when nodes anchor only one tie
$f_{cu} = 0.65\ f_c'$ when nodes anchor more than one tie

and resistance factors are

$\phi = 0.7$ for struts and nodes
$\phi = 0.9$ for ties

13.2 HAND-CALCULATION EXAMPLE OF STM

Two hand-calculation cases are covered in this section; the first case is a Hammerhead Pier originally reported by Fu et al. (2005), and the second case is a pier-supported footing covered in the final report of NCHRP Project 20-07 Task 217 (Martin and Sanders 2007).

13.2.1 Hammerhead Pier No. 49 of Thomas Jefferson Bridge, Maryland

A simple model is designed by hand to demonstrate the procedures for STM (Fu et al. 2005). The same structure will be seen in Section 13.64 under Case Study 4—Pier Cap 2. Also a finite element method analysis will be done for the same pier cap where loading was increased to see the formation of cracks.

13.2.1.1 Data

Material strength: $f_c' = 24.13$ MPa (3.5 ksi); $f_y = 275.8$ MPa (40 ksi)
Load from each girder: $P = 1289.92$ kN (290 kip)
Strength reduction factor for struts, ties, and nodes: $\phi = 0.75$ (ACI 9.3.2.6)

Through nodes 2 and 3 reactions are transferred to the pier. At each point $2P = 580$ kip (2579.84 kN) load is transferred. The 3D of the pier and the cap is 1524 mm (60″) thick (Figure 13.9).

Putting $b = 1.524$ m (60″) and $\beta_s = 1.0$ (ACI A.3.2.1)
The length of bearing area required $L_{\text{bearing}} = 2P/\phi(0.85\beta_s f_c')b = 110$ mm (4.332″)

13.2.1.2 Determination of member forces

$P1 = P/\sin(59.93°) = 335.100$ kip (1490.525 kN)

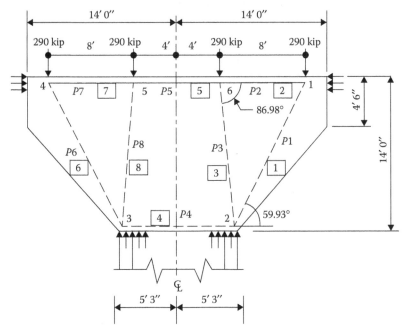

Figure 13.9 Strut-and-tie model developed for the Hammerhead Pier.

$P6 = P1 = 335.100$ kip (1490.525 kN)
$P2 = P1\cos(59.93°) = 145.307$ kip (646.326 kN)
$P7 = P2 = 145.307$ kip (646.326 kN)
$P3 = P/\sin(86.98°) = 290.403$ kip (1292.713 kN)
$P8 = P3 = 290.403$ kip (1291.713 kN)
$P4 = P1 \cos(59.93°) - P3 \cos(86.98°) = 152.605$ kip (678.787 kN)
$P5 = P2 - P3 \cos(86.98°) = 130.007$ kip (578.271 kN)

13.2.1.3 Design of the tie

Members 2, 5, 7 (Figure 13.9) (ACI A.2.6 and A.4.1)

Required area of steel for members 2 and 7: $A_{st\_2} = P2/\phi f_y = 3125$ mm² (4.844 in²)
Required area of steel for member 5: $A_{st\_5} = P5/\phi f_y = 2795$ mm² (4.334 in²)

Minimum reinforcement (ACI 11.8.5)

From geometry of the pier cap: Depth $d = 4267.2$ mm (14'); Width $b = 1524$ mm (60")
$A_{stmin} = 0.04(f_c'/f_y)bd = 22{,}761$ mm² (35.28 in²)

Minimum reinforcement for crack control (ACI 11.8.4)

According to ACI 318 11.8.4, closed stirrups or ties of area A_h parallel to A_s shall be provided.
   To simplify, assume

$N_{uc} = 0$, $A_n = 0$ in²
$A_{st} = \max(A_{st2}, A_{sts}, A_{stmin}, A_{st}) = 35.28$ in² (22,761 mm²)
$A_h = 0.5(A_{st} - A_n) = 17.64$ in² (11,381 mm²)
8 # 5 in 10 layers @ 305 mm (12″) c/c

Determination of the required depth to satisfy the stress limits at nodes 1 or 4 and to check the anchorage:

For nodal zone anchoring one tie $\qquad$ $\beta_n = 0.8$ (ACI A.5.2)
$f_{cu} = 0.85\beta_n f_c'$ $\qquad\qquad\qquad$ $\phi f_{cu} = 1.785$ ksi (12.31 MPa)
Required depth $\qquad\qquad\qquad$ $d_{reqd} = P2/(\phi f_{cu} b) = 35$ mm (1.375″)

13.2.1.4 Design of the strut

Members 1, 3, 6, 8 (ACI A.2.6 and A.3.2); $\beta_{s\_bottle} = 0.75$
   By providing four two-legged no. 5 rebars as stirrups at 305 mm (12″) c/c, which is also required for crack control and calculated earlier in accordance to ACI 11.8.4,

$A_{s2} = 1548$ mm² (2.4 in²) $\qquad$ $s_2 = 305$ mm (12″)
$\gamma_2 = 86.98°$ $\qquad\qquad\qquad$ $(A_{s2}/bs_2)\sin(\gamma_2) = 0.003$

The stress in these bottle-shaped members will be limited to $\phi f_{cu} = \phi 0.85\beta_{s\_bottle} f_c'$

$\phi f_{cu} = 1673.437$ psi (11.538 MPa)
Required depth for members 1 and 6: $d_{1\_strut} = P1/(\phi f_{cu} b) = 79$ mm (3.129″)
Required width for members 3 and 8: $d_{3\_strut} = P3/(\phi f_{cu} b) = 73$ mm (2.892″)

Member 4 (ACI A.2.6 and A.3.2)

   This member is considered prism strut: $\beta_{s\_prism} = 1$
   The stress in prism-shaped members will be limited to

$\phi f_{cu\_prism} = \phi 0.85\beta_{s\_prism} f_c' = 2231$ psi (15.383 MPa)

Required width for member 4: $D_{4\_strut} = P1/(\phi f_{cu\_prism} b) = 64$ mm (2.503″)

13.2.2 Representative pile-supported footing

This hand-calculation example is illustrated in the final report of NCHRP Project 20-07 Task 217 (Martin and Sanders 2007). Figure 13.10a depicts a 3.3 m × 3.3 m × 0.9 m (11′ × 11′ × 3′) footing supported by nine piles arranged in a 3 × 3 pattern. The total factored loading, including the pile cap and the soil overburden, was 5164 kN (1161 kip) with a transverse moment of 521 kN-m (384 kip-ft). The concrete used in the original design had an f_c' of 20.7 MPa (3000 psi), and the steel used was grade 60

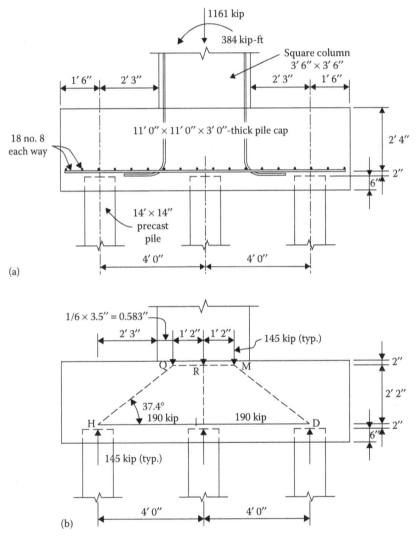

Figure 13.10 (a) Details of the existing footing. (b) Section through the centerline of footing. (Continued)

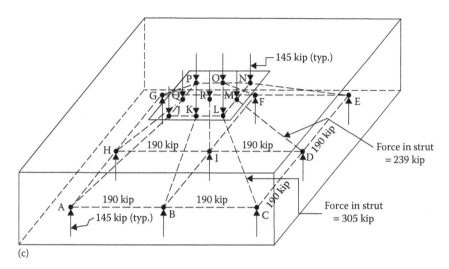

(c)

Figure 13.10 (Continued) (c) 3D STM truss resulting from the flow of forces and resulting member forces. (Data from Martin Jr., B.T. and Sanders, D.H., "Analysis and Design of Hammerhead Pier Using Strut and Tie Method," Final Report-Project 20-07_Task 217, National Cooperative Highway Research Program, Transportation Research Board, Washington, DC, November, 2007.)

(f_y = 413.7 MPa). The original reinforcement was determined by checking the moment capacity at the face of the column as well as one-way and two-way shear at the critical sections.

Assume typical load for each nine-load location is from the factored loading 574 kN (= 5164 kN/9) plus load based on moment 71 kN (the moment divided by the section modulus of the pile group, 521/7.3 = 71 kN or 384/24 = 16 kip) for a total of 645 kN (145 kip) (see Figure 13.10b and c).

13.2.2.1 Check the capacity of the ties

The required area of reinforcement (A_{st}), is

$$A_{st} = \frac{P_u}{\phi f_y} = \frac{190}{0.9 \times 60} = 3.52 \text{ in}^2 (2270.96 \text{ mm}^2)$$

The factored flexural resistance required to resist 1.2 M_{cr}

$$M_{cr} \geq 1.2 \times \frac{bh^2}{6} 0.36\sqrt{f_c'} = 1.2 \times \frac{12 \times 11 \times 36^2}{6} 0.36\sqrt{3} = 21,334 \text{ kip-in}$$

$$= 1778 \text{ kip-ft} (2412.34 \text{ kN-m})$$

The minimum area of flexural reinforcement corresponding to 1.2M_{cr} can then be determined.

$$A_{s,min} = \frac{1778}{2.33 \times 0.9 \times 60} = 14.13 \text{ in}^2(9116.11 \text{ mm}^2)$$

With three parallel ties, the area of reinforcement required to ensure the factored flexural resistance is at least $1.2M_{cr}$, which is equal to $14.13/3 = 4.71 \text{ in}^2 (3038.7 \text{ mm}^2)$.

The amount of reinforcement required to resist 1.33 times the factored loads is

$$A_{st} = 1.33 \times 3.52 = 4.68 \text{ in}^2 (3019.35 \text{ mm}^2)$$

The amount of reinforcement required to resist 1.33 times the factored loads is less than the amount required to resist $1.2M_{cr}$; therefore, this smaller amount will be checked against the amount provided by the original design of the footing. There are presently 18 no. 8 bars provided in the lower mat in each direction. This equals $18 \times 0.79 = 14.22 \text{ in}^2 (9174.18 \text{ mm}^2)$. This results in $14.22/3 = 4.74 \text{ in}^2 (3058.06 \text{ mm}^2)$ per tie zone. This reinforcement is distributed across the full width of the footing and not the limits of the nodes. Even though the total amount of reinforcement is greater than the $4.68 \text{ in}^2 (3019.35 \text{ mm}^2)$ required, it is not placed within the region defined by the nodes and therefore does not meet the requirements of STM.

13.2.2.2 Check the capacity of struts

Take the representative strut AJ (Figure 13.10c).

Using node A as representative of all the corner nodes, the area of the vertical projection of the strut may be calculated as

$$A_{srut} = \frac{25.45 \times 6.88}{2} + 2 \times 25.45 + \left(\frac{25.45 + 19.8}{2}\right) \times 2$$
$$= 87.55 + 50.90 + 45.25 = 1.185 \times 10^5 \text{ mm}^2(183.7 \text{ in}^2)$$

Because this value is a vertical section of the strut, the cross-sectional area perpendicular to the axis of the strut can be calculated by $\cos(28.4°) \times 183.70 = 1.0426 \times 10^5 \text{ mm}^2 (161.6 \text{ in}^2)$.

The limiting compressive stress f_{cu} in the strut depends on the principal strain, ε_1, in the concrete surrounding the tension ties.

The tensile strain in tie AB is

$$\varepsilon_s = \frac{P_u}{A_{st}E_s} = \frac{190}{4.74 \times 29,000} = 1.382 \times 10^{-3}$$

In accordance with AASHTO LRFD C5.6.3.3.3, the strain will be approximately equal to $1.382 \times 10^{-3}/2 = 0.691 \times 10^{-3}$ at the midpoint of the strut. Using the angle between the plane of the tension ties and the diagonal strut of 28.4°, the principal strain e_1 can be determined using the following:

$$\varepsilon_1 = \varepsilon_s + (\varepsilon_s + 0.002)\cot^2\alpha_s$$
$$= 0.691 \times 10^{-3} + (0.691 \times 10^{-3} + 0.002)\cot^2 28.4°$$
$$= 9.9 \times 10^{-3}$$

and the limiting compressive stress f_{cu}, the nominal resistance P_n, and then the factored resistance P_r, in the strut are

$$f_{cu} = \frac{f_c'}{0.8 + 170\varepsilon_1} = \frac{3}{0.8 + 170 \times 9.90 \times 10^{-3}} = 1.21\,\text{ksi} \leq 0.85 \times 3$$
$$= 2.55\,\text{ksi}\,(17.58\,\text{MPa})$$

$$P_n = f_{cu}A_{cs} = 1.21 \times 161.6 = 195.5\,\text{kip}\,(869.63\,\text{kN})$$

$$P_r = \phi P_n = 0.7 \times 195.5 = 137\,\text{kips}\,(609.4\,\text{kN})$$

As this is less than the factored load in the strut of 305 kip (1356.7 kN), the strut capacity is inadequate. To meet the strength requirement of the strut, the depth of the footing would need to be increased by approximately 355.6 mm (14″). This increase in depth would decrease the load in the strut and increase the area of the strut due to the change in the geometry of the STM.

13.2.2.3 Check nodal zone stress limits

The CCC nodal zone at the column–cap interface has a stress of

$$f_c = \frac{9 \times 145}{42 \times 42} = 0.74\,\text{ksi}\,(5.1\,\text{Mpa})$$

This value is below the nodal stress limit for a CCC node of

$$0.85\phi f_c' = 0.85 \times 0.70 \times 3 = 1.78\,\text{ksi}\,(12.27\,\text{MPa})$$

The stress in the CTT nodal zone immediately above the piles is

$$f_c = \frac{145}{14 \times 14} = 0.74\,\text{ksi}\,(5.1\,\text{MPa})$$

As the CTT nodal zones immediately above the piles have tension ties in at least two directions, the nodal zone stress limit is

$$0.65\phi f_c' = 0.65 \times 0.70 \times 3 = 1.36\,\text{ksi}\,(9.38\,\text{Mpa}) > 0.74\,\text{ksi}\,(5.1\,\text{MPa})$$

13.2.2.4 Check the detailing for the anchorage of the ties

The no. 8 bars are required to develop a force of 190 kip (845.16 kN) at the inner face of the piles. The original plans called for no hooks or any other anchorage device. The stress in the no. 8 bars at the inner faces of the piles is

$$f_s = \frac{190}{6 \times 0.79} = 40 \, \text{ksi} (275.80 \, \text{MPa})$$

In accordance with AASHTO LRFD paragraph 5.11.2.1.1, the basic tension development length of a no. 8 bar, l_d, is 868.68 mm (34.2″). The development length can be reduced as a function of the amount of stress in the bar; hence (40/60) × 34.2 = 584.2 mm (23″). Because a development length of 609.6 mm (24″) is provided, the original anchorage details are acceptable.

In this section a very simple structure was analyzed to demonstrate the STM method. But when the structure is more complicated with larger numbers of members or when the structure is indeterminate, STM goes beyond the limits of hand calculation. Since the procedure is based on trial and error to get the optimum STM model, a computer program will be necessary. CAST (computer-aided strut-and-tie), a state-of-the-art program developed by Kuchma (2005) sponsored by the National Science Foundation, is a very useful tool with a user-friendly graphics interface.

The following case studies will demonstrate the usage of STMs in the transportation-related field. All cases can be simulated by using planar STM models. The first four cases were solved earlier by hand calculations (Fu et al. 2005) and later by CAST for verification. The fifth case is for an integral pier bent, which was covered in the final report of NCHRP Project 20-07 Task 217 (Martin and Sanders 2007) and then solved by CAST for this chapter.

13.3 2D ILLUSTRATED EXAMPLE I—ABUTMENT ON PILE

An abutment on piles is widely used in bridges, and one under construction can be seen in Figure 13.11a and b (Fu et al. 2005). For the case study, the abutment considered is 10.06 m (33′) long, 0.91 m (3′) wide, and 0.91 m (3′) deep. Eleven prestressed concrete deck beams bearing on elastomeric pads are supported at an interval of 0.91 m (3′) along the length of the abutment. The concrete slabs span 15.24 m (50′) and transfer 107.61 kip (478.67 kN) factored load on each elastomeric pad. The abutment is supported on six piles spaced at 1.83 m (6′) on center. With this geometry, where depth is half the distance between the supports, this abutment is a special deep beam where Bernoulli's region does not exist and there is a disturbed region throughout. AASHTO states that Bernoulli's region does not exist when the depth-to-span ratio exceeds two-fifth. This beam exceeds that limit. According to one of the criteria of St. Venant's principle, D-regions are

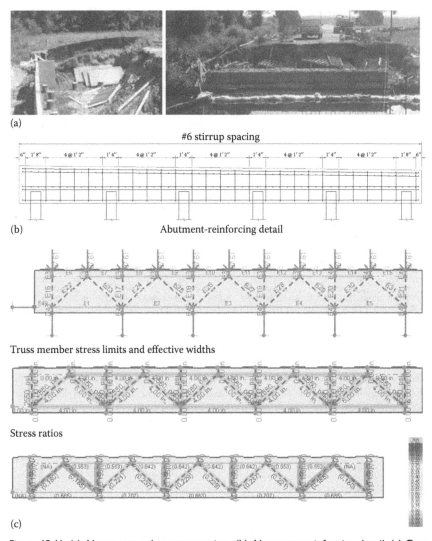

Figure 13.11 (a) Abutment under construction. (b) Abutment reinforcing detail. (c) Case study 1—Truss model and results using CAST program.

those parts of a structure within a distance equal to the beam depth of the member from the concentrated force (load or reaction).

13.3.1 General properties

D-region thickness = 914.4 mm (36″).
Concrete cylinder strength = 4000 psi (27.58 MPa)
Nonprestressed reinforcement yield strength = 60,000 psi (413.69 MPa)

The truss model and the results obtained from CAST are presented in Figure 13.11c. Based on the calculation by the CAST program, maximum compression in the diagonal strut is 101.87 kip (453.14 kN) and in the vertical strut is 107.61 kip (478.67 kN). Maximum tension in the top tie is 31.76 kip (141.28 kN) and in the bottom tie is 50.87 kip (226.28 kN). Size of the upper nodes is determined by the size of bearing, and the size of the lower nodes is decided by the sizes of piles. Rebar sizes and arrangements are finalized after a few iterations. Bearing reinforcement details in the width direction can be determined by a simple truss model in the horizontal direction. The abutment is 914.4 mm (3′) wide, and the strut section 914.4 mm × 152.4 mm (36″ × 6″) provides the required strength for the struts. For ties, three no. 6 bars can provide the required strength. However, code-specified minimum reinforcement must be provided to prevent temperature-, creep-, and shrinkage-related issues.

13.4 2D ILLUSTRATED EXAMPLE 2—WALLED PIER

Another common structure found in the transportation field is a solid shaft bridge pier on a mat foundation shown in Figure 13.12a (Fu et al. 2005). This case study is done for a 5.49-m (18′) high by 0.91-m (3′) wide wall on a mat foundation. Four girders are resting on the wall, and each girder reaction is 215.22 kip (957.35 kN). St. Venant's principle states: "The localized effects caused by any load acting on the body will dissipate or smooth out within regions that are sufficiently away from the location of the load." Elevation of the structure is shown in Figure 13.12b.

Based on the same principle, an STM model is developed for the walled pier and presented in Figure 13.12c. The inclined angle q can either be obtained from a stress trajectory plot or be assumed to vary from 65° for $l/d = 1°$–55° for $l/d = 2.0$, where l is the wall length and d is the height. A reasonable path at a 2-to-1 slope is created here to flow the concentrated loads from the top of the wall toward the mat foundation. Maximum strut force is 128.9 kip (573.38 kN), and maximum tie force is 50.22 kip (223.39 kN), which are in the same range of Case Study 1, and a similar strut width and reinforcement will be sufficient. Again, for this case, minimum steel per code provisions applicable to the wall have to be provided.

13.5 2D ILLUSTRATED EXAMPLE 3—CRANE BEAM

A conservative estimate of the resistance of a concrete structure may be obtained by the application of the lower-bound theorem of plasticity. If sufficient ductility is present in the system, a STM fulfills the conditions for the application of the lower-bound theory. The lower-bound theorem requires identifying at least one plausible load path and ensuring that no portion of the load path is overstressed.

This case study pertains to the gantry crane beam at the Maryland Port Authority Harbor as shown in Figure 13.13a (Fu 1994; Fu et al. 2005). The beam section is 1.83 m (6′) deep by 0.61 m (2′) wide and has five spans, each 1.83 m (6′). 135# gantry rail on continuous base plate (1/2″ or 12.7 mm thick by 24″ or 609.6 mm wide), anchored with the beam and the whole assembly, is encased except for the top 25.4 mm (1″) of the rail

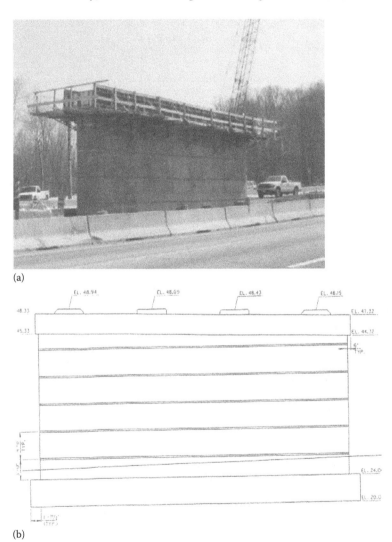

(a)

(b)

Figure 13.12   (a) Solid shaft bridge pier on a mat foundation under construction. (b) Walled pier. (Continued)

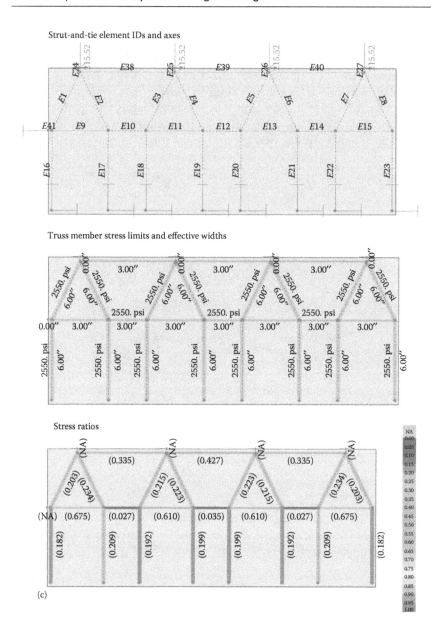

Figure 13.12 (Continued) (c) Case study 2—Truss model and results using CAST program.

for wheel movement. A schematic sketch of the structure can be seen in Figure 13.13b.

Five-span continuous beam models are built with five different configurations to simulate the stress trajectories for the moving wheel loads of the crane. Five configurations represent the first wheel placed at 0, L/5,

(a)

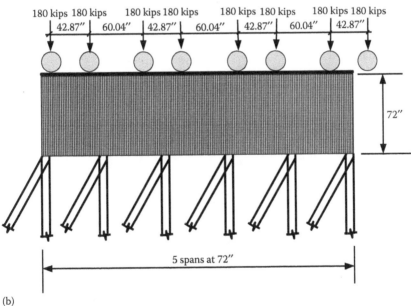

(b)

Figure 13.13 (a) Gantry crane beam at the Maryland Port Authority Harbor. (b) Schematic sketch of a gantry crane beam. (Continued)

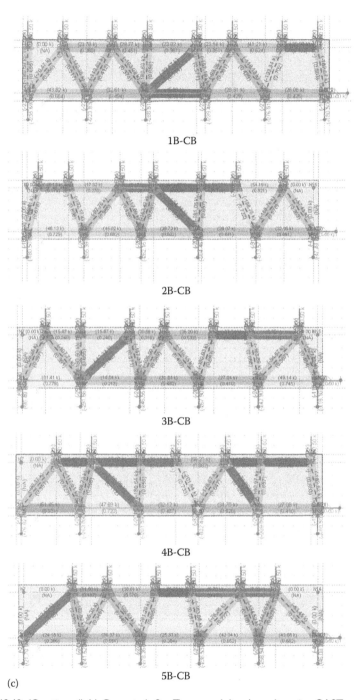

(c)

Figure 13.13 (Continued) (c) Case study 3—Truss model and results using CAST program.

$2L/5$, $3L/5$, and $4L/5$ from the end support, and other wheels follow the location of the wheel spacing. As shown in Figure 13.13b, crane loads are applied at the top of the deep beam, and the self-weight of the deep beam is considered as loads to the deck. The crane load consists of eight wheels, each 180.5 kip (802.90 kN) (factored). The envelope results for each case are tabulated in the study report to the Maryland Port Authority (Fu 1994).

Results from the CAST for all the five configurations are shown in Figure 13.13c. For Case No. 4B, the maximum tension force is 61.45 kip (273.34 kN) and the maximum compression force is 201.84 kip (897.83 kN). Beam thickness is 609.6 mm (24″). Based on wheel contact width and height of rail, the width of the strut will be 254 mm (10″) minimum; hence the strut section considered is 254 mm × 609.6 mm (10″ × 24″). Reinforcements of four no. 6 rebars are provided at the top and bottom for the tie members. Truss forces and stress interaction (actual/allowable) ratios are well below unity for all the members.

After achieving the solution for the members, a detailed nodal analysis is performed. With 254-mm (10″) width struts, the nodes at the bottom ends of the most heavily loaded members were overstressed. A few iterations were necessary to optimize the strut width (ranging from 254 mm [10″] to 304.8 mm [12″]) so that the stress triangles within the nodal zone get reoriented and meet the strength requirement of the code-specified limit of the nodal zone.

The stress fields in struts and ties are idealized to be uniaxial, whereas the stress fields in nodal zones are biaxial. These conditions cause stress discontinuity at the interface of the strut and node stress fields and at the interface of the tie and node stress fields. The stress discontinuity also occurs along the longitudinal boundary of the strut or tie stress fields if the selected stress distribution across the effective width is uniformly distributed. For 2D structures, the interface between two different stress fields is commonly referred to as the line of stress discontinuity. Although the term *line* is used, the stress discontinuity actually occurs on a surface perpendicular to the plane of the structures, across the D-region thickness. For this reason, reinforcement is required at the nodal locations perpendicular to the plane of the structures. This reinforcement can be seen in Figure 13.11b provided for the case 1 example.

13.6 2D/3D ILLUSTRATED EXAMPLE 4—HAMMERHEAD PIER OF THOMAS JEFFERSON BRIDGE

This structure is located in St. Mary's and Calvert counties in Southern Maryland (Fu et al. 2005). It was completed and put into service in 1977. During an inspection in 1979, cracks were observed in the deepwater piers. These piers developed cracks from the corner of the girder base plate and were propagated for great lengths. The scope of this case study is to highlight the application of a newer-generation STM, which was not in practice at the time of the original design. Thus, these piers were not designed with adequate reinforcement and remedial post-tensioning was required.

Depth-to-span ratios vary from 1 to 2 and girders are transferring loads very close to the support edge, making these Hammerheads ideal candidates for STM applications.

1. Pier cap 1 (Figure 13.14a). Length 8.53 m (28′), width 1.22 m (4′), depth at the end 1.07 m (3′-6″), and at the pier face 2.74 m (9′), four loads at 250 kip (1112.06 kN), each placed on the top of the cap. The first load is 0.61 m (2′) from the left end, and the rest are at 2.44 m (8′) intervals. The last load is 0.61 m (2′) from the right end.
2. Pier cap 2 (Figure 13.14b). Length 8.53 m (28′), width 1.524 m (5′), depth at the end 1.37 m (4′-6″), and at the pier face 4.27 m (14′), four loads at 290 kip (1289.98 kN), each placed on the top of the cap. The locations of loads are the same for Pier cap 1.
3. Pier cap 3 (Figure 13.14c). Length 8.53 m (28′), width 1.83 m (6′), depth at the end 1.83 m (6′), and at the pier face 8.53 m (28′), four loads at 550 kip (2446.52 kN), each placed on the top of the cap. The locations of loads are the same for Pier cap 1.

As per this case study 7.5 in² (4838.7 mm²) reinforcement at the top tie level provided acceptable strength for all three Hammerheads. However, a minimum requirement of reinforcement for crack control needs to be provided in accordance with ACI 318. The STM results can be seen in Figure 13.14d.

There could be numerous reasons for the cracks to develop. Shrinkage, stress concentration, or some erection condition may be a few of them. During STM analysis, the presence of cracks was not considered, but the existence of the crack will redistribute the stress flow. The choice of load path is limited by the deformation capacity of the beam, and a situation may arise when, due to the presence of the crack, a structure is unable to undergo the force distribution to reach the assumed load path. In connection with the crack, the common retrofit is post-tensioning. In the STM, the external post-tensioning can be efficiently modeled as external load. All forces acting on the anchorage zone shall be considered in the selection of an STM, which should follow a path from the anchorages to the end of the anchorage zone.

A finite element analysis was done for pier cap 2, using ANSYS. The SOLID65 elements (3D-reinforced concrete solid) were used. The physical model can be seen in Figure 13.14e. In the original analysis, girder reaction was 290 kip

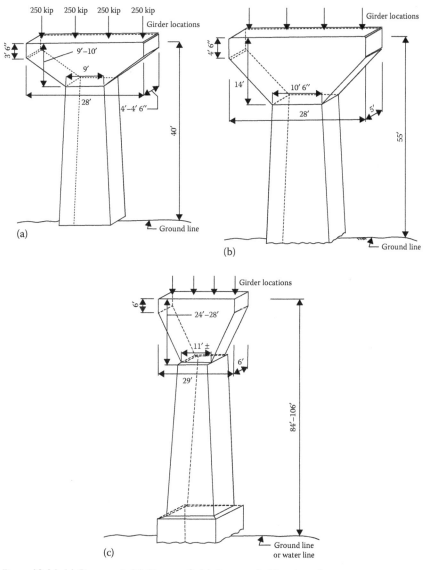

Figure 13.14 (a) Pier cap 1. (b) Pier cap 2. (c) Pier cap 3. (Continued)

(1289.98 kN), but then the load was increased to see the crack formation. Cracks were observed at about 750 kip (3336.17 kN) for the girder reaction. From the stress contour S_x (lateral stress, Figure 13.14f), the tension zone can be identified where reinforcement shall be provided as tie members or stirrups. The S_y (vertical stress) contours can be seen in Figure 13.14g. In this figure, the formation of struts is clearly visible.

Strut-and-tie element IDs and axes

Truss member stress limits and effective widths

Stress ratios

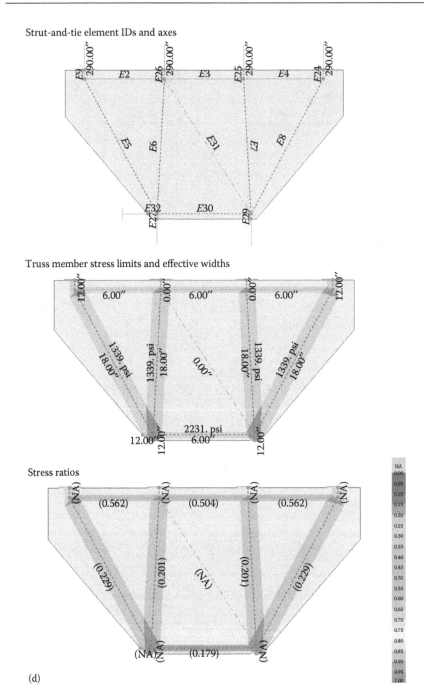

(d)

Figure 13.14 (Continued) (d) Case study 3—Truss model and results using CAST program. (Continued)

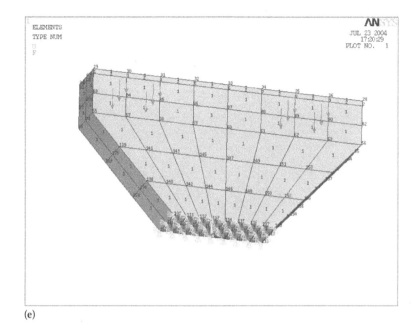

(e)

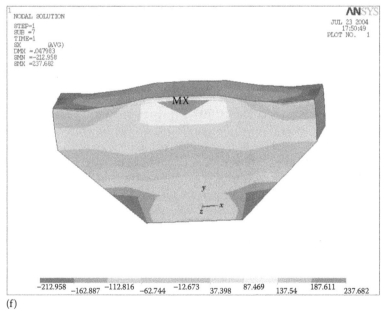

(f)

Figure 13.14 (Continued) (e) Case study 4—ANSYS model. (f) Case study 4—ANSYS model (S_x—lateral stress). (Continued)

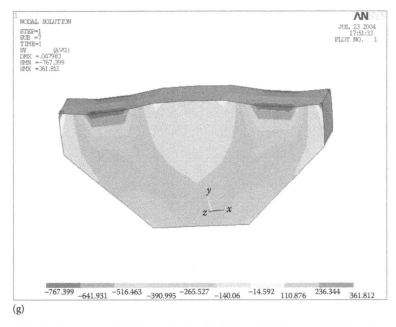

(g)

Figure 13.14 (Continued) (g) Case study 4—ANSYS model (S_y—vertical stress).

13.7 2D ILLUSTRATED EXAMPLE 5—INTEGRAL BENT CAP

This case study is illustrated in the final report of NCHRP Project 20-07 Task 217 (Martin and Sanders 2007). A three-span rigid-frame structure has the configuration shown in Figure 13.15a. The superstructure consists of a four-cell cast-in-place box girder carrying a 12-m (40′) roadway. The box girders are fully supported during casting and are integral with the bent caps. The superstructure geometry is shown in Figure 13.15b, and the geometry of the bent is shown in Figure 13.15c. The bent cap concrete has an f_c' of 27.6 MPa (4 ksi), and the mild reinforcing is grade 60 (f_y = 413.7 MPa). The reinforcing for the integral cap in bent 3 is designed using AASHTO LRFD strut-and-tie provisions and HL-93 live loading applied to the spans as shown in Figure 13.15d.

In this example, there are two live loading cases on the same STM model, except varied load magnitudes. The first case places the live load on the cantilever to maximize the negative moment. This is illustrated in Figure 13.15e, and the resulting forces on the STM are shown in Figure 13.15f. The second case loads the middle of the bent with live load to maximize the positive moment in the cap. This is illustrated in Figure 13.15g, and the resulting loading on the STM is shown in Figure 13.15h.

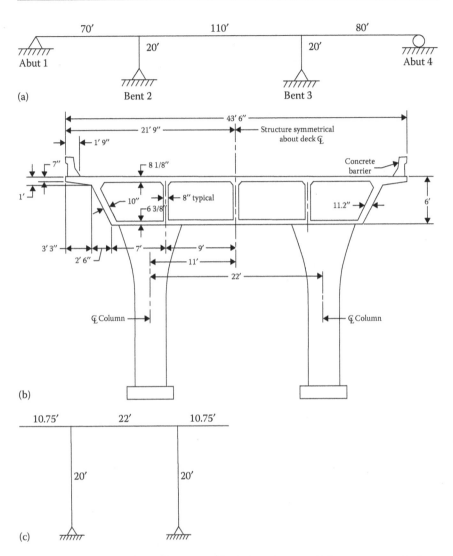

Figure 13.15 (a) Rigid-frame geometry. (b) Cross section of the structure. (c) Bent geometry. (Continued)

13.8 ALTERNATE COMPATIBILITY STM AND 2D ILLUSTRATED EXAMPLE 6— CRACKED DEEP BENT CAP

Alternate to the previously demonstrated STM examples, Scott et al. (2012) proposed another type of model called compatibility STM (C-STM), where shear resistance in structural concrete elements is resisted by a combination

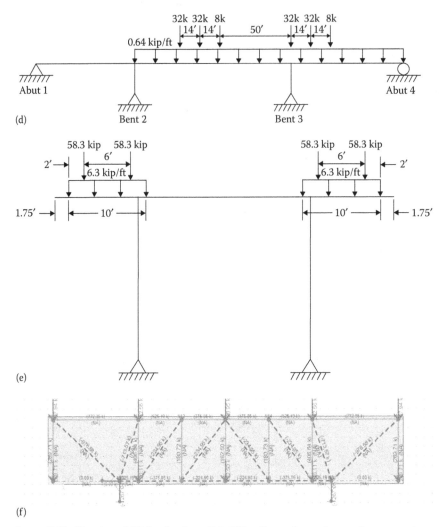

Figure 13.15 (Continued) (d) Application of HL-93 loading to determine maximum reactions. (e) Live load configuration to maximize cantilever moments. (f) Factored loading resulting from maximizing of cantilever moments. (Continued)

of truss and arch action. Arch action refers to the compressive stress field that forms the main corner-to-corner diagonal concrete strut from an applied load, whereas the truss action specifically pertains to the shear mechanism that engages the transverse reinforcement through *smeared* diagonal concrete struts resembling a truss. The contribution of each mechanism was apportioned according to the longitudinal and transverse reinforcement ratios.

Numerical integration schemes were considered to model the discrete crack patterns for reinforced concrete beams. The truss model geometry is

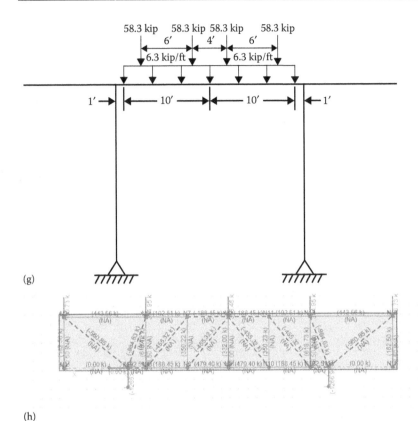

(g)

(h)

Figure 13.15 (Continued) (g) Factored live load configuration to maximize positive moments. (h) Loading resulting from maximizing of positive moments. (Data from Martin Jr., B.T. and Sanders, D.H., "Analysis and Design of Hammerhead Pier Using Strut and Tie Method," Final Report-Project 20-07_Task 217, National Cooperative Highway Research Program, Transportation Research Board, Washington, DC, November, 2007.)

defined by first locating the node coordinates. The horizontal coordinates of the boundary nodes is either defined (1) by an applied load or bearing support or (2) at the intersecting lines of thrust from the beam and column members. The transverse tension ties in the truss mechanism are then located according to the selected numerical truss (single-point Gauss quadrature).

Each member in the C-STM is comprised of two elements that model the individual behavior of steel and concrete in that member. The two elements are constrained together to give the combined steel–concrete response. After assigning axial rigidities of steel and concrete elements and defining nonlinear constitutive material relations, the C-STM could be applied to any nonlinear structural analysis software. The C-STM can be modeled

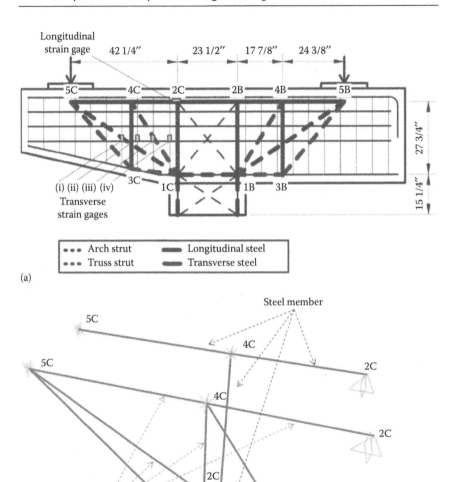

Figure 13.16 (a) C-STM model of a deep bent cap by Scott et al. (2000). (Data from Scott, R.M et al., *ACI Structural Journal*, 109, 635–644, 2012.) (b) C-STM model of a deep bent cap by SAP2000. (Data from SAP2000, "Integrated Software for Structural Analysis & Design," Computer and Structures, Inc., Berkeley, CA, 2007.)

using separate trusses with nodes constrained together to give the combined steel–concrete member response. This is most easily simulated by duplicating the assigned nodes in the out-of-plane axis to form two separate trusses and constraining the degree of freedom for each of the duplicate

Table 13.1 Variables of the deep bent cap example

L mm (in)	j_d mm (in)	L/j_d	ρ_T	ρ_L	η	A_s mm² (in²)	A'_s mm² (in²)	b_w mm (in)	d mm (in)	k_d mm (in)	d' mm (in)	α	N_h	A_{sh} mm² (in²)
1073 (42.25)	705 (27.76)	1.52	0.0030	0.0073	0.55	4813 (7.46)	4051 (6.28)	838 (33)	787 (31)	193 (7.6)	83 (3.25)	33.3°	4	396 (0.614)

L, the beam length; b_w, the beam width; d, the effective depth of the beam from the extreme concrete compression fiber to the centroid of the tension steel; d', distance from extreme compression fiber to the centroid of longitudinal compression reinforcement; s, the stirrup spacing; j, the internal lever arm coefficient, which, in lieu of a more precise analysis, may be taken as $j = 0.9$; A_s, the area of longitudinal tension reinforcement; A'_s, the area of compression reinforcement; A_{sh}, the area of one set of stirrups; ρ_L, the volumetric ratio of longitudinal steel to concrete, $\rho_L = A_L/b_w d$, where A_L is the area of longitudinal reinforcement contributing to the tension tie; ρ_T the volumetric ratio of transverse steel to concrete over one hoop spacing $\rho_T = A_{sh}/b_w s$; η, the arch breadth scalar, used to apportion the contribution of arch-and-truss action defined as a function of the longitudinal and transverse reinforcement and the member's span–depth ratio, given by

$$\eta = \frac{\rho_L f_y}{\rho_L f_y + \rho_T f_{yh} j \cot^2 \alpha}$$

k, the elastic compression zone coefficient

$$k = \sqrt{\left(\rho_L + \rho'_L\right)^2 n^2 + 2\left(\rho_L + \frac{\rho'_L d'}{d}\right)n + \left(\rho_L + \rho'_L\right)r};}$$

N_h, integer part of active hoops in truss mechanism $N_h = \mathrm{int}(L/s - 1)$; α, the corner-to-corner diagonal angle.

Table 13.2 Axial rigidity assignments of the SAP2000 model

Member	Steel		Concrete		Comments
	E_s ksi (GPa)	A_s mm² (in²)	E_c ksi (GPa)	A_c mm² (in²)	
2-4 4-5	$E_s = 29{,}000\ (200)$	$A_s = 4813\ (7.46)$	$E_c = 4490\ (30{,}960)$	$b_w kd = 161{,}740\ (250.7)$	Tension chord
1-3	$E_s = 29{,}000\ (200)$	$A'_s = 4051\ (6.28)$	$\psi_e E_c = 3680\ (25{,}373)$	$b_w kd = 161{,}740\ (250.7)$	Compression chord
3-4	$E_s = 29{,}000\ (200)$	$N_h A_{sh} = 1584\ (2.46)$	$E_c = 4490\ (30{,}960)$	$(4c + 2d)N_h s = 165{,}350\ (256.3)$	Active hoop steel including tension stiffening
1-5	—	—	$E_c = 4490\ (30{,}960)$	$\dfrac{0.375\eta b_w\, jd}{\cos\alpha} = 144{,}645\ (224.2)$	Concrete strut in arch mechanism
1-4	—	—	$E_c = 4490\ (30{,}960)$	$\dfrac{0.5(1-\eta)b_w\, jd}{\sqrt{0.423+\tan^2\alpha}} = 145{,}390\ (220.7)$	Concrete strut in truss mechanism
3-5	—	—	$E_c = 4490\ (30{,}960)$	$\dfrac{0.5(1-\eta)b_w\, jd}{\sqrt{0.577+\tan^2\alpha}} = 131{,}100\ (203.2)$	Concrete strut in truss mechanism

A_c, the area of concrete in compression; b_w, the beam width; d, the effective depth of the beam from the extreme concrete compression fiber to the centroid of the tension steel; s, the stirrup spacing; j, the internal lever arm coefficient, which, in lieu of a more precise analysis, may be taken as $j = 0.9$; E_c, Young's modulus for concrete; E, Young's modulus for steel; A_s, the area of longitudinal tension reinforcement; A'_s, the area of compression reinforcement; A_{sh}, the area of one set of stirrups; k, the elastic compression zone coefficient

$$k = \sqrt{\left(\rho_L + \rho'_L\right)^2 n^2 + 2\left(\rho_L + \rho'_L\frac{d'}{d}\right)n} + (\rho_L + \rho'_L)n;$$

ψ_e, strain compatibility coefficient, in lieu of a more precise analysis, 0.6 is recommended; N_h, integer part of active hoops in truss mechanism $N_h = \mathrm{int}(L/s-1)$; α, the corner-to-corner diagonal angle.

nodes. The steel and concrete elements are then drawn with pinned-end connections between the appropriate node points.

An example of a cracked deep reinforced concrete bent cap by C-STM is illustrated here. Figure 13.16a represents the applied C-STM, where the suffies "C" and "B" refer to the tapered cantilever and beam ends, respectively. A finite element model of the tapered cantilever was established in SAP2000 (2007) to be analyzed, as illustrated in Figure 13.16b. Table 13.1 shows physical and material variables of the deep bent cap example. Based on those variables, axial rigidities are calculated and listed in Table 13.2, which are then assigned to the SAP2000 (2007) model.

Chapter 14

Stability

14.1 BASICS OF STRUCTURAL STABILITY

Structural stability is the ability of a structure to resist loading. Loss of such ability, so-called instability, is a state in which the structure is no longer in equilibrium with change in the geometry of a structure or structural component under loads. One phenomenon of structural failure led by instability is excessive structural displacements or component deformations. The underlying causes are the loss of stiffness in some particular degrees of freedom due to geometric and/or material constitutional reasons, that is, geometric and material nonlinearities.

According to the principle of minimum total potential energy, a structure is in equilibrium when the total energy no longer changes or the first-order derivative of the total energy to displacements equals to zero. As illustrated in Section 3.2.1, Equation 3.1 (or Equation 12.4 where $\partial \Pi / \delta d = 0$), which leads to the establishment of global equilibrium equation 3.3, reveals any possible state that makes the total energy minimal or maximal (locally or globally). Further, the value of the second-order derivative tells the trend of total energy changes as shown in Figure 14.1 and Equations 14.1 through 14.3. The engineering purpose of stability analyses is to find any practical solution for Equation 3.3, or a state, that meets Equation 14.2 or 14.3.

$$\frac{\delta^2 \Pi}{\delta d^2} > 0 \quad \text{The solution of Equation 3.3 is structurally stable} \quad (14.1)$$

$$\frac{\delta^2 \Pi}{\delta d^2} = 0 \quad \text{The solution of Equation 3.3 is in a state of unknown} \quad (14.2)$$

$$\frac{\delta^2 \Pi}{\delta d^2} < 0 \quad \text{The solution of Equation 3.3 is structurally unstable} \quad (14.3)$$

From the perspective of stiffness matrix analysis in the global equilibrium formulation (Equation 3.3 where $[K_0 + K_\sigma + K_L]da = F$), Equations 14.2

435

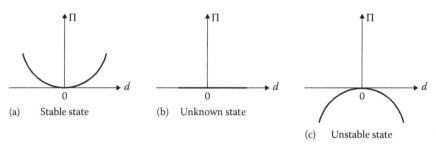

Figure 14.1 (a–c) States of structural equilibrium.

and 14.3 are equivalent to any diagonal element being zero and being less than zero, respectively. Based on the making of the global stiffness and its changing from positive to zero or even negative, the instability of a structure can be in the following three categories:

1. Buckling. Scenarios where the stiffness change due to the large displacement is ignored ($K_L = 0$), and when evaluating elastic matrix D in Equation 3.12, a constant Young's modulus E is assumed, that is, small displacements and elastic material. Only the stiffness of initial stress K_σ is considered. Therefore, buckling is an elastic stability problem in which the stiffness due to geometric change is ignored. When buckling happens, the structure suddenly changes to an unstable or unknown state. As a point clearly divides the structural states from stable to unstable, buckling is often referred to as bifurcation buckling and the loads to reach this point are called critical loads. A column or beam under compression as shown in Figure 14.2a is a typical buckling problem. By solving general eigenproblem as shown

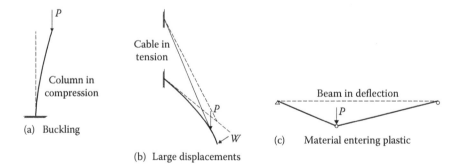

Figure 14.2 (a–c) Categories of structural instability.

in Equation 14.4, the critical load factor (eigenvalue) and displacement patterns of buckling (eigenvector) can be obtained.

$$|K_0 + \lambda K_\sigma| = 0 \tag{14.4}$$

2. Excessive displacements. Scenarios where the stiffness changes due to both initial stress and large displacements are considered and when evaluating elastic matrix D in Equation 3.12, a constant E is assumed, that is, large displacements and elastic material. The development of excessive displacements is gradual. The bifurcation point as in category (1) does not exist. Cable-stayed structures under certain load combinations as shown in Figure 14.2b can develop excessive displacements. For long-span bridges, as large displacements are more significant than inelastic material, excessive displacements under certain load combinations should be investigated.

3. Collapse. Scenarios that are the same as category (2), but when evaluating elastic matrix D in Equation 3.12, the tangent at the current strain position on material constitutive curves is used instead of a constant E, that is, inelastic material. Similar to that of category (2), no bifurcation point exists in the equilibrium changes from stable to unstable. Figure 14.2c shows a simple example of structural collapse due to inelastic material. Depending on the material property, collapse could happen before large displacements develop. As this type of instability is due to material entering the inelastic stage, the ultimate load leading to collapse or structural failure is often called limited state capacity or ultimate collapse capacity. It is common to conduct limited state capacity analyses for middle- and short-span bridges. For particular types of structures, such as PC/RC girder bridges, stiffness changes due to initial stress and large displacement can be simply ignored so as to simplify the iterations. This type of instability is not covered in this book.

14.2 BUCKLING

Buckling means loss of the stability of an equilibrium configuration, without fracture or separation of the material or at least prior to it (Cook et al. 2002). In general, there are two types of buckling: bifurcation buckling and snap-through buckling. Bifurcation buckling is the type of buckling based on the elementary column theory where a straight prebuckling configuration under a critical load P_{cr} is no longer in a stable state of equilibrium and may also be in a different buckled configuration. As shown in Figure 14.3, the primary path is following the original load–displacement curve and its extension. Also shown in the same figure, the secondary path

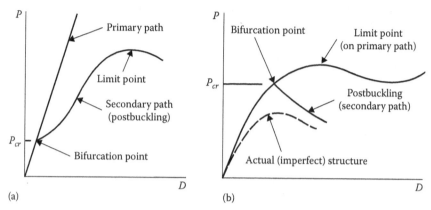

Figure 14.3 Possible load versus displacement behavior of thin-walled structures. (a) Linear prebuckling path and rise postbuckling path. (b) Nonlinear prebuckling path and drop postbuckling path. (From Cook, R.D. et al.: *Concepts and Applications of Finite Element Analysis*, 4th edition, New York, 2002. Copyright Wiley-VCH Verlag GmbH & Co. KGaA. Reproduced with permission.)

is the alternative path that originates when the critical load is reached. The two paths intersect at the bifurcation point. Once past the bifurcation point, the primary path is unstable. It is possible that mathematically the structure follows the primary path, whereas the real structure follows the secondary path. If the secondary path has a positive derivative (rises), the structure has postbuckling strength (Figure 14.3a). A limit point is a maximum on a load–displacement curve, but this point is not a bifurcation point because there is no immediate adjacent equilibrium configuration. When a limit-point load is reached under increasing load, snap-through buckling occurs, as the structure assumes a new configuration. A collapse load is the maximum load a structure can sustain without gross deformation. It may be greater or less than the computed bifurcation buckling load as shown in Figure 14.3.

Linear perturbation analyses can be performed from time to time during a fully nonlinear analysis by including the linear perturbation steps between the general response steps. The linear perturbation response has no effect as the general analysis is continued. If geometric nonlinearity is included in the general analysis on which a linear perturbation study is based, stress stiffening or softening effects and load stiffening effects are included in the linear perturbation analysis.

The loads for which the stiffness matrix becomes singular are searched by an eigenvalue buckling problem. Equation 14.4 has nontrivial solutions where K_0 is the tangent stiffness matrix when the loads are applied, and K_σ is the initial stress stiffness. Eigenvalue buckling is generally used to estimate the critical buckling loads of stiff structures, for example, structures carrying their loads primarily by axial or membrane action. Even when the response of a structure is nonlinear prior to collapse, a general eigenvalue or linear buckling analysis can provide useful estimates of collapse mode shapes. Generally speaking, eigenvalue analysis is a straightforward problem. However, some structures have many buckling modes with closely spaced eigenvalues, which can cause numerical problems. In these cases it often helps to apply enough preload, just below the buckling load, before performing the eigenvalue extraction. In many cases a series of closely spaced eigenvalues indicate that the structure is imperfection sensitive.

In mathematics, an eigenvalue of Equation 14.4 indicates that at least one diagonal element in the sum matrix becomes zero when K_σ is amplified by that time. In structures, it means the critical point has been reached if applied load has been multiplied by a factor of eigenvalue. In engineering, it is meaningful only when its associated load is clearly defined. For example, when K_σ is due to all structural weights, the first eigenvalue (λ) predicts that the structure will lose its stability if all structural weights are equally multiplied by a factor of λ. If an analysis is to know how many times a live load will cause buckling, K_0 and K_σ in Equation 14.4 should be adjusted accordingly. To accurately predict the buckling load, a special-purpose finite element analysis (FEA) package, which can sum K_σ at one stage due to certain loads into K_0 and compute K_σ at another stage due to another load, should be employed. Taking a cable-stayed bridge as an example, K_0 in Equation 14.4 should be able to include all the initial stresses accumulated from the first construction stage until the deck is superimposed, and K_σ in Equation 14.4 counts for only one particularly extreme live load. Therefore, the eigenvalue may predict a meaningful engineering safety factor.

14.2.1 Linear buckling of a steel plate

14.2.1.1 Formulation of plate buckling

In this section, plate buckling theory is discussed. The von Karman large deflection equations for flat isotropic plates with in-plane loading were modified to account for anisotropy by Rostovtsev (1940), and later the effect of initial imperfections were included resulting in the following simultaneous equations, which are considered the most general equations currently available for solving plate buckling problems (Murray 1984):

$$D_x \frac{\partial^4 \omega}{\partial x^4} + 2H \frac{\partial^4 \omega}{\partial x^2 \partial z^2} + D_z \frac{\partial^4 \omega}{\partial z^4} = \frac{\partial^2 \phi}{\partial z^2} \frac{\partial^2 (y + \omega)}{\partial x^2} + \frac{\partial^2 \phi}{\partial x^2} \frac{\partial^2 (y + \omega)}{\partial z^2}$$

$$-2 \frac{\partial^2 \phi}{\partial x \partial z} \frac{\partial^2 (y + \omega)}{\partial x \partial z} + q \qquad (14.5)$$

$$\frac{1}{t_z E_z}\frac{\partial^4 \phi}{\partial x^4} + 2\left(\frac{1}{K_{xz}} - \frac{v_x}{t_x E_x} - \frac{v_z}{t_z E_z}\right)\frac{\partial^4 \phi}{\partial x^2 \partial z^2} + \frac{1}{t_x E_x}\frac{\partial^4 \phi}{\partial z^4} = \frac{\partial^2 y}{\partial z^2}\frac{\partial^2 \omega}{\partial x^2}$$

$$+2\frac{\partial^2 y}{\partial x \partial z}\frac{\partial^2 \omega}{\partial x \partial z} - \frac{\partial^2 y}{\partial x^2}\frac{\partial^2}{\partial z^2} - \frac{\partial^2 \omega}{\partial z^2}\frac{\partial^2 \omega}{\partial x^2} + \left(\frac{\partial^2 \omega}{\partial x \partial z}\right)^2 \qquad (14.6)$$

where:
ω is the lateral deflection
ϕ is the stress function
q is the load intensity on the plate
D_x, D_z are plate stiffnesses
E_x, E_z are moduli of elasticity
t_x, t_z are the thicknesses of plate
v_x, v_z are the Poisson's ratios

14.2.1.2 Solving plate and box girder buckling problem

The high bending moments and shearing forces for long-span bridges may consider the use of fabricated plate and box girders. In their simplest form, plate and box girders can be considered as an assemblage of webs and flanges. To reduce the self-weight of these girders, slender plate sections are employed. Hence the local buckling and postbuckling reserve the strength of plates, they are important design criteria. For the efficient use of thin plates, flanges and webs in a box girder are often reinforced with stiffeners. However, there are some difficulties that are usually encountered by the designers of plated structures (Ryall et al. 2000):

- The engineer's simple "plane sections remain plane" theory of bending is no longer adequate, even for linear elastic analysis.
- Nonlinear elastic behavior caused by the buckling of plates can be of great importance and must be allowed for.
- Because of this complex nonlinear elastic behavior as well as stress concentration problems, some yielding may occur at loads that are quite low in relation to ultimate collapse loads. While such yielding may not be of great significance with regard to rigidity and strength, it means that simple maximum stress criteria are no longer sufficient.
- Because of the buckling problem in plates and stiffened panels, complete plastification is far from being realized at collapse. Hence simple plastic criteria are also not sufficient.
- Complex interactions occur between flanges, webs, and diaphragms, and the pattern of this interaction can change as the level of load increases.

To demonstrate the linear buckling problem, a rectangular plate is compressed in its middle plane by forces uniformly distributed along the sides $x = 0$ and $x = a$, as shown in Figure 14.4.

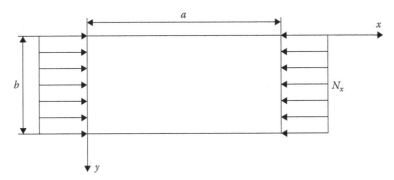

Figure 14.4 Simply supported rectangular plate uniformly compressed in one direction.

Thus, the expression for the critical value of the compressive force $N_{x,cr}$ can be simplified as

$$N_{x,cr} = \frac{\pi^2 D}{a^2} \left(m + \frac{1}{m} \frac{a^2}{b^2} \right)^2 \tag{14.7}$$

Equation 14.7 with $m = 1$ can be represented in the form

$$N_{x,cr} = k \frac{\pi^2 D}{a^2} \tag{14.8}$$

where k is a factor depending on the ratio a/b and is shown in Figure 14.5 by the curve marked $m = 1$. The critical value of the compressive stress σ_{cr} is then given by

$$\sigma_{cr} = \frac{N_{x,cr}}{h} = \frac{k\pi^2 E}{12(1 - v^2)} \frac{h^2}{b^2} \tag{14.9}$$

where
    h is the thickness of the plate
    a is the length
    b is the width
    m is the number of half-waves in which the plate buckles have been determined

14.2.2 Linear buckling of steel members

14.2.2.1 Buckling of steel structure members

Steel members in compression in a truss structure have to be analyzed for buckling loads. Usually buckling becomes a governing criterion in structures like arched bridges, guyed towers, the top chord of a pony truss, or any other unbraced compression member.

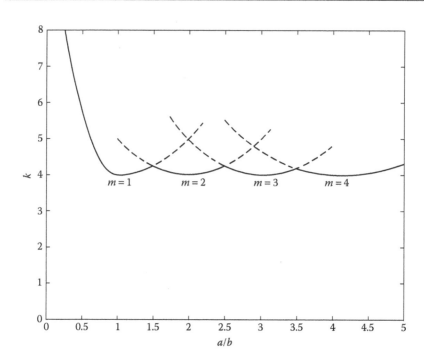

Figure 14.5 Buckling stress coefficients for uniaxially compressed plate. (Data from Ryall, M.J. et al., *Manual of Bridge Engineering*, Thomas Telford Publishing, London, 2000.)

In this chapter, the pony truss, a half-through bridge truss that has its deck between the top and bottom chords and has no top lateral bracing, is used as an example. A pony truss can be idealized as a continuous beam with intermittent spring support (Figure 14.6). The stiffness of these spring supports will depend on the vertical and diagonal members of the truss and floor beams. A method for solving the buckling of a continuous beam on elastic foundation was suggested by Timoshenko (1936).

Many classical methods were developed for solving the buckling problem, but most of them are based on the idealization of a bridge as a continuous beam on elastic foundation. In this chapter, the method of finding a buckling load of a pony truss bridge as suggested by Timoshenko (1936) is illustrated. Another effective method, which gave comparable results but is not listed here, was established by the Structural Stability Research Council (SSRC) Guide (Galambos 1998). A case study of a 27-m (90′) pony truss is considered, and the results are compared with those based on an ANSYS numerical model.

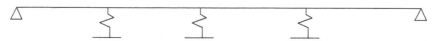

Figure 14.6 Pony truss idealized as a continuous beam on spring support.

14.2.2.2 Buckling analysis of a pony truss by Timoshenko's method

Length of vertical members $= l$ (as shown in Figure 14.7)
Modulus of elasticity $= E$
Moment of inertia of vertical members $= I_v$
Length of floor beam $= d$ (as shown in Figure 14.7)
Moment of inertia of floor beam $= I_b$
Length of diagonal members $= a$
Moment of inertia of diagonals $= I_d$
Length of each panel $= 2c$
Total length of all top chord members $= L$
Moment of inertia of top chord members $= I_t$

1. Calculate the modulus of equivalent elastic foundation
   a. Vertical members

$$A = \frac{l^3}{3EI_v} \qquad B = \frac{l^2 d}{2EI_b} \qquad R_{01} = \frac{1}{A+B}$$

   b. Diagonal members

$$A = \frac{a^3}{3EI_v} \qquad B = \frac{l^2 d}{2EI_b} \qquad R_{02} = \frac{2}{A+B}$$

   c. Considering all parts

$$I_{eq} = I_v + 2I_d \left(\frac{l}{a}\right)^3$$

$$A = \frac{l^3}{3EI_{eq}} \qquad B = \frac{l^2 d}{2EI_b} \qquad R_{02} = \frac{1}{A+B}$$

2. Calculating the buckling load
   Calculate b

$$b = \frac{R_0}{c} \qquad P_e = \frac{\pi^2 EI_t}{L}$$

Calculate $bL^4/16EI_t$ and find out $1/m$ from the following table by interpolating

$bL^4/16EI_t$	0	5	10	15	22.8	56.5	100	162.8	200	300	500	1000
$1/m$		0.696	0.524	0.443	0.396	0.363	0.324	0.29	0.259	0.246	0.225	0.204	0.174

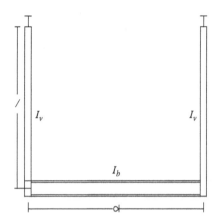

Figure 14.7 Floor beam, vertical members, and diagonal members of a pony truss bridge.

Calculate the buckling load as

$$P_{cr} = m^2 P_e$$

14.2.2.3 Case study of pony truss by Timoshenko's method

The following bridge has been considered for the case study:

Length of vertical members $= l = 120''$ (3048 mm)
Modulus of elasticity $= E = 29,000$ ksi (199,955 MPa)
Moment of inertia of vertical members $= I_v = 9.906$ mm $\times 10^7$ mm (238 in⁴)
Length of floor beam $= d = 6756$ mm (266'')
Moment of inertia of floor beam $= I_b = 1.361$ mm $\times 10^9$ mm (3270 in⁴)
Length of diagonal members $= a = 3810$ mm (150'')
Moment of inertia of diagonals $= I_d = 9.906$ mm $\times 10^7$ mm (238 in⁴)
Length of each panel $= 2c = 180''$ (4572 mm)
Total length of all top chord members $= L = 29,041$ mm (1143.36'')
Moment of inertia of top chord members $= I_t = 2.219$ mm $\times 10^8$ mm
   (533 in⁴)

1. Calculate the modulus of equivalent elastic foundation
   a. Vertical members

$$A = \frac{l^3}{3EI_v} = 0.0835 \text{ in/kip}$$

$$B = \frac{l^2 d}{2EI_b} = 0.025 \text{ in/kip}$$

$$R_{01} = \frac{1}{A+B} = 9.22 \text{ kip/in}$$

b. Diagonal members

$$A = \frac{a^3}{3EI_v} = 0.163 \text{ in/kip}$$

$$B = \frac{l^2 d}{2EI_b} = 0.037 \text{ in/kip}$$

$$R_{02} = \frac{2}{A+B} = 9.98 \text{ kip/in}$$

c. Considering all parts

$$I_{eq} = I_v + 2I_d \left(\frac{l}{a}\right)^3 = 481.712 \text{ in}^4$$

$$A = \frac{l^3}{3EI_{eq}} = 0.041 \text{ in/kip}$$

$$B = \frac{l^2 d}{2EI_b} = 0.025 \text{ in/kip}$$

$$R_{02} = \frac{1}{A+B} = 15.099 \text{ kip/in}$$

2. Calculating the buckling load
Calculate b

$$b = \frac{R_0}{c} = 0.08388$$

$$P_e = \frac{\pi^2 EI_t}{L} = 116.6966 \text{ kip}$$

$$\frac{bL^4}{16EI_t} = 579.658$$

Hence from the table: $1/m = 0.198$

$$P_{cr} = m^2 P_e = 2988.66 \text{ kip } (13,293.6 \text{ kN})$$

$b^{\mu}/16EI_t = 579.658$
From Timoshenko Table 2.9, $1/m = 0.198$
Then, $m = 5.061$
$P_{cr}/P_e = m^2 = 25.61$
$P_{cr} = m^2 P_e = 2988.66$ kip (13,293.6 kN) $> 2.12 P_{max} = 866.2956$ kip (3853.3 N)
$P_{max} = 408.63$

The P_{cr} calculated here, 2988.66 kip, is far above $1.5P_{max}$ and even greatly exceeds $2.12P_{max}$ allowed by AASHTO (2013). It can be concluded that the response in a linear analysis step is the linear perturbation response about the *base state*. The base state is the current state of the model at the end of the last general analysis step prior to the linear perturbation step. If the first step of an analysis is a perturbation step, the base state is determined from the initial conditions.

14.3 FEM APPROACH OF STABILITY ANALYSIS

A technique of seeding the finite element mesh with an initial displacement field is employed in this study to initiate out-of-plane deformations of the flat compression panels. In this technique, the finite element mesh is subjected to a linearized buckling analysis to obtain the first buckling mode. The displacement field associated with this lowest mode is then superimposed on the finite element model as a seed imperfection for use in the incremental nonlinear analysis.

As previously discussed in Chapter 3, for stiffness analysis, K_T, the total tangential stiffness matrix is the sum of three terms: (1) K_0, the usual, small displacements stiffness matrix; (2) K_σ, initial stress matrix or geometric matrix; and (3) K_L, the initial displacement matrix or large displacement matrix. For short-span bridges, if the large deformation is ignored, the total tangential stiffness will have only K_0, the elastic, small displacement stiffness matrix, and K_σ, the initial stress stiffness matrix. For a long-span cable-stayed bridge, as the axial forces along the pylon and the girder are in compression, K_σ will reduce K_T. If the loads that cause the initial stress, usually the structural weight and cable stressing, keep increasing, a critical point will be reached, at which the determinant of the total stiffness matrix is zero.

Such a bifurcation stability problem can be solved as an eigenvalue problem (Tang 1976; Ermopoulos 1992). In actual situations, however, it rarely happens due to the flaws in building the structure. K_L should also be considered, and the full Newton–Raphson process is required. In some typical situations, it is easy to understand. For example, the transverse stability due to the live load of a vertically stayed cable bridge under transverse wind loads will be enhanced after the deck moves laterally away from the centerline. Not only the tension, the positive K_σ but also the laterally sloped geometry K_L of the cables will enhance the lateral stiffness.

Again, the stability analysis of a long-span cable-stayed bridge can be combined with its nonlinear analysis. The analysis of a long-span cable-stayed bridge with a main span of 1088 m, however, shows that the statically geometrical nonlinear stability analysis is not sufficient. The total tangential stiffness, with K_L included, hardly reaches zero. This suggests that aerodynamic stability analysis and the geometric plus material nonlinear analyses are required (Ren 1999). When material nonlinearity is considered, a uniaxial representation of the bilinear elastic and perfectly plastic steel constitutive law is employed. The von Mises yield criterion, which is considered most suitable for structural steels, can be selected to extrapolate a yield surface in three-dimensional (3D) principal stress space.

A full nonlinear stability analysis provides greater accuracy by incrementally increasing load application until a structure becomes unstable. This condition of instability is achieved when a small increase in the load level causes a very large change in displacement. Nonlinear stability analysis is a static method that accounts for material and geometric nonlinearities, load perturbations, geometric imperfections, and gaps. Either a small destabilizing load or an initial imperfection is necessary to initiate the solution of a desired buckling mode.

A nonlinear analysis requires incremental load steps in an explicit or implicit manner. At the end of each increment, the structure geometry changes and possibly the material is nonlinear or the material has yielded. An explicit nonlinear analysis performs the incremental procedure, and at the end of each increment updates the stiffness matrix based on the geometry changes and material changes (if applicable). An implicit nonlinear analysis does the same thing but uses Newton–Raphson iterations to enforce equilibrium, which is the primary difference between the two types of analyses. Either explicit or implicit nonlinear static analysis can be used. However, for nonlinear stability analysis, the implicit method is preferred.

14.4 3D ILLUSTRATED EXAMPLE WITH LINEAR BUCKLING ANALYSIS OF A PONY TRUSS, PENNSYLVANIA

This example is to verify the hand calculation of a pony truss bridge shown in Section 14.2.2 by eigenvalue buckling analysis. Eigenvalue buckling analysis done by ANSYS predicts the theoretical buckling loads of an ideal elastic structure by performing classical Euler buckling analysis. Eigenvalues are computed for the given structure with the given boundary conditions and loading. The cross section and the perspective view of the bridge are shown in Figures 14.7 and 14.8, respectively.

Figure 14.8 Pony truss bridge.

Solving a 3D model on ANSYS shown in Figure 14.9, the following eigenvalue results were obtained:

Set	Time/Freq
1	0.31468E+07
2	0.34171E+07
3	0.34276E+07
4	0.34995E+07
5	0.37006E+07

Hence the buckling load = 3146.8 kip (13,997 kN), which is close to 2988.66 kip (13,293.6 kN) as calculated from Timoshenko's method. Figures 14.10 and 14.11 show the different buckling modes.

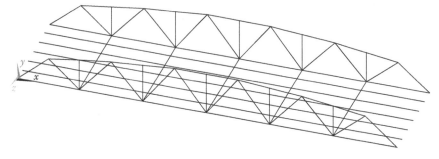

Figure 14.9 ANSYS model of the pony truss bridge.

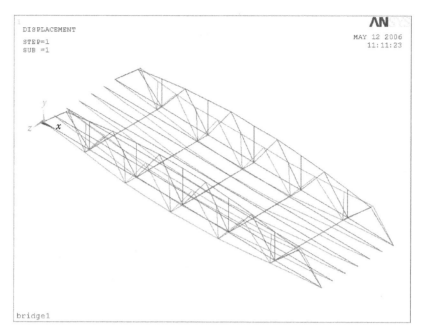

Figure 14.10 First mode of the pony truss bridge buckling.

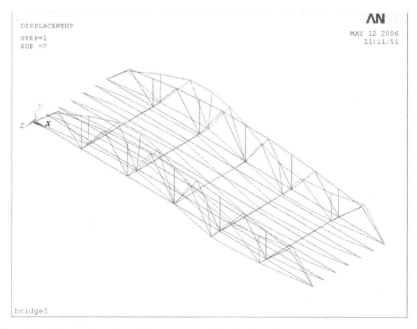

Figure 14.11 Second mode of the pony truss bridge buckling.

14.5 3D ILLUSTRATED EXAMPLE WITH LINEAR BUCKLING ANALYSIS OF A STANDARD SIMPLE ARCH RIB

This example is to demonstrate the basic stability analysis—linear, or elastic, buckling analysis. In this example, a prismatic single arch rib with a span of 50 m is fixed at both ends. The geometry of the rib axis is a parabolic curve with a chord height-to-span length ratio of 0.3. Moment inertia on both in-plane and out of plane are equivalent. Both weight-equivalent cross-sectional area and stiffness-equivalent cross-sectional area are the same too. The material is concrete.

Figures 14.12 through 14.14 show the first three modes of linear buckling analysis, respectively. The ratios of critical loads to the applied loads, the structural weight, are 408.516, 1046.208, and 1259.367, respectively. Table 14.1 shows the comparison between VBDS (Wang and Fu 2005), a special-purpose bridge FEA package, and a theoretical formula (Li 1996). Values in the table are converted from critical load factors (λ).

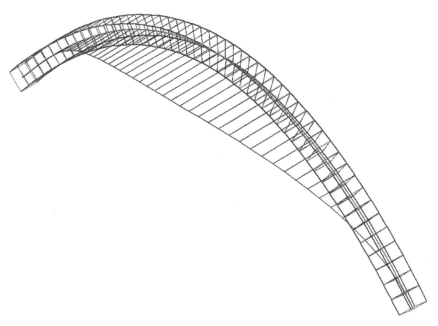

Figure 14.12 The first mode of a simple arch bridge bulking, out of plane ($\lambda = 408.516$).

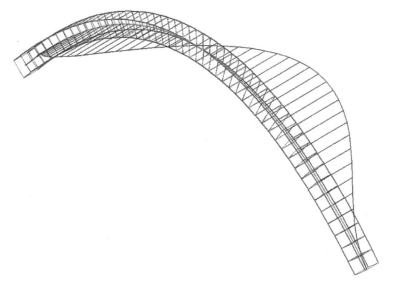

Figure 14.13 The second mode of a simple arch bridge bulking, out of plane ($\lambda = 1046.208$).

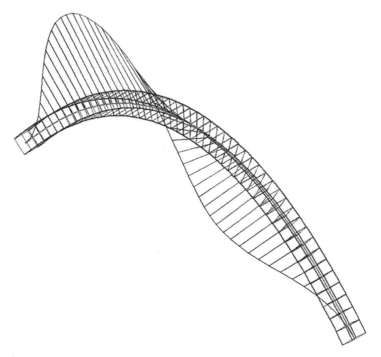

Figure 14.14 The third mode of a simple arch bridge buckling, in-plane ($\lambda = 1259.367$).

Table 14.1 Comparison of the buckling load of a simple arch bridge (kN/m)

Category	VBDS	Li's theoretical method
First mode (lateral)	10,213	10,780
Third mode (plane)	31,484	32,200

14.6 3D ILLUSTRATED EXAMPLE WITH LINEAR BUCKLING ANALYSIS OF A PROPOSED TIED-ARCH BRIDGE—LINYI, PEOPLE'S REPUBLIC OF CHINA

In this example, a tied-arch bridge with steel tube concrete ribs is used to illustrate selections of load cases to form the initial stress stiffness matrix so as to conduct meaningful linear buckling analyses.

Figure 14.15 shows the main dimensions of the bridge. The ribs are steel tubes filled with concrete, the hangers are high-strength steel strands, the tie girders are post-tensioned concrete girders, and the lateral wind bracing beams are hollow steel tubes. The construction sequences include the following three main stages:

1. Stage 1. Cast tie girders and lateral beams with temporary supports at each hanger location and post-tensing tie girders.
2. Stage 2. Install ribs and lateral wind bracing beams, then install hangers, and fill the rib concrete and first-time jacking hangers.
3. Stage 3. Build deck and its attachments and jack the hangers and finalize their jacking stress levels.

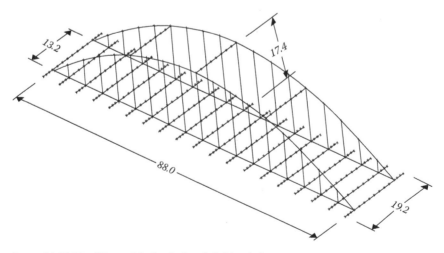

Figure 14.15 The 3D model of a tied-arch bridge (m).

The load cases contain the following:

1. Structural weights of all installed components at stage 1
2. Structural weights of all newly installed components at stage 2
3. Hanger tuning in stage 2
4. Superimposed deck loads in stage 3
5. Hanger tuning in stage 3

All other FEA-related properties are not listed here.

This example includes many analyses such as stage changing, hanger tuning, and live loading. The stability-related analyses include (1) finding the live loads that make the compression on top of one rib maximal and (2) comparing buckling load factors regarding different acting loads and whether or not accumulated initial stresses are considered.

One live load that causes compression on top of one rib maximal is analyzed. As a 3D model and influence surface loading are used, the lateral distribution of live loads is clearly displayed by the axial force distribution and structure displacements as shown in Figures 14.16 and 14.17, respectively. Figures 14.18 through 14.20 show the first three modes of buckling considering only the extreme live loads in the initial stress stiffness matrix (K_σ) and including total accumulated initial stress stiffness (from stages 1 to 3) in K_0. The buckling pattern (eigenvectors) are all out of plane, indicating that the ribs have a much higher stiffness in vertical plane than in

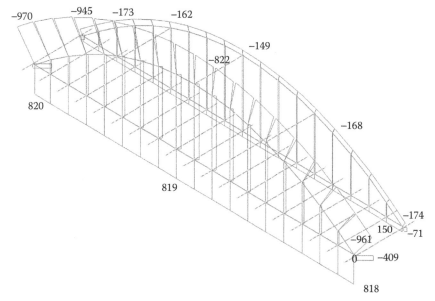

Figure 14.16 Axial forces (kN) under live loads that cause the compression on the top of one rib maximal.

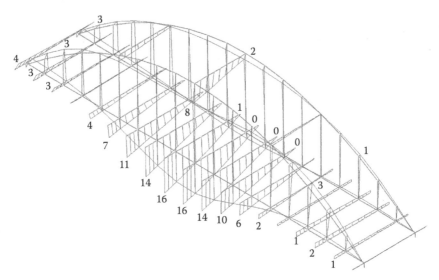

Figure 14.17 The correspondent displacements (mm) under live loads that cause the compression on the top of one rib maximal.

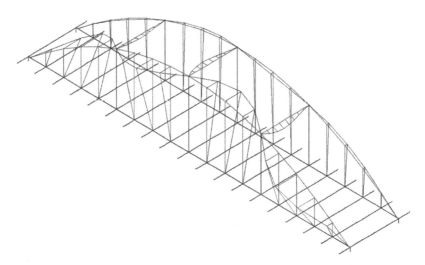

Figure 14.18 The first mode of buckling (out of plane) considering only the extreme live loads in initial stress stiffness matrix (K_σ) and including total accumulated initial stress stiffness (from stages 1 to 3) in K_0.

lateral plane. The corresponding critical load factors (λ) are 51.00, 62.4, and 95.28, respectively. The critical load factor 51.00 of the first mode, for example, means the arch bridge would enter the first bifurcated point when the live loads are increased by 51 times. Note that it is increased by 51 times, not 50 times, as the initial stress is already accumulated in K_0.

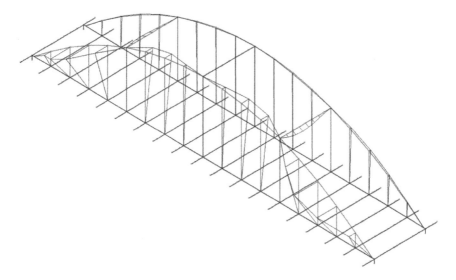

Figure 14.19 The second mode of buckling (out of plane) considering only the extreme live loads in initial stress stiffness matrix (K_σ) and including total accumulated initial stress stiffness (from stages 1 to 3) in K_0.

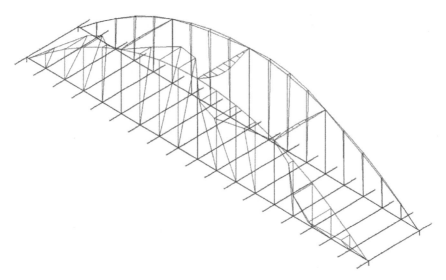

Figure 14.20 The third mode of buckling (out of plane) considering only the extreme live loads in initial stress stiffness matrix (K_σ) and including total accumulated initial stress stiffness (from stages 1 to 3) in K_0.

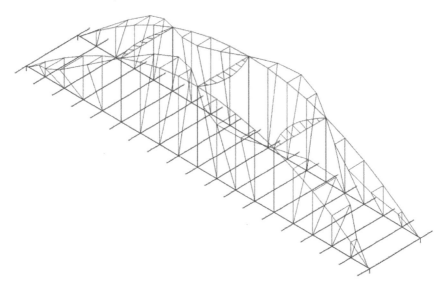

Figure 14.21 The first mode of buckling (out of plane) considering all loads accumulated up to the current stage in initial stress stiffness matrix (K_σ) and including total accumulated initial stress stiffness (from stages 1 to 3) in K_0.

Figure 14.21 shows the first mode of buckling considering all loads accumulated up to stage 3 in the initial stress stiffness matrix (K_σ) and including total accumulated initial stress stiffness (from stages 1 to 3) in K_0. The buckling pattern is out of plane. The corresponding critical load factor (λ) is 4.23, much lower than that when considering only live loads, indicating that the critical load factors in bifurcated buckling analysis is engineering meaningful only when the acting loads are clearly defined. Also, the nature of acting loads is shown by the difference of Figure 14.21 from Figures 14.18 through 14.20.

Because most bridges are built in many stages, whenever the initial stress stiffness is evaluated in any stage, the initial stress should be accumulated from the first stage to the stage prior to (or upto) the current stage. Also, the initial stress stiffness should be able to include the linear stiffness matrix (as the so-called initial stress considered). Moreover, it has to be able to pick a particular load case as the acting load case in buckling analysis. Further, to be more practical, the analyzed extreme live loads should be able to be saved as load cases. To simplify, (1) when computing K_0 of Equation 14.4, stiffness due to accumulated initial stress should be able to be included, (2) when computing K_σ of Equation 14.4, the acting loads should be able to be selected among many different dead and live load cases, and (3) analyzed extreme live loads should be able to be treated as a regular load case, which are very important and practical features when initial stress problems such as buckling or stability are regarded.

14.7 3D ILLUSTRATED EXAMPLE WITH NONLINEAR STABILITY ANALYSIS OF A CABLE-STAYED BRIDGE, JIANGSU, PEOPLE'S REPUBLIC OF CHINA

For the demonstration of nonlinear stability analysis, the same cable-stayed bridge, Sutong Bridge, Jiangsu, China, in Chapter 11 is taken as an example to illustrate issues. The typical cross section of the steel box girder, the concrete pylon, and elevation profile are shown in Figures 12.23, 12.26, and 12.29, respectively. As described in Chapter 11, the steel girder and the pylon are modeled as a 3D frame, the diaphragm at the anchor position is modeled as a rigid body, and the cable is modeled as a 3D truss. Totally, the model is meshed with 1032 elements and 1035 nodes. VBDS is employed in the analysis. ANSYS is also employed for checking some analyses. Several different loading patterns are taken in the stability analysis of this bridge. Table 14.2 lists the load patterns and critical load of the stability analysis. In the six loading patterns, only the increment of the construction load, which includes a 100-ton crane at the end of the girder and a uniform load of 1 ton/m at the maximum single-cantilever stage, shows the coupling of bending in vertical and lateral directions. Figure 14.22 shows the vertical and lateral displacements when the construction loads increase to

Table 14.2 Loading patterns and the critical loads in stability analysis

Loading patterns	Description	Critical case
At S_0, increase V step by step	To search the live load safety factor without wind interfering at service stage	When the live loads are increased by 40 times of the normal live load, the vertical displacements at the center of the main span abruptly reached 42 and 13 m at the top of the pylon. The structure, however, still maintains some degree of stiffness. No lateral displacement significantly increased.
At S_0, increase S step by step	To search the whole structural weight safety factor without wind interference at service stage	At about three times of S, the displacements increase abruptly. No lateral displacement significantly increased.
At S_1 plus W, increase C step by step	To search the construction load safety factor with wind interference at maximum dual-cantilever stage	When increased to 240 times of C, the displacements increase abruptly. No lateral displacement significantly increased.
At S_1, increase W step by step	To search the static wind load safety factor at maximum dual-cantilever stage	Still remains in elastic even at 50 times of W, while the lateral displacement at the end of the girder reaches to 7 m.

(Continued)

Table 14.2 (Continued) Loading patterns and the critical loads in stability analysis

Loading patterns	Description	Critical case
At S_2 plus W, increase C step by step	To search the construction load safety factor with wind interfering at maximum single-cantilever stage	At 46 times of C, the vertical displacement at the end of the girder increased to over 100 m accompanied with 42 m of lateral displacements (Figure 14.22).
At S_2, increase W step by step	To search the static wind load safety factor at maximum single-cantilever stage without the consideration of the construction load	At 48 times of W, the lateral displacement at the end of the girder increased to over 100 m.

S_0, the ideal state at the service stage (the structural weight, cable tuning, and the superimposed dead load); S_1, the state at the maximum dual-cantilever stage (the structural weight and the cable tuning); S_2, the state at the maximum single-cantilever stage (the structural weight and the cable tuning); S, the whole structure weight plus superimposed dead load; V, the live loads that cause the maximum vertical displacement at the center of the main span; C, a 100-ton crane at one or two ends of the cantilever and 1 ton/meter of the other construction load; W, the lateral wind load.

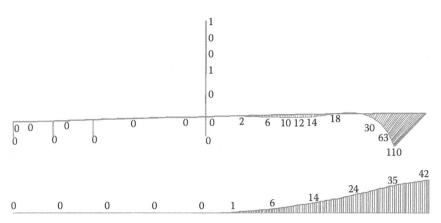

Figure 14.22 The vertical (top) and the lateral (bottom) displacements (m) of the girder when the construction loads are increased by 46 times of the normal construction loads at the maximum single-cantilever stage.

46 times the earlier construction load, while the lateral wind load remains unchanged. The stability analysis also shows that the structure at the stage when its main span is ready to close is more vulnerable than at the stage when its side span reaches the second auxiliary pier. Although the results of these six loading patterns show that the structure has sufficient stability against live loads, wind load, construction load, and the structural weight, the full nonlinear ultimate analysis (Ren 1999), in which the material nonlinearity is also considered, and the aerodynamic stability analysis are required.

Chapter 15

Redundancy analysis

15.1 BASICS OF BRIDGE REDUNDANCY

Redundancy is the quality of a bridge to perform as designed in a damaged state because of the presence of multiple load paths. Conversely, nonredundancy is the lack of alternate load paths, meaning the failure of a single primary load-carrying member would result in the failure of the entire structure. Three types of redundancy, load path, structural, and internal redundancies, have been identified much earlier. Recently, the FHWA provides a new definition for these three types of redundancy in the *FHWA Bridge Design Handbook* (FHWA/NSBA/HDR 2012), and they are summarized in Table 15.1. In general, redundancy issue should exist for all types of bridges. However, of all bridge construction materials, only steel bridge members may have such designation as fracture critical, and with regard to the topic of structural redundancy, the nonredundant steel members are the fracture critical members (FCMs). FCMs are those in axial tension or tension components of bending members whose failure would result in the failure of the structure. These elements are labeled as such on the contract drawings and are subjected to more stringent design, testing, and inspection criteria than those that are part of a redundant system (Fu and Schelling 1989, 1994; Fu 2000). Caltrans (2004) made a list of members or components, including but not limited to the following, identified as FCMs:

- Tension ties in arch bridges
- Tension members in truss bridges
- Tension flanges and webs in two-girder bridges
- Tension flanges and webs in single or double box girder bridges
- Tension flanges and webs in floor beams or cross girders
- Tension braces in the cross frame of horizontally curved girder bridges
- Attachments welded to an FCM when their dimension exceeds 100 mm (4″) in the direction parallel to the calculated tensile stress in the FCM
- Tension components of bent caps
- Splice plates of an FCM

Table 15.1 Types of redundancy

Type	Description
Load path redundancy	A member is considered load path redundant if an alternative and sufficient load path is determined to exist. Load path redundancy is the type of redundancy that designers consider when they count parallel girders or load paths. However, merely determining that alternate load paths exist is not enough. The alternative load paths must have sufficient capacity to carry the load redistributed to them from an adjacent failed member. If the additional redistributed load fails, progressive failure of the alternative load path occurs, and the members could in fact be fracture critical. In determining the sufficiency of alternative load paths, all elements present (primary and secondary members) should be considered.
Structural redundancy	A member is considered structurally redundant if its boundary conditions or supports are such that the failure of the member merely changes the boundary or support conditions but does not result in the collapse of the superstructure. Again, the member with modified support conditions must be sufficient to carry loads in its new configuration. For example, the failure of the negative-moment region of a two-span continuous girder is not critical to the survival of the superstructure if the positive-moment region is sufficient to carry the load as a simply supported girder.
Internal redundancy	A member is considered internally redundant if an alternative and sufficient load path exists within the member itself such as the multiple plies of riveted steel member.

Source: FHWA/NSBA/HDR, "Steel Bridge Design Handbook FHWA-IF-12-052—Vol. 9: Redundancy," Federal Highway Administration, USDOT, November 2012, http://www.fhwa.dot.gov/bridge/steel/pubs/if12052/volume09.pdf."

Moreover, Caltrans made a comprehensive flowchart for identifying FCMs of complex steel bridges in Figure 15.1.

The definition of a narrow plate girder (PG) system varies slightly from that used in stability discussions when focusing on redundancy. Whereas the system could contain any number of closely spaced girders in stability discussions, twin girder systems alone constitute a narrow system in the context of redundancy. This is due to the fact that only two primary elements exist to transfer load. If one of these fails, the second would be unable to support the entire weight of the structure, resulting in collapse. Other elements of the bridge, particularly the deck, could be able to carry additional loads encountered due to a nonredundant member failure and prevent collapse, which has been seen in the past. This built-in redundancy is difficult to predict, however, and is not explicitly recognized in the design. As such, for typical PG bridges, a minimum of three girders are required to provide alternate load paths and be considered system redundant.

To a lesser degree, studies of concrete bridge redundancy were made assuming cracking concrete, yield reinforcement, or reaching ultimate moment and shear capacity of the longitudinal or transverse beams (Imhof et al. 2004).

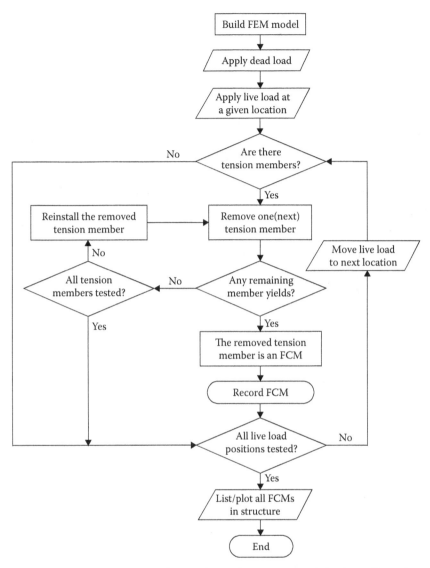

Figure 15.1 Flowchart for identifying FCMs of complex steel bridges. (Data from Caltrans, "Memo to Designers 12-2: Guidelines for Identification of Steel Bridge Members," August 2004.)

Another type of redundancy is the structural behavior under dynamic loads, such as earthquake loading or blast loading. The effect of blast loading is more localized than earthquake's global effect. The ability to sustain local damage without total collapse (structural integrity) is a key similarity between seismic-resistant and blast-resistant designs (NIST 2001). In general, the term *progressive collapse* has constantly been used in the redundancy analysis. As stated in ASCE 7-10 (2010), *progressive collapse* is defined as the spread of an initial local failure from element to element, eventually resulting in the collapse of an entire structure or disproportionately large part of it. Progressive collapse due to earthquake loading will be discussed more in Chapter 17—Dynamic/Earthquake Analysis.

To achieve targeted integrity during blast, the redundancy of the gravity load-carrying structural system takes center stage in tackling the issue of progressive collapse. This is not explicitly addressed in any code. However, ASCE 7-10 (2010) implies a desired alternate load path in the event one or more beams and/or columns of a building fail as a result of a blast. The structure should be able to remain stable by redistributing the gravity loads to other members and subsequently to the foundation through an alternate load path, while keeping building damage somewhat proportional to the initial failure.

For performance-based designs, factors considered include life safety issues, progressive collapse mechanisms, ductility of certain critical components, and redundancy of the whole structure. Blast load damages structures through propagating spherical pressure waves, which can be simulated by a series of equivalent loads. Performance of bridge elements under equivalent static loads can be considered as reasonably similar to that under the original dynamic blast loads. For the evaluation of the existing bridges under blast loading, the structural performance levels, the immediate occupancy (IO) level, life safety (LS) level, and the collapse prevention (CP) level, adopted by FEMA (1998) for the seismic evaluation of buildings, are used here. More details about these three levels will be discussed in Chapter 17.

15.2 PRINCIPLE AND MODELING OF BRIDGE REDUNDANCY ANALYSIS

The emphasis of this chapter is to illustrate how to conduct a bridge redundancy analysis. NCHRP Report 403 (1998) proposed a series of tables for system factors to be used in the design and evaluation equations for common-type bridges. The system factor tables developed in the NCHRP study are applicable to standard prestressed concrete and steel bridges. Bridges with configurations that are not covered by the tables have to be checked by performing a detailed incremental structural analysis. A steel truss bridge was mentioned specifically in this report to illustrate how the direct analysis can be applied in practice. This approach is allowed by Penn DOT Design Manual Part IV, Section 3—Loads and Load Factors (Penn DOT 2000). Commentary for extreme event IV states that

> For this extreme event, a 3D analysis is required. The objective of this analysis is survival of the bridge (i.e., the bridge may have large permanent deflections, but it has not collapsed).

Thus, a three-dimensional (3D) nonlinear model of the truss bridge is recommended for the structural analysis. In this chapter, the safety analysis is conducted as follows:

1. Member failure check, ultimate capacity check, and functionality check. It is proposed that the check of member failure be performed using a 3D elastic analysis of the structural system [ANSYS (2012) or SAP2000 (2007)]. Member capacity is calculated using AASHTO member strength formulas (AASHTO 2012, 2013). Penn DOT load combination extreme event III will be applied on a linear elastic structural model.
2. Damaged condition check. It is proposed that the check be performed using ANSYS (2012) or SAP2000 (2007) to analyze the damaged structure on a structural model. ANSYS (2012) or SAP2000 (2007) also may be applied using several degrading models to simulate the incremental analyses. Penn DOT load combination extreme event IV will be applied on a nonlinear elastic structural model.

The intention is to prove that although this bridge geometry does not satisfy redundancy criteria, the conservatives of the member design ensure that enough system safety is still available. Note that extreme events III and IV described here can be replaced by any extreme cases described in other codes.

15.2.1 Analysis cases

When possible, alternate load paths should be included in the design. Though this is not always an option, special consideration is warranted during the design of nonredundant structures. Due to the criticality of the primary load-carrying members, attention should be paid to fatigue, and effort should be made to eliminate detrimental details when possible. Sophisticated analyses have been performed in the past with some effectiveness to determine if two-girder systems are truly nonredundant or not, to account for the membrane action of the deck and to determine load-shedding properties of secondary members. These analyses are rather grueling and are not suggested as part of a typical design to avoid the penalties associated with the use of nonredundant members and FCMs.

Before 1998, there was some discussion but little guidance on the assessment of redundancy. The AASHTO LRFD Specifications (AASHTO 2013) specifications for the design of highway bridges recognize the importance of redundancy and require its consideration when designing steel bridge members. The specifications state that a structure is nonredundant when the failure of a single element could cause collapse.

The AASHTO LRFD specifications (AASHTO 2013) and Penn DOT Design Manual Part IV (Penn DOT 2000) proposed a format explaining how redundancy can be included in the design process by using load factor modifiers η_R, where this redundancy factor ≥ 1.05 for nonredundant members, $= 1.00$ for conventional levels of redundancy, and ≥ 0.95 for exceptional levels of redundancy.

In 1998 NCHRP Report 403 was published, entitled "Redundancy in Highway Bridge Superstructures" (NCHRP 1998). A clear guideline for a redundancy check was given. The limit states that should be checked to ensure adequate bridge redundancy and system safety are defined as

1. Member failure. A traditional check of individual member safety using elastic analysis and nominal member capacity.
2. Ultimate limit state. The ultimate capacity of the intact bridge system. It corresponds to the formation of a collapse mechanism for bridges.
3. Functional limit state. A maximum acceptable live load displacement in a main longitudinal member equal to the span length/100.
4. Damaged condition limit state. The ultimate capacity of the bridge system after damage to one main load-carrying element.

Penn DOT Design Manual Part IV (Penn DOT 2000) has an even more explicit statement on the checking of redundancy for truss bridges, which are as follows:

1. Provision of a third line of trusses where possible
2. Use of stitched built-up components, which are designed to support the entire component load with any one element assumed to be broken and for which joints and splices have been designed to transmit component loads with any one element of the component assumed to be broken (based on load combination extreme event III)
3. Demonstration through 3D analysis that failure of any tension component, or other components designated by the department, of a two-truss system will not cause the collapse of the entire structure (based on load combination extreme event IV)

A series of analysis cases were defined to match the proper analysis methodology with the appropriate truss configuration.

15.2.2 Finite element modeling

For steel bridge redundancy analysis, two levels of analysis should be made. First level is the 2D or 3D linear analysis to identify FCMs, as shown in Figure 15.1. Second level is the 3D nonlinear analysis to check the performance under loading. In the 3D nonlinear analysis, steel plastic behavior is described by bilinear kinematic hardening material model.

15.3 3D EXAMPLE WITH REDUNDANCY ANALYSIS OF A PONY TRUSS, PENNSYLVANIA

This illustrated example is to demonstrate the redundancy capacity check using nonlinear finite element analysis. A truss configuration was selected to represent the 142'-8 1/2" (43.5-m) Foxstop Road Bridge, shown in Figures 15.2 and 15.3, where all designs are fabricated using grade 50 (345-MPa) steel. The 3D model in ANSYS is shown in Figure 15.4. Following the flowchart shown in Figure 15.1, the first step is to identify FCMs.

After the finite element model is made by the first-level linear analysis, the FCMs are identified. Seventeen (17) FCMs are identified per truss panel (where A as the left truss panel and B as the right truss panel): nine on truss A are bottom chord members (elements 1 through 9) and eight are diagonal members (elements 28, 31, 34, 37, 40, 43, 46, and 49). Due to symmetry and simplification, only three bottom chord members (L1L3, L5L7, and L9L11) and two diagonal member-cut cases (U2L3 and U4L5) and one uncut case were analyzed for each load case. Specifically, a series of code checks are required for these bottom chord and diagonal members of the truss bridge in the uncut and cut conditions, which are shown in Table 15.2. Figure 15.5a shows the semi-box section for the bottom chord with two channels, and Figure 15.5b shows the wide flange section for the diagonal members, both shown in ANSYS plastic section designation prepared to the second-level nonlinear analysis.

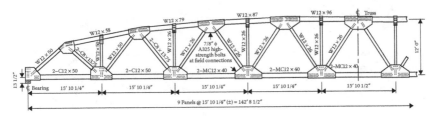

Figure 15.2 USB 43.5-m (142'-8 1/2") Foxstop Road Bridge detail.

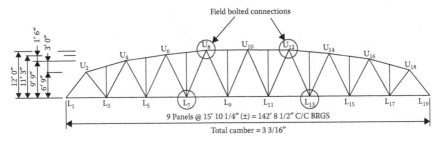

Figure 15.3 USB 43.5-m (142'-8 1/2") Foxstop Road Bridge elevation view of truss.

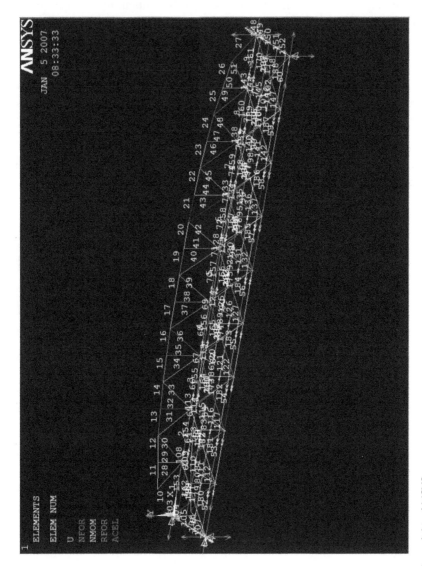

Figure 15.4 3D model in ANSYS.

Table 15.2 Redundancy analysis

Design alternates	Structural elements	Code checks
All basic uncut designs	Main truss	Dead load conditions
	Members	Live load maxima
	Gusset plates	Tension allowables
	Connections	Column compression
		Fatigue (stress range)
All basic cut designs	Cross beams	Nonfatigue
	supports	Secondary stresses
		Half-truss stability
		Redundancy
		Deflections

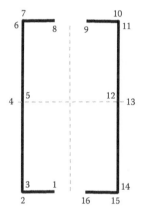

Local coordinate number	Local coordinate
1	$(-dx,-d2,0)$
2	$(-bf1,-d2,tf)$
3	$(-bf1-0.00001,-d2,0)$
4	$(-bf1-0.00001,0,tw)$
5	$(-bf1,0,0)$
6	$(-bf1,d2,tw)$
7	$(bf1-0.00001,d2,0)$
8	$(-dx,d2,tf)$
9	$(dx,d2,0)$
10	$(bf1,d2,tf)$
11	$(bf1+0.00001,d2,0)$
12	$(bf1+0.00001,0,tw)$
13	$(bf1,0,0)$
14	$(bf1,-d2,tw)$
15	$(bf1,-d2-0.00001,0)$
16	$(dx,-d2-0.00001,tf)$

(a) Bottom chord double-channel section

Local coordinate number	Local coordinate
1	$(-d2,bf2,0)$
2	$(-d2,-bf2,tf)$
3	$(-d2,0,0)$
4	$(d2,0,tw)$
5	$(d2,bf2,0)$
6	$(d2,-bf2,tf)$

(b) Diagonal wide flange section

Figure 15.5 (a, b) Plastic section definition of FCMs.

To maximize the live loading effect, live loading positions are dependent on the locations of cutting members. If the cutting member is close to the center of the bridge, the truck is positioned at the centerline. Both truck and lane loads are such transversely positioned that truss panel A will be more heavily loaded.

15.3.1 Loading cases

The full dead load (DL) and live loading plus impact (LL + I) were applied as specified by AASHTO LRFD (AASHTO 2013) and Penn DOT Design Manual (Penn DOT 2000) for extreme cases III and IV. They are summarized as follows:

1. All DL intensities were computed from the actual weights of the individual components of the bridge, and their load factors are listed in Table 15.3.
2. (LL + I) was determined by applying a full PHL-93 or P-82 live load longitudinally to obtain maximum tension and compression effects for all members. The impact and distribution factors specified by AASHTO were utilized. The Penn DOT vehicular live loading on the roadways of bridges or incidental structures, designated as PHL-93 (similar to AASHTO HL-93, except higher-design tandem), shall consist of a combination of the following:
   a. Design truck (HS-20 as shown in Figure 15.6a) or design tandem (two axles of 31.25 kip or 139 kN)
   b. Design lane load (0.64 kip/ft or 9.3 kN/m) and the P-82 permit truck is shown in Figure 15.6b. Note that the loading HL-93 or PHL-93 used here can be replaced by any other design vehicles of any code and P-82 can be replaced by permit vehicles of any jurisdiction.

Two load combinations are considered in either extreme event III (no cut) or extreme event IV (cut) case. Computer-run cases in relation to element-cut and load cases are listed in Table 15.4. Load factors in relation to load combinations and load cases are shown in Table 15.5.

Table 15.3 Dead load factors for extreme events III and IV

Load factors		Extreme event III	Extreme event IV
γ_{DC}	Maximum	1.25	1.05
	Minimum	0.90	0.95
γ_{DW}	Maximum	1.50	1.05
	Minimum	0.65	0.90

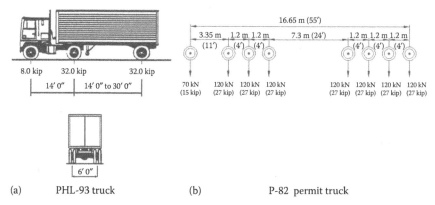

(a)   PHL-93 truck   (b)   P-82 permit truck

Figure 15.6 Penn DOT (a) PHL-93 and (b) P-82 permit truck configuration.

15.3.2 Results

A series of analysis cases, which are defined in Table 15.4, were developed to assess the appropriate AASHTO code requirement (AASHTO 2012, 2013) as applied to each member bridge configuration and failure mode. Analysis case 1 can be obtained by either TRAP (BEST Center 2006) or ANSYS program. The results of this analysis case are not covered in this book. Analysis cases 2 and 3, which are in the scope of the redundancy analysis, have to be obtained by 3D analysis, and the ANSYS program is used. Specifically, a total of 24 ANSYS runs with 24 analysis cases were investigated for the redundancy analysis (Table 15.6), each case requiring the application of multiple loadings for the 188 finite elements, which compose each bridge configuration.

Contained within this section is a summary of results of the ANSYS analysis of the Foxstop Road Bridge.

15.3.2.1 Extreme event III

The 3D frame analysis uses the entire truss–deck system assemblage in determining the stress in two plane trusses and floor beams. A review of four load cases with no element cut reveals the following:

1. The maximum and average stresses due to bending for all dead and maximum live load combinations are investigated as specified by AASHTO LRFD specifications (AASHTO 2013) and Penn DOT Design Manual Part IV (Penn DOT 2000).
2. The level of secondary stresses is generally low, and predominant stresses are axial stresses on the truss panels.

Table 15.4 Summary of analysis cases

Analysis case		Methodology				Load cases						Structural elements				
		Analysis methodology		Bridge member evaluation			AASHTO Live loading					Truss members		Floor system		
No.	Description	2D truss (TRAP)	3D frame (ANSYS)	Plain truss	All members	Dead loading	Nonfatigue	Fatigue	Sidewalk	Wind	Primary stresses	Secondary stresses	Floor beams	Stringers	Connections
1	Primary stress analysis and code check of plane truss for the following:														
	(a) PHL-93	X		X		X	X	X			X				X
	(b) P-82	X													X
2	Secondary stress analysis and code check for truss and floor system for the following:														
	(a) PHL-93		X		X	X	X					X	X	X	X
	(b) P-82		X		X	X	X					X	X	X	X
3	Redundancy stress analysis and code check for truss and floor system for one bottom chord channel cut														
	(a) PHL-93		X		X	X	X				X	X	X	X	X
	(b) P-82		X		X	X	X				X	X	X	X	X

Table 15.5 Load factors in relation to load combinations and load cases

Load combination with element-cut status	Load cases	Load factors				
		γ_{DC}	γ_{DW}	γ_{LL} truck	γ_{LL} lane	γ_{PL}
Extreme event III (no cut)	1	1.25	1.5	1.3×1.33	1.3	–
	2	1.25	1.5	–	–	1.1×1.33
	3	1.25	1.5	1.3×1.33	1.3	–
	4	1.25	1.5	–	–	1.1×1.33
Extreme event IV (cut element)	1	1.05	1.05	1.15×1.33	1.15	–
	2	1.05	1.05	–	–	1.05×1.33
	3	1.05	1.05	1.15×1.33	1.15	–
	4	1.05	1.05	–	–	1.05×1.33

Table 15.6 Computer-run cases in relation to element-cut and load cases (one lane case)

Cases of element cut	Load cases	Computer-run cases
No cut	Each with load cases	P82CLnocut, P82L3nocut, PHL93CLnocut, PHL93L3nocut
Cut 1 (first bottom chord)	1. PHL-93 truck at midspan and PHL-93 lane all over	P82CLcut1, P82L3cut1, PHL93CLcut1, PHL93L3cut1
Cut 3 (third bottom chord)	2. P-82 truck at midspan	P82CLcut3, P82L3cut3, PHL93CLcut3, PHL93L3cut3
Cut 5 (fifth bottom chord)	3. PHL-93 truck at 1/3 L and PHL-93 lane all over	P82CLcut5, P82L3cut5, PHL93CLcut5, PHL93L3cut5
Cut 28 (first diagonal chord)	4. P-82 truck at 1/3 L	P82CLcut28, P82L3cut28, PHL93CLcut28, PHL93L3cut28
Cut 31 (second diagonal chord)		P82CLcut31, P82L3cut31, PHL93CLcut31, PHL93L3cut31

3. No member exhibits a combined (axial and bending) stress that exceeds the allowable given in AASHTO specifications or Penn DOT Design Manual.

4. Noncomposite action is conservatively considered for the deck system. For 3D analysis, floor beams are considered as part of the frame action. The model showed that the stresses of floor beams (elements 103–152) are under the allowables and they are not the governing cases compared to the truss members.

5. No yielding is found in any element of these four load cases.

The results of these analyses are summarized in Table 15.7 under the title of "no cut" in the column to identify an element cut. By investigation the most critical stresses for a "no cut" case are –37.96 ksi (261.7 MPa) at element 62 under load 2-a: 1.25Dc + 1.5DW + 1.3*1.33P-82 (at L/3). Because it is considered as extreme event III, the allowable stress is the yield stress of the section, which is 50 ksi (345 MPa) in this case. Also, the worst vertical displacements are 104 mm (4.1″) under load case 1-a. Displacement of 104 mm (4.1″) of the 43.5-m (142′-8 1/2″) Foxstop Road Bridge corresponding to L/418 is adequate for the extreme event III limit state.

15.3.2.2 *Extreme event IV*

As stated in Section 15.3.2, the test of redundancy required a 3D frame analysis of the entire system under conditions of severing a single bottom chord of any twin-channel bottom chord member while sustaining the full AASHTO (or Penn DOT) dead and live loads applied to attain the maximum stresses. Or the severed members may be identified as diagonal tension members as listed in Table 15.6. The results of these analyses also are given in Table 15.7 for the various member-cut conditions. A summary of these results follows:

1. Five (5) *cut* cases are identified, and their maximum stresses and elastic and inelastic strains are summarized in Table 15.7. With four load cases, there are 20 runs in total. Among these 20 runs, one case for bottom chord cut case (load case 1-b-cut 1) and four cases for diagonal cut cases (load cases 1-b-cut 28, 2-b-cut 28, 3-b-cut 28, 4-b-cut 28) have members plastified. After plastification, stresses stay at the level of yield stress, and strains may still grow to their respective maximum strains under the current loading condition.
2. Ductility ratio listed under column (7) is defined as the maximum strain divided by the yield strain (0.001724). Resulting from all analyses, the maximum ductility ratio identified is 4.45, which is associated with "load case 2-b with element 28 cut."
3. The same "load case 2-b with element 28 cut" gives vertical (y-direction) displacements of 95 mm (3.75″). Displacement of 95 mm (3.75″) of the 43.5-m (142′-8 1/2″) Foxstop Road Bridge corresponds to L/457, which is also considered adequate for the extreme event IV limit state.

Table 15.7 Summary of governing stresses, strains, and displacements for the two extreme load cases truss model

Load case (1)	Element cut (2)	Maximum elastic stress (ksi) (3) Tension	Compression	Element of maximum stress (4) Tension	Compression	Elastic strain (5) Tension	Compression	Plastic strain (6) Tension	Compression	Displacement at x-direction (in.) (8)	Displacement at y-direction (in.) (9)	Displacement at z-direction (in.) (10)
1-a: 1.25Dc + 1.5DW + 1.3*1.33P-82 (at CL) (no cut)	No cut	35.45	-37.87	113	73	0.001223	-0.001306	NA	NA	0.440	-4.10	-0.859
	1	50	-50	108	108	0.001724	-0.001724	0.000446	-0.000450	-0.382	-3.51	-0.821
	3	36.47	-36.58	113	113	0.001257	-0.001261	NA	NA	0.373	-3.474	0.745
	5	32.46	-32.67	113	113	0.001119	-0.001127	NA	NA	0.372	-3.499	-0.777
1-b: 0.05Dc + 1.05DW + 1.15*1.33P-82 (at CL) (cut)	28	50	-50	28	11	0.001724	-0.001724	0.00360	-0.001672	-0.406	-3.646	-1.487
	31	48.132	-43.23	31	30	0.001660	-0.001491	NA	NA	0.374	-3.512	-0.750
2-a: 1.25Dc + 1.5DW + 1.3*1.33P-82 (at L/3) (no cut)	No cut	35.70	-37.96	126	62	0.001231	-0.001309	NA	NA	0.454	-3.977	-0.833
	1	49.18	-49.126	108	108	0.001696	-0.001694	NA	NA	0.383	-3.385	-0.777
	3	32.38	-32.43	113	113	0.001117	-0.001118	NA	NA	0.386	-3.366	-0.724
2-b: 1.05Dc + 1.15*1.33P-82 (at L/3) (cut)	5	32.29	-32.7	128	128	0.001113	-0.001128	NA	NA	0.384	-3.385	-0.747
	28	50	-50	11,28	11	0.001724	-0.001724	0.005962	-0.004389	0.391	-3.750	-2.040
	31	43.44	-43.46	31	12	0.001498	-0.001499	NA	NA	0.385	-3.406	-0749

(Continued)

Table 15.7 (Continued) Summary of governing stresses, strains, and displacements for the two extreme load cases truss model

Load case (1)	Element cut (2)	Maximum elastic stress (ksi) (3)		Element of maximum stress (4)		Elastic strain (5)		Plastic strain (6)		Displacement at x-direction (in.) (8)	Displacement at y-direction (in.) (9)	Displacement at z-direction (in.) (10)
		Tension	Compression	Tension	Compression	Tension	Compression	Tension	Compression			
3-a: 1.25Dc + 1.5DW + 1.3 Lane + 1.3*1.33PHL93 (at CL) (no cut)	No cut	26.67	−30.35	113,117	20	0.000920	−0.001047	NA	NA	0.365	−3.38	−0.649
	1	45.26	−45.26	108	108	0.001561	−0.001561	NA	NA	−0.311	−2.860	−0.616
	3	27.49	−27.53	113	113	0.000948	−0.000949	NA	NA	0.305	−2.834	0.555
	5	25.07	−25.53	123	71	0.000865	−0.000880	NA	NA	0.305	−2.854	−0.585
3-b: 1.05Dc + 1.05DW + 1.15Lane + 1.15*1.33PHL93 (at CL) (cut)	28	50	−50	28	10	0.001724	−0.001724	0.001918	−0.00018	−0.322	−2.94	−0.998
	31	36.72	−34.68	31	12	0.001266	−0.001196	NA	NA	0.306	−2.863	−0.5655
4-a: 1.25Dc + 1.5DW + 1.3 Lane + 1.3*1.33PHL93 (at L/3) (no cut)	No cut	26.84	−31.53	79	66	0.000926	−0.001087	NA	NA	0.377	−3.233	−0.632
	1	43.29	−43.38	108	108	0.001493	−0.001496	NA	NA	0.315	−2.726	−0.587
	3	29.17	−29.26	113	113	0.001006	−0.001009	NA	NA	0.317	−2.707	−0.544
	5	22.98	−26.49	128	66	0.000792	−0.000914	NA	NA	0.316	−2.721	−0.56
4-b: 1.05Dc + 1.05DW + 1.15Lane + 1.15*1.33PHL93 (at L/3) (cut)	28	50	−50	28	10,12	0.001724	−0.001724	0.002468	−0.000485	0.321	−2.859	−1.149
	31	39.34	−36.77	31	12	0.001357	−0.001268	NA	NA	0.317	−2.744	−0.568

Table 15.8 Summary of maximum strains and ductility factors for the two extreme load cases

Load case	Element cut	Maximum strain		Ductility factor
		Tension	Compression	
1-a: 1.25Dc + 1.5DW + 1.3*1.33P-82 (at CL) (no cut)	1	0.002170	−0.002174	1.26
1-b: 0.05Dc + 1.05DW + 1.15*1.33P-82 (at CL) (cut)	28	0.005324	−0.003396	3.088
2-a: 1.25Dc + 1.5DW + 1.3*1.33P-82 (at L/3) (no cut)	28	0.007686	−0.006113	4.45
2-b: 0.05Dc + 1.05DW + 1.15*1.33P-82 (at L/3) (cut)				
3-a: 1.25Dc + 1.5DW + 1.3Lane + 1.3*1.33PHL93 (at CL) (no cut)	28	0.003642	−0.001904	2.11
3-b: 1.05Dc + 1.05DW + 1.15Lane + 1.15*1.33PHL93 (at CL) (cut)				
4-a: 1.25Dc + 1.5DW + 1.3Lane + 1.3*1.33PHL93 (at L/3) (no cut)	28	0.004192	−0.002209	2.43
4-b: 1.05Dc + 1.05DW + 1.15Lane + 1.15*1.33PHL93 (at L/3) (cut)				

4. Only a limited number of elements yield by cutting any FCM. This means the structure would shake down after a few members yielded. A nonlinear analysis program is capable of redistributing the load after any member plastifies. The sum of the elastic and plastic strains yields a ductility ratio of 4.45, shown in Table 15.8.

15.4 3D REDUNDANCY ANALYSIS UNDER BLAST LOADING OF A PC BEAM BRIDGE, MARYLAND

This example demonstrates the analysis under equivalent blast load on a prestressed concrete beam bridge formed by 3D frame elements with plastic hinges assigned at specific locations. The bridge was designed using AASHTO's *Standard Specifications for Highway Bridges* for an HS-20-44 live load. A representative prestressed concrete beam span is simply supported, 18.3 m (60′) in length and 12.1 m (39′-8″) wide. There are six AASHTO type III beams, spaced 2.2 m (7′-2″) center to center. Figure 15.7 shows the bridge's typical half-section, with symmetry occurring at the centerline.

The bridge deck is 178 mm (7″) thick, which includes a 13-mm (1/2″) monolithic wearing surface. The AASHTO type III beam cross section

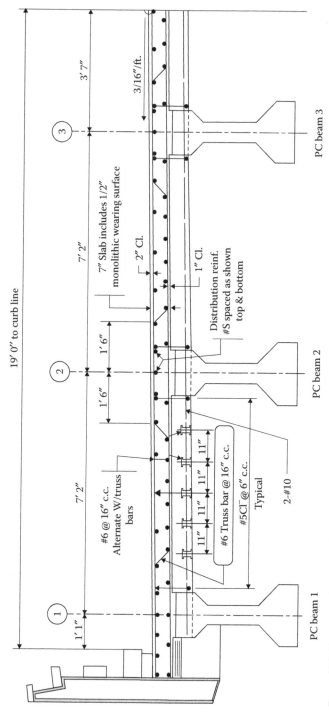

Figure 15.7 Prestressed concrete beam typical half-section.

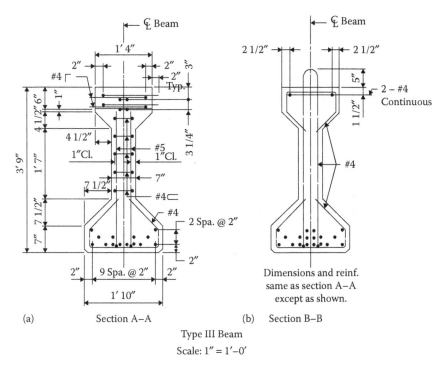

Figure 15.8 (a, b) AASHTO type III beam cross sections.

dimensions and prestressing tendon layout are shown in Figure 15.8. Section A–A corresponds to the end of the bridge, while Section B–B is the beam cross section at the bridge midspan. The prestressed beam concrete has a minimum 28-day compressive strength of 5000 psi (34.5 MPa) and a minimum compressive strength of 4000 psi (27.6 MPa) when jacking. The prestressing tendons are number 7 wire strand, with 1/2″ (13 mm) diameter and a cross-sectional area of 0.153 in² (98.7 mm²). The capacity of the wire is 270 ksi (1861.7 MPa).

15.4.1 Bridge model

SAP2000 (2007) is used to create a model of the prestressed concrete beam bridge. As this example is concerned with the response of the deck as well as the beams, the deck is also modeled using frame elements. By defining the deck as frame elements, nonlinear hinges (or plastic hinges) can be assigned, so the deck will exhibit nonlinear plastic behavior. To properly model the bridge deck and account for transverse and longitudinal stiffness, a grid of frame elements is created. The deck frame elements are

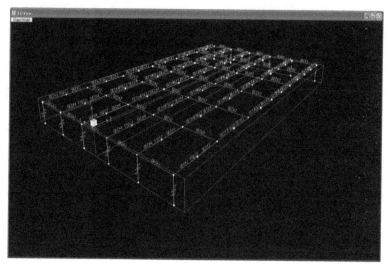

Figure 15.9 Grid of deck frame elements for PC beam bridge.

connected to joints along the prestressed concrete beam centerlines. Four frame elements are defined, two in the transverse direction and two in the longitudinal direction.

The width of the deck sections are calculated via the beam tributary area. The distance between joints along the centerline of the concrete beams in the longitudinal direction is designated as b and equal to 3.05 m (10′). The spacing between the concrete beams in the transverse direction is designated as s and equal to 2.18 m (7′-2″). The overhang distance between the exterior concrete beams and the edge of the bridge is designated as o and equal to 0.58 m (1′-11″). Figure 15.9 shows the deck *grid* model.

Figure 15.8 shows that the AASHTO type III beams have 20 prestressing tendons. The tendons are modeled as truss elements. Four of these tendons are deflected strands that vary along the beam length. The remaining 16 strands are straight through the beam length. The four deflected tendons are modeled together as one at their centroid location, with a cross-sectional area equal to four times the area of one tendon, or 0.612 in² (395 mm²). The straight tendons are also modeled as one tendon group, with a cross-sectional area equal to 2.448 in² (1579 mm²). The two top *tendons* are used to resist tension on the *top* of the beam at release and are not modeled in this analysis. Figure 15.10 illustrates the tendon layout in SAP2000 (2007).

15.4.2 Attack scenarios

The five attack scenarios for the PC beam bridge are restated in Table 15.9. Each scenario is characterized by a charge weight of TNT and location

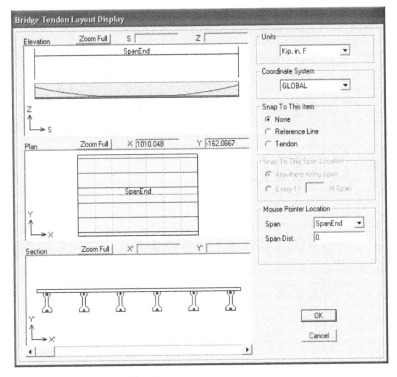

Figure 15.10 Prestressing tendon layout.

Table 15.9 Attack scenarios for PC beam bridge

Attack scenario	Charge weight (lb TNT)	Blast location along bridge length (ft)
1	674	34
2	1009	44
3	437	13
4	2911	36
5	1821	21

along the bridge's 18.3 m (60′) length. The charge weights and locations were assigned based on their probability distributions and were randomly generated to several scenarios (Mahoney 2007).

The static equivalent loads of each attack scenario can be calculated. Table 15.10 displays these calculations for attack scenario 1 as an example. Having the distance between the blast and each bridge joint (*D*) and the angle of blast (θ_i) calculated, the program AT-Blast (ARA 2004) based

Table 15.10 Attack scenario 1 (674-lb TNT) static equivalent load calculations

| | Joint no. | Coordinates relative to origin | | Z (ft) | D (ft) | θ_i (°) | Pressure (psi) | Tributary area (ft²) | Load on joint (K) |
		x	y	X (ft)						
PC beam 1	1	0	17.92	38	4	39	84	59.19	27.50	234
	2	10	17.92	30	4	30	82	119.43	55.00	946
	3	20	17.92	23	4	23	80	**233.64**	55.00	1850
	4	30	17.92	18	4	19	78	**312.75**	55.00	2477
	5	40	17.92	19	4	19	78	**312.75**	55.00	2477
	6	50	17.92	24	4	24	81	197.09	55.00	1561
	7	60	17.92	32	4	32	83	100.12	27.50	396
PC beam 2	1	0	10.75	36	4	36	84	73.51	35.83	379
	2	10	10.75	26	4	27	81	150.43	71.67	1552
	3	20	10.75	18	4	18	77	**321.06**	71.67	3313
	4	30	10.75	11	4	12	71	**784.48**	71.67	8096
	5	40	10.75	12	4	13	72	**668.49**	71.67	6899
	6	50	10.75	19	4	20	78	**306.30**	71.67	3161
	7	60	10.75	28	4	28	82	137.49	35.83	709
PC beam 3	1	0	3.58	34	4	34	83	86.82	35.83	448
	2	10	3.58	24	4	25	81	173.95	71.67	1795
	3	20	3.58	14	4	15	75	**472.39**	71.67	4875
	4	30	3.58	5	4	7	53	**7869.88**	71.67	81217
	5	40	3.58	7	4	8	60	**3487.37**	71.67	35990
	6	50	3.58	16	4	17	76	**365.83**	71.67	3775
	7	60	3.58	26	4	27	81	150.43	35.83	776

Group										
PC beam 4	1	0	−3.58	34	4	34	83	86.82	35.83	448
	2	10	−3.58	24	4	25	81	173.95	71.67	1795
	3	20	−3.58	14	4	15	75	**472.39**	71.67	4875
	4	30	−3.58	5	4	7	53	**7869.88**	71.67	81217
	5	40	−3.58	7	4	8	60	**3487.37**	71.67	35990
	6	50	−3.58	16	4	17	76	**365.83**	71.67	3775
	7	60	−3.53	26	4	27	81	150.43	35.83	776
PC beam 5	1	0	−10.75	36	4	36	84	73.51	35.83	379
	2	10	−10.75	26	4	27	81	150.43	71.67	1552
	3	20	−10.75	18	4	18	77	**321.06**	71.67	3313
	4	30	−10.75	11	4	12	71	**784.48**	71.67	8096
	5	40	−10.75	12	4	13	72	**668.49**	71.67	6899
	6	50	−10.75	19	4	20	78	**306.30**	71.67	3161
	7	60	−10.75	28	4	28	82	137.49	35.83	709
PC beam 6	1	0	−17.92	38	4	39	84	59.19	27.50	234
	2	10	−17.92	30	4	30	82	119.43	55.00	946
	3	20	−17.92	23	4	23	80	**233.64**	55.00	1850
	4	30	−17.92	18	4	19	78	**312.75**	55.00	2477
	5	40	−17.92	19	4	19	78	**312.75**	55.00	2477
	6	50	−17.92	24	4	24	81	197.09	55.00	1561
	7	60	−17.92	32	4	32	83	100.12	27.50	396

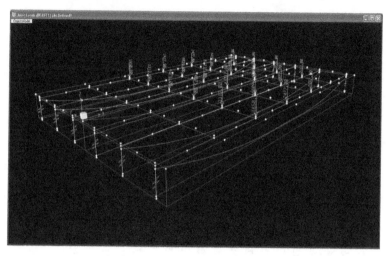

Figure 15.11 Attack scenario 1 (674-lb TNT) static equivalent joint loads.

on an open-air hemispherical explosion calculates the static pressure at each joint. The influence surface is the surface area expanding radially from the explosion centroid. As the blast magnitude increases, the influence surface increases. Trial and error aided in deciding that the blast loads may be cut off at pressures less than 200 psi (1.38 MPa). Using the tributary area method, the pressure is resolved into joint loads. In Table 15.10, the pressures that appear in bold are greater than or equal to 200 psi (1.38 MPa), so the corresponding joint loads of these pressures are applied to the PC beam bridge model. Figure 15.11 shows the static equivalent joint loads for attack scenario 1 applied to the prestressed concrete beam bridge model.

15.4.3 Analyze structural response

The nonlinear static analysis output shows the performance of structural members' plastic hinges with nodes color-coded showing the hinge's state on the moment–rotation or force–deformation curve. The analysis generated responses for multiple steps. Figure 15.12 demonstrates the final response step for one of the attack scenarios, which reveals that the PC beam bridge experiences total failure in every attack scenario. Therefore, the bridge under attack has no additional redundancy and will have to be replaced. This result simplifies the quantification of damaged areas by performance levels. Table 15.11 summarizes the structural damage under IO, LS, and CP. With all attack scenarios under the category of ($\geq$CP), there is no redundancy left for this type of bridges.

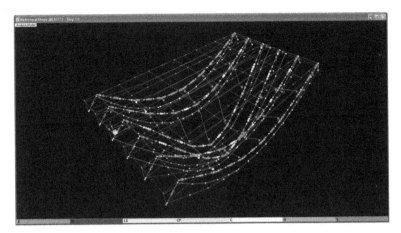

Figure 15.12 Attack scenario 1 response (step 11).

Table 15.11 PC beam bridge structural damage costs

Attack scenario (i)	Damaged area by performance level (ft²)		
	≥IO	≥LS	≥CP
1	–	–	2380
2	–	–	2380
3	–	–	2380
4	–	–	2380
5	–	–	2380

15.5 3D ANALYSIS UNDER BLAST LOADING OF A STEEL PLATE GIRDER BRIDGE, MARYLAND

This example demonstrates the analysis under equivalent blast load on a three-span continuous steel PG bridge formed by 3D frame elements with plastic hinges assigned at specific locations. The three-span bridge was designed using AASHTO's *Standard Specifications for Highway Bridges* for an HS-20-44 live load. Each span of the bridge is 61.6 m (202') long, totaling 184.7 m (606'), with 22 equally spaced diaphragms per span. The concrete deck is 165 mm (6.5") thick and 11.68 m (38'-4") wide. There are two steel built-up PGs and five rolled beam stringers. The PGs are spaced 8.53 m (28') center to center. Between the PGs, the interior stringers are spaced 2.13 m (7') apart. The exterior stringers are 1.4 m (4'-7") center to center from the PGs. Figure 15.13 shows the bridge's typical section and girder/stringer numbering scheme.

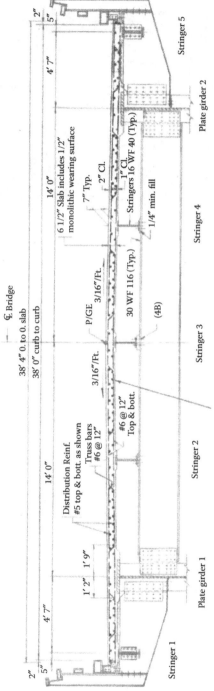

Figure 15.13 Continuous steel PG bridge typical section.

All structural steel sections are A36 (248-MPa) carbon steel, and the concrete deck is lightweight concrete. The stringers are W16 × 40 rolled beams, and the PG sections vary along the bridge length. There are three different PG sections, each having a constant web plate depth of 120″ (3048 mm). The web thickness varies along the PG from 9.5 mm (3/8″) to 11 mm (7/16″). The PG flange plates are 762 mm (30″) wide, with a thickness ranging from 8 mm (1 5/16″) to 57 mm (2 1/4″).

15.5.1 Bridge model

A model of the three-span continuous PG bridge is created by SAP2000 (2007). As this example is concerned with the response of the major structural elements (e.g., PGs, stringers, deck), the deck is modeled using frame elements. By defining the deck as frame elements, nonlinear hinges can be assigned at the ends, so the deck will exhibit nonlinear plastic behavior. To properly model the bridge deck and show transverse and longitudinal stiffness, a grid of frame elements is created. The deck frame elements are connected to the joints along the PG and stringer centerlines. The modeling details are similar to the PC bridge example discussed in Section 15.4.1.

The width of the deck sections are calculated via the girder tributary area. The distance between joints along the centerline of the PGs and stringers in the longitudinal direction is designated as b and equal to 2.6 m (8.5′). The distance between the PGs and exterior stringers is designated as s_1 and equal to 1.4 m (4′-7″). The spacing between the PGs and interior stringers in the transverse direction is designated as s_2 and equal to 2.1 m (7′). The overhang distance between the exterior stringers and the bridge edge is designated as o and equal to 0.2 m (7″).

Deck section 1 is along the end of the bridge in the transverse direction. The width of this section is equal to half the distance between joints along the PGs and stringers, or $b/2$. Deck section 2 is also in the transverse direction, with a width equal to the spacing between joints, or b. Deck section 3 falls in the longitudinal direction along the exterior stringers, so the width is defined as the overhang distance plus half the beam spacing, or $(o + s_1/2)$. Deck section 4 elements, also in the longitudinal direction, are along the PGs, with a width equal to $(s_1 + s_2)/2$. Deck section 5 elements are in the longitudinal direction along the interior stringers, with a width equal to their center-to-center spacing, or s_2. Figure 15.14 shows a plan view of the deck grid at the left end of the bridge model.

15.5.2 Attack scenarios

The five attack scenarios for the steel PG bridge are restated in Table 15.12. Each scenario is characterized by a charge weight of TNT and location

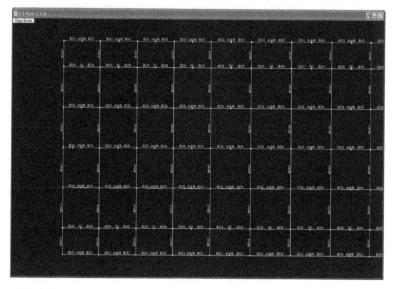

Figure 15.14 Grid of deck frame elements for PG Bridge.

Table 15.12 Attack scenarios for steel PG bridge

Attack scenario	Charge weight (lb TNT)	Blast location along bridge length (ft)
1	674	347
2	1009	444
3	437	130
4	2911	361
5	1821	209

along the bridge's 184.7 m (606′) length. As in the previous example, charge weights and locations were assigned based on their probability distributions and were randomly generated to several scenarios (Mahoney 2007).

The static equivalent loads of each attack scenario are calculated. Table 15.13 displays these calculations for attack scenario 1 as an example. The blast load on each joint is calculated the same way as the previous example. In Table 15.13, the pressures that appear in bold are greater than or equal to 200 psi (1.38 MPa), so the corresponding joint loads of these pressures are applied to the PG bridge model. Figure 15.15 shows the static equivalent joint loads for attack scenario 1 applied to the PG bridge model.

Table 15.13 Attack scenario 1 (674-lb TNT) static equivalent load calculations

	Joint no.	Coordinates relative to origin		X (ft)	Z (ft)	D (ft)	θ_i (°)	Pressure (psi)	Tributary area (ft²)	Load on joint (K)
		x	y							
Stringer 1	35	312.75	18.58	39	4	39	84	59.19	27.98	239
	36	322.5	18.58	31	4	31	83	107.86	27.98	435
	37	332.25	18.58	24	4	24	80	205.34	27.98	827
	38	342	18.58	19	4	20	78	306.30	27.98	1234
	39	351.75	18.58	19	4	20	78	306.30	27.98	1234
	40	361.5	18.58	24	4	24	80	205.34	27.98	827
	41	371.25	18.58	31	4	31	83	107.86	27.98	435
	42	381	18.58	39	4	39	84	59.19	25.00	213
Plate Girder 1	35	312.75	14	37	4	37	84	68.22	56.45	555
	36	322.5	14	28	4	29	82	128.05	56.45	1041
	37	332.25	14	20	4	21	79	290.81	56.45	2364
	38	342	14	15	4	15	75	472.39	56.45	3840
	39	351.75	14	15	4	15	75	472.39	56.45	3840
	40	361.5	14	20	4	21	79	290.81	56.45	2364
	41	371.25	14	28	4	28	82	137.49	56.45	1118
	42	381	14	37	4	37	84	68.22	50.43	495

(Continued)

Table 15.13 (Continued) Attack scenario 1 (674-lb TNT) static equivalent load calculations

| | | Coordinates relative to origin | | | | | | | | |
Joint no.		x	y	X (ft)	Z (ft)	D (ft)	θ_i (°)	Pressure (psi)	Tributary area (ft²)	Load on joint (K)
35	Stringer 2	312.75	7	35	4	35	83	81.09	68.25	797
36		322.5	7	25	4	26	81	161.43	68.25	1587
37		332.25	7	16	4	17	76	365.83	68.25	3595
38		342	7	9	4	9	65	1833.63	68.25	18021
39		351.75	7	8	4	9	65	1833.63	68.25	18021
40		361.5	7	16	4	17	76	365.83	68.25	3595
41		371.25	7	25	4	26	81	161.43	68.25	1587
42		381	7	35	4	35	83	81.09	60.97	712
35	Stringer 3	312.75	0	34	4	34	83	86.82	68.25	853
36		322.5	0	25	4	25	81	173.85	68.25	1709
37		332.25	0	15	4	15	75	472.39	68.25	4643
38		342	0	5	4	6	51	11320.26	68.25	111256
39		351.75	0	5	4	6	50	12000.89	68.25	117945
40		361.5	0	15	4	15	75	472.39	68.25	4643
41		371.25	0	24	4	25	81	173.95	68.25	1710
42		381	0	34	4	34	83	86.82	60.97	762

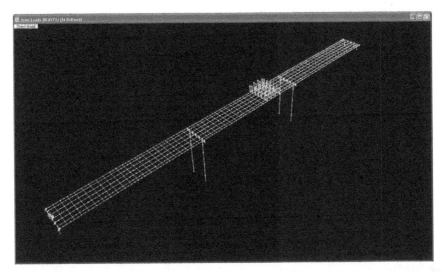

Figure 15.15 Attack scenario 1 (674-lb TNT) static equivalent joint loads.

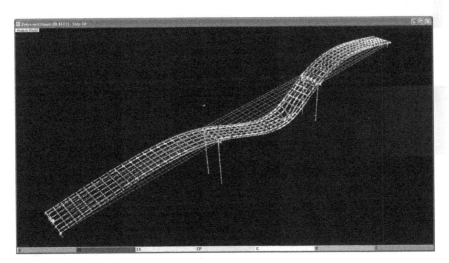

Figure 15.16 Attack scenario 1 response (step 10).

15.5.3 Analyze structural response

The nonlinear static analysis output shows the performance of the structural members' plastic hinges with nodes color-coded showing the hinge's state on the moment–rotation or force–deformation curve. As revealed in the final response steps for one of the attack scenarios (Figure 15.16),

Table 15.14 SG bridge structural damage costs

Attack scenario (i)	Damaged area by performance level (ft²)		
	≥IO	≥LS	≥CP
1	3310	540	3930
2	2861	1962	4303
3	2902	1353	4273
4	1926	1341	3578
5	2041	930	4439

the steel girder (PG) bridge experiences damage in all three performance levels but still has some redundancy left. Table 15.14 summarizes the structural damage under IO, LS, and CP. With all attack scenarios under the category of (≥CP), there is some redundancy left for this type of bridges.

Chapter 16

Integral bridges

16.1 BASICS OF INTEGRAL BRIDGES

16.1.1 Introduction

An integral bridge is a jointless bridge with no bearing at the connection point where the superstructure and substructure are framed together. Therefore, integral bridges are categorized as rigid-frame structures because they eliminate expansion joints. Integral bridges include integral abutment bridges (IABs) as well as integral piers. As most discussions focus on IABs, this section will briefly mention integral piers before moving onto IABs.

One type of integral bridge is the integral pier, which involves building a monolithic or framing-in joint at the pier. There are a number of ways to form an integral pier (Sisman and Fu 2004). A common method in concrete construction is a frame-type structural system, namely, cast-in-place concrete box girder bridges and was also carried over to steel I-girders framing into a concrete pier cap or diaphragm. More recently, a number of versions of integral piers have been developed, which involve steel plate girder construction to improve substructure layouts, eliminate detrimental effects of a skewed substructure, or enhance bridge performance under seismic loads. Some of the new concepts use steel framing-in caps, which integrate with the concrete columns (Figure 16.1); others are various versions of traditional concrete caps with varying structural boundary conditions. Whichever method, integral piers concealed within the boundaries of superstructure lines are definite enhancements to the aesthetic value of a bridge, whether it is in an urban setting or on a country road.

The main type of integral bridges is the IAB. An integral abutment is a stub abutment on a single row of flexible piles and constructed without joints. These bridges allow for expansion and contraction through movement at the abutments. In the conventional design of the superstructure, bridges are idealized as a continuous beam with simply supported ends. Figure 16.2 shows possible configurations for a typical four-span highway

Figure 16.1 Integral pier on ramp FR-A over SR6060, Pittsburgh, Pennsylvania.

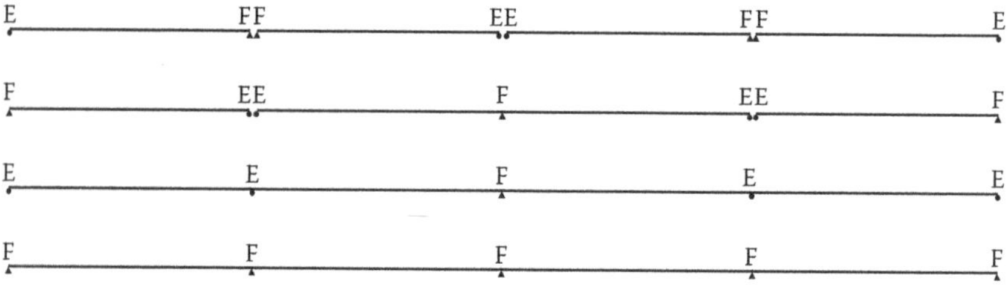

E   FF   EE   FF   E

F   EE   F   EE   F

E   E   F   E   E

F   F   F   F   F

Figure 16.2 Possible configurations for a typical four-span highway bridges.

bridge. However, unlike traditional bridges that sit on bearings with heavy abutments, integral bridges can be formed by casting the deck integrally with short abutments supported on a single row of flexible piles (Figure 16.3) to take care of the longitudinal thermal movement of the bridge.

IABs are designed without any expansion joints in the bridge deck. These bridges are generally designed with stiffness and flexibility spread throughout the structure–soil system so that all supports accommodate the thermal and braking loads. They are single- or multiple-span bridges that have their superstructure cast integrally with their substructure. Generally, these bridges include capped pile stub abutments. Piers for IABs may be constructed either integrally with or independently of the superstructure. Integral or semi-integral bridges are defined as single- or multiple-span continuous bridges with rigid, nonintegral foundations and movement systems primarily composed of integral-end diaphragms, compressible backfill, and movable bearings in a horizontal joint at the superstructure–abutment interface (Wasserman and Walker 1996).

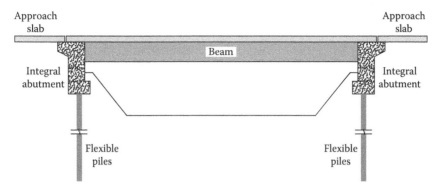

Figure 16.3 Integral abutment with flexible piles.

16.1.2 Types of integral abutment

There are several different basic types of abutments, which are conventional, semi-integral (Figure 16.4), and integral abutments (Figure 16.5). Integral abutments are a class of abutments in which the superstructure is integrally connected to the abutment and the abutment foundation. Generally the girders are set on an abutment cap, and a closure pour is cast, which encases the ends of the girders such that the girder ends are embedded several inches or more into the abutment concrete (FHWA 2012). Integral abutments are typically founded on a single line of vertical steel H-piles, although integral details have occasionally been used with piles, drilled shafts, and spread footings. Based on surveys, more than half of integral abutments in the United States are built with their piles oriented for weak axis bending to minimize the stresses in the abutments.

Semi-integral abutments are different from integral abutments in that integral abutments have no intentional moment relief detail (hinge) anywhere between the superstructure and the abutment foundation (Figure 16.5). With semi-integral abutments, however, the superstructure is integrally connected to the abutment backwall, but the abutment backwall is isolated from the abutment cap by means of certain hinge detail (Figure 16.4). The superstructures for semi-integral bridges are generally supported on bearings as with a conventional structure, thus allowing longitudinal translation. In this case the backwall is separated from the abutment stem, yet the beam ends are encased in the backwall as in an IAB. Semi-integral abutments offer some of the advantages of fully integral abutments such as elimination of expansion joints and a robust-end diaphragm detail for the superstructure, while also reducing the moment demand on the piles by providing a reliable hinge detail that allows the piles to behave in a free-head rather than fixed-head manner (FHWA 2012).

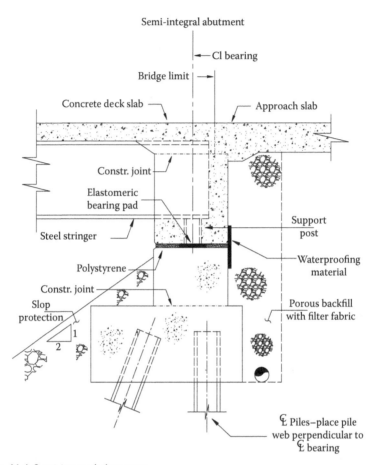

Figure 16.4 Semi-integral abutment.

An important step in integral and semi-integral abutment design is to figure out all the loads and calculate horizontal forces and moments in the foundation elements. The following loads are considered in the integral abutment design:

- Abutment cap self-weight
- Abutment backwall self-weight
- Abutment wingwall self-weight
- Miscellaneous dead loads (bearing seats, lateral restraints, etc.)
- Superstructure dead load
- Approach slab dead load
- Lateral soil pressure on the backwall (active and passive)
- Lateral soil pressure on the wingwalls
- Longitudinal applied forces (in select cases, depending on the nature of the bearings provided between the superstructure and the abutment)

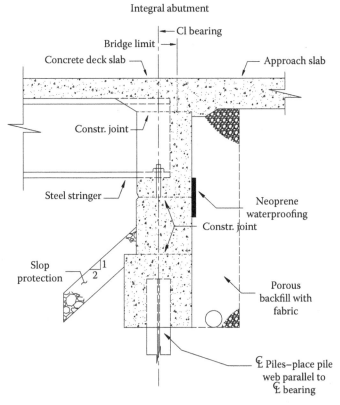

Integral abutment

Cl bearing

Bridge limit

Concrete deck slab

Approach slab

Constr. joint

Steel stringer

Neoprene
waterproofing

Constr. joint

Slop
protection

1
2

Porous
backfill with
fabric

℄ Piles–place pile
web parallel to
℄ bearing

Figure 16.5 Full-integral abutment.

- Induced forces due to longitudinal movements (most important thermal movements)
- Seismic loads

In addition to the primary effects due to dead load, live load, and so on, integral bridges are subjected to secondary effects due to (1) creep and shrinkage, (2) thermal gradients, (3) differential settlement and differential deflections, (4) pavement-relief pressures when moisture and sustained high temperatures trigger pavement growth, and (5) soil–pile interaction (Arockiasamy et al. 2004).

16.2 PRINCIPLE AND ANALYSIS OF IABs

The difference of modeling integral bridges and other types of bridges is handling the soil–structure interaction. Analysis methodologies range widely from simple to comprehensive analyses. A more comprehensive analysis is usually combined with a nonlinear soil–structure interaction analysis of the foundation elements. A simplified way to approach this is to separate the foundation analysis from the rest of the structure and consider the foundation elements independently. For the case of pile or drilled shaft foundations, this lateral analysis would be accomplished via a laterally loaded pile analysis, often facilitated by a standardized computer model based on a p–y curve analysis of the lateral response of the soil until the laterally loaded pile analysis and the structural analysis converge.

A more rigorous approach to a comprehensive analysis might involve the modeling of the soil response directly in the structural analysis model. This step eliminates the tedious iterations of exchanging information manually between the geotechnical and the structural analysis models, but the resulting soil–structure interaction model can become fairly complex (FHWA 2012). Often a simple two-dimensional (2D) model is a sufficiently comprehensive approach to the soil–structure interaction analysis. For a skewed or curved bridge, a full three-dimensional (3D) analysis may be warranted.

In many integral abutments with foundations on steel piles, longitudinal movements of the bridge will cause sufficiently high internal loads so that the plastic moment capacity of the pile is exceeded. In those cases, the common assumption is to allow a plastic hinge to form during the analysis, which provides significant moment relief for any movements above those that cause yielding of the piles. Some designers have pointed out that allowing a plastic hinge at the pile–abutment interface while simultaneously sizing the pile to prevent even a nominal overstress in terms of bending–axial interaction lower in the pile represents an inconsistent design approach, but to date there have been no known significant in-service problems for piles designed in this fashion.

16.2.1 Force analysis

In this section, a sample design calculation to attain moment and shears of a semi-integral abutment is provided for better understanding of the loading due to earth pressure. Figure 16.6a shows the cross section of the superstructure, and Figure 16.6b shows the elevation view of the semi-integral abutment. The calculation steps for the backwall moments and shears are shown here:

Earth pressure resultant per unit width

$$w = \frac{1}{2} K_p \left(H_{\text{backwall}} \right)^2 \tag{16.1a}$$

For unit weight of soil $\gamma = 145$ pcf (2325 kg/m³), $K_p = 4$, assuming the use of Expanded Polystyrene (EPS) material behind backwall, and backwall height $H_{\text{backwall}} = 1.93$ m (6.33′), calculated $w = 11.6$ klf (169.2 kN/m). This can be assumed a distributed line load applied along the abutment.

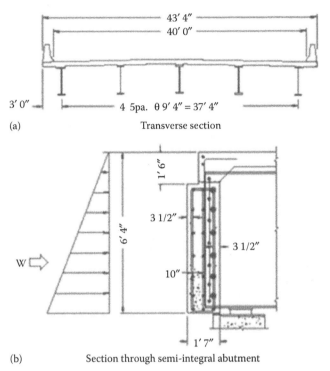

(a) Transverse section

(b) Section through semi-integral abutment

Figure 16.6 (a, b) Sample design of a semi-integral abutment.

Beam/girder spacing along the skew can be calculated with

$$L = \frac{S_{Beam}}{\cos 30°} \qquad (16.1b)$$

With beam spacing $S = 2.84$ m ($9.33'$) and bridge skew angle of $30°$, $L = 3.28$ m ($10.77'$).

Then, by assuming continuous beam along the abutment with supports at girder lines, the moments, shears, and reaction under triangular earth pressure can be calculated as

$$\text{Max positive moment: } M_{pos} = 0.08wl^2 = 0.08(11.6 \text{ klf})(10.77 \text{ ft})^2$$
$$= 107.6 \text{ ft-kip}(145.9 \text{ kN-m})$$

$$\text{Max negative moment: } M_{neg} = 0.10wl^2 = 0.10(11.6 \text{ klf})(10.77 \text{ ft})^2$$
$$= 134.6 \text{ ft-kip}(182.5 \text{ kN-m})$$

Max shear: $V_{\text{max}} = 0.6wl = 0.6(11.6 \text{ klf})(10.77 \text{ ft})$
$$= 75.0 \text{ kip}(101.7 \text{ kN})$$

Max reaction at girder: $R_{\text{max}} = 1.1wl = 1.1(11.6 \text{ klf})(10.77 \text{ ft})$
$$= 137.4 \text{ kip}(186.3 \text{ kN})$$

16.3 MODELING OF IABs

When an IAB is analyzed, 2D and 3D models using the finite element method (FEM) can be built. Three types of soil modeling are used: (1) equivalent cantilever finite element model, (2) soil spring finite element model, and (3) soil continuum finite element model.

16.3.1 Equivalent cantilever finite element model

For piles used in the IAB design, there are two pile design alternatives, (1) conventional elastic design approach and (2) inelastic design approach, which address the following three AASHTO specification design criteria (Greimann 1989):

1. Capacity of the pile as structural member (Case A)
2. Capacity of the pile to transfer the load to the ground (Case B)
3. Capacity of the ground to support the load (Case C)

In Case A, a pile embedded in soil can be analytically modeled as an equivalent beam–column structural member without transverse loads between the member ends and with a base fixed at a specific soil depth. There can be either a fixed or pinned head based on the rotational restraint at the pile head. Figure 16.7 shows an idealized fixed-headed pile for both (a) an actual system and (b) the corresponding equivalent cantilever system. The total length l of the equivalent cantilever equals the sum of the length l_u above the ground and the length l_e from the soil surface to the fixed base of the equivalent cantilever. The pile length, l_e, that defined whether the pile behaves as a rigid or flexible pile is given as (Greimann 1989)

$$l_e = 4\left(\sqrt[4]{\frac{EI}{k_b}} \right) \tag{16.2}$$

where:
> E, I is the modulus of elastic and moment of inertia with respect to the plane of bending of the pile
>
> k_b is the horizontal stiffness of the soil

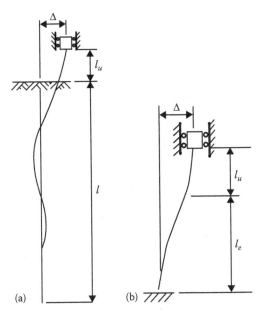

Figure 16.7 Cantilever idealization of a fixed-headed pile. (a) Actual system. (b) Equivalent system.

For nonuniform soil conditions, an equivalent uniform lateral soil stiffness parameter, k_e, is used to evaluate the length l_e as

$$k_e = \frac{3}{l_0^3} \int_0^{l_0} k_h(x)(l_0 - x_1)^2 \, dx \qquad (16.3)$$

where x_1 is the depth below the abutment. As length l_0 is a function of k_e, interaction is needed for the calculation of k_e.

AASHTO Cases B and C assume that the lateral displacement of the pile can affect the capacity of the pile to transfer load to the ground through vertical friction along the embedment length in Case B, but should not affect the end-bearing resistance of flexible piles, nor the capacity of the ground to support the load in Case C. Details of these two cases (Greimann 1989) are not discussed here as the equivalent cantilever finite element model is a simplified method with approximation compared to the next two methods.

16.3.2 Soil spring finite element model

This modeling technique represents the soil around a pile as a Winkler foundation with distributed springs and dashpots (for dynamic analysis only) that are constant or frequency dependent or with lumped springs concentrated at a finite number of nodes. In the modeling process, while the bridge deck, abutment walls, girders, and cross members at the piers are idealized using four-node shell elements, piles and remaining cross members are modeled as beam elements. The soil backfill and the piles, fixed at their base, support the abutments.

16.3.2.1 Soil spring and p–y curve

To allow the stiffness of the deck–girder connection to be varied, spring tied elements are employed at their interface. Nonlinear spring elements model the soil backfill as well as the soil around the piles. A set of p–y curves may be generated using the modified Ramberg–Osgood model as shown in Figure 16.8 for different types of soil, particularly very stiff clay, loose sand, and dense sand. Similar curves for f–z (load–slip) and q–z (pile tip load–settlement) are also generated using the same modified Ramberg–Osgood model. Based on Greimann and Wolde-Tinsae (1988), the modified Ramberg–Osgood model can be used to approximate the p–y, f–z, and q–z soil displacement–resistance curves as follows:

$$P = \frac{k_h y}{\sqrt[n]{1 + |y/y_u|^n}} \qquad (16.4a)$$

where:

$$y_u = \frac{P_u}{k_h} \qquad (16.4b)$$

where:
  k_h is the initial lateral stiffness
  P is the generalized soil resistance
  P_u is the ultimate lateral soil resistance

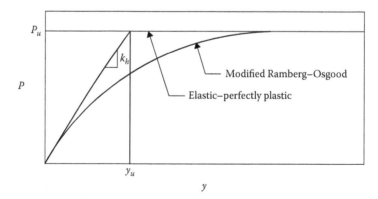

Figure 16.8 The modified Ramberg–Osgood curve for a typical P–y curve.

n is the shape parameter
y is the lateral displacement of the pile
y_u is the ultimate lateral displacement

Alternatively, the guidelines by the American Petroleum Institute (API) (1993) are used to develop the P–y curves, which represent the stiffness for the nonlinear springs substituting the soil around the piles. The P–y relationship is a hyperbolic tangent curve defined as follows:

$$P = AP_u\tanh\left[\frac{kz}{AP_u}y\right]$$ (16.5a)

where:
P_u is the ultimate bearing capacity
k is the parameter defined by φ angle of internal friction
z is the depth in the soil
y is the lateral displacement of the pile
A is the parameter that varies with soil depth in case of static loading
according to Equation 16.5a

$$A = 3.0 - 0.8\frac{X}{D} \geq 0.9$$ (16.5b)

where:
X is the soil depth
D is the average pile length

16.3.2.2 Soil behind the abutment

The soil–structure interaction is modeled by attaching linear springs at the selected nodes of the abutment and piles. The springs simulate the effect of the abutment fill on the bridge. As shown in Figure 16.9, the number of soil springs behind the abutment depends on the size of the tributary area each spring represents.

Using the design curves by National Cooperative Highway Research Program (NCHRP, Barker 1991), passive and active earth pressure effects behind the abutment can be modeled for the soil with the corresponding unit weight and φ angle of internal friction.

16.3.2.3 Soil around piles

Figure 16.10 shows the soil–pile interaction where the soil is idealized by three sets of springs: lateral springs k_h, vertical springs k_v, and a point spring k_q. Table 16.1 lists the parameters for soil spring (Greimann and Wolde-Tinsae 1988).

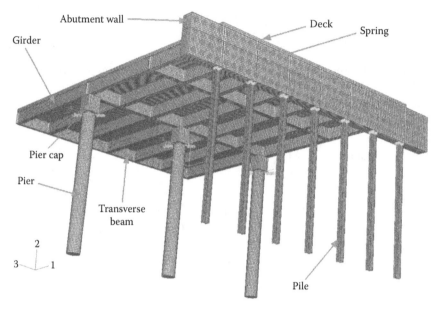

Figure 16.9 Rendering finite element model of an integral abutment bridge. (Data from Shah, B.R., "3D Finite Element Analysis of Integral Abutment Bridges Subjected to Thermal Loading," MS Thesis, Kansas State University, Manhattan, KS, 2007.)

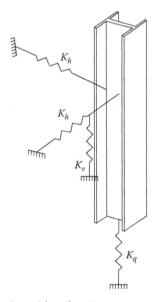

Figure 16.10 Soil–pile interaction with soil springs.

Table 16.1 Parameters for soil springs

Parameter	Case	
	Clay	Sand
Lateral springs		
P_u	$9c_uB$	$3\gamma Bk_px$
k_h	$67c_u$	n_hx
Vertical springs		
f_{max} (H-piles), (klf)	The least of $2(d + b_f)c_u$ $2(d + 2b_f)c_a$ $2(dc_u + b_fc_a)$	$0.02N[2(d + 2b_f)]$
f_{max} (others), (klf)	The lesser of: l_gc_a l_gc_u	$0.04Nl_g$
k_v	$10f_{max}/z_c$	$10q_{max}/z_c$
Point spring		
q_{max} (ksf)	$9c_u$	$8N_{corr}$
k_q	$10q_{max}/z_c$	$10q_{max}/z_c$

B = pile width;
b_f = flange width of H-pile (ft);
c_a = adhesion between soil and pile = αc_u (psf);
c_u = undrained cohesion of the clay soil = $97.0N + 114.0$ (psf);
d = section depth of H-pile or diameter of pipe pile (ft);
J = 200 for loose sand, 600 for medium sand, 1500 for dense sand;
l_g = gross perimeter of the pile (ft);

$$k_p = \tan^2\left(45° + \frac{\varphi}{2}\right);$$

N = average standard penetration blow count;
N_{corr} = corrected standard penetration test (SPT) blow count at the depth of pile tip
     = N (uncorrected) if $N \leq 15$;
     = $15 + 0.5\ (N - 15)$ if $N > 15$;
n_h = constant of subgrade reaction = $J\gamma/1.35$;
x = depth from the soil surface;
z_c = relative displacement required to develop f_{max} or q_{max}
   = 0.4″ (0.033′) for sand;
   = 0.2″ (0.021′) for clay;
α = shear strength reduction factor;
γ = effective unit soil weight;
φ = angle of internal friction.

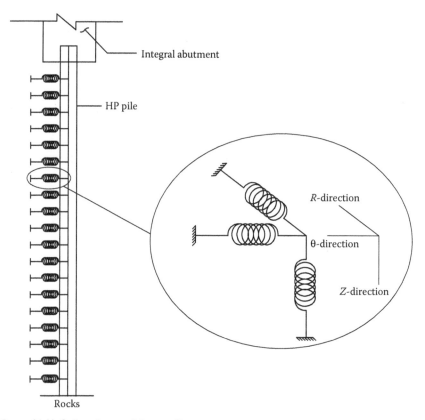

Integral abutment

HP pile

R-direction

θ-direction

Z-direction

Rocks

Figure 16.11 Soil spring model on a pile.

   Three types of soil resistance–displacement models can describe soil characteristics (Greimann and Wolde-Tinsae 1988): lateral resistance–displacement
(*p–y*) curves; longitudinal load–slip (*f–z*) curves; and pile tip load–settlement
(*q–z*) curves. The *p–y* curves represent the relationship between the lateral
soil pressure against the pile (force per unit length of the pile) and the corresponding lateral pile displacement. The *f–z* curves describe the relationship between skin friction (force per unit length of the pile) and the relative
vertical displacement between the pile and the soil. The *q–z* curves describe
the relationship between the bearing stress at the pile tip and the pile tip
settlement. The total pile tip force is *q* times the effective pile tip area. All
three types of curves assume the soil behavior to be nonlinear and can be
developed from basic soil parameters where the *p–y* curve is the most prominent one, so it is called the *p–y* method. Figure 16.11 shows the spring
model of a steel pile.

16.3.3 Soil continuum finite element model

In this approach a 2D or 3D finite element model of the superstructure and substructure, including surrounding soil, is built. Both pile and soil can be modeled into a 3D finite element model using eight-node solid continuum elements with a nonlinear response (Khodair and Hassiotis 2013). While an elastic–plastic response was adopted for the pile elements, the Mohr–Coulomb model with strain hardening idealized the nonlinear soil response. A surface-to-surface contact algorithm was employed to model the sand–pile interaction. To model the tangential contact, the friction coefficient for the interaction between pile and soil materials was calculated.

In the 3D finite element model shown in Khodair and Hassiotis (2013), the pile and soil were modeled using 3D eight-node solid continuum elements. Three boundary conditions were imposed in the finite element model: (1) the pile is fixed at the bottom to model the embedment of the piles, (2) all degrees of freedom associated with the exterior surface of the sand surrounding the piles are restrained to model the confinement of the galvanized steel sleeves by crushed stone backfill (which may not be the case for others), and (3) guided fixation at the top of the pile is modeled by tying the nodes at the top surface of the pile to a defined reference point located in the centroid of the cross section of the pile at its top to simulate the embedment of the piles into the abutments. The steel piles were modeled using an elastic–perfectly plastic model. The soil was modeled using a strain hardening model implementing Mohr–Coulomb failure criterion. The soil–pile interaction was simulated by adopting tangential and normal contact behavior in the model. Master and slave surfaces were defined in the model such that the exterior surface of the pile was used to model the master surface and the interior surface of the sand was used for the slave surface.

For a predrilled hole case, such as a drilled shaft, special treatment has to be made where interface elements allow relative movement between the structural elements and the contact soil. More detailed description of the modeling technique is covered by an illustrated example in Section 16.5.

16.4 ILLUSTRATED EXAMPLE OF A STEEL GIRDER BRIDGE IN SOIL SPRING FINITE ELEMENT MODEL

An IAB described in a PhD dissertation at the University of Maryland (Thanasattayawibul 2006) is used as a case study in this chapter. The cross section of the bridge is shown in Figure 16.12. The bridge consists of a 178-mm (7″)-thick concrete slab that is supported by six girders. There is a 0.6-m (2′) overhang on each side of the bridge. There are 11 piles supporting

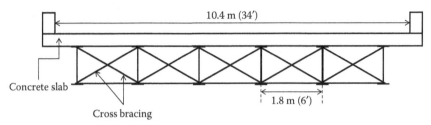

Figure 16.12 Cross section of the illustrated bridge.

the bridge abutment oriented for strong axis bending. Girders, cross bracing, and piles are beams of W30X132, L6X6, and HP10X42, respectively. The piles are placed such that the bending occurs around their strong axis. The pile length in this example is 12.5 m (41'), of which 0.3 m (1') is within the abutment. The total width of the abutment is equal to the width of the bridge, 10.4 m (34') as shown in Figure 16.12.

16.4.1 Structure

The model is analyzed using the ANSYS program. The shell element type that is chosen for the slabs, girders, and piles is SHELL 43, a four-node plastic shell. The element has plasticity, creep, stress stiffening, large deflection, and large strain capabilities. The element has six degrees of freedom at each node: translations in the nodal x-, y-, and z-directions and rotations about the nodal x-, y-, and z-axes. Cross bracings are modeled using beam elements of type BEAM 4, a 3D elastic beam. BEAM 4 is a uniaxial element with tension, compression, torsion, and bending capabilities. The element has six degrees of freedom at each node: translations in the nodal x-, y-, and z-directions and rotations about the nodal x-, y-, and z-axes. Abutments are modeled using solid elements of type SOLID 45. The element has plasticity, creep, swelling, stress stiffening, large deflection, and large strain capabilities. The element is defined by eight nodes having three degrees of freedom at each node: translations in the nodal x-, y-, and z-directions. Multipoint constraint elements, MPC184, with rigid beam option are used to connect all elements together. MPC184 comprises a general class of multipoint constraint elements that implement kinematic constraints using Lagrange multipliers. A rigid beam option has six degrees of freedom at each node: translations and rotations in x-, y-, and z-directions.

   As stated, the concrete slab is modeled using shell elements, and a node is placed at each end of the typical section, along the centerline of each girder, along each end of the girders' top flange, and at a point halfway between girders. Beam elements are used to model the cross bracings with the same nodes at the intersection of webs and flanges. The layout of nodes for the concrete slab, girders, and cross bracings is shown in Figure 16.13.

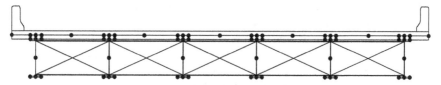

Figure 16.13 Superstructure node distribution of the IAB example.

16.4.2 Soil

Soil is modeled using spring elements, COMBIN39, a spring between a node and ground. The spring is a unidirectional element with nonlinear generalized force–deflection capability. The element has three degrees of freedom at each node: translations in the nodal x-, y-, and z-directions, with their properties as described in Section 16.3.2. There are three types of springs used in the model. The first type represents the displacement in lateral and longitudinal directions and consists of two springs. Both springs are at the center of the web. They are modeled at each layer of the nodes along the pile starting at one layer below the bottom of the abutment and continuing all the way to one layer above the tip of the pile. The second type of spring represents friction along the pile. It consists of a single spring at each node along the web of each pile starting one layer below the bottom of the abutment and ending one layer above the tip of the pile. The third and final type of spring is the tip spring that represents the settlement in the pile and consists of seven springs at each node at the tip of the pile. This spring representation of the tip of each pile allows for uniform resistance to pile settlement and is used in the analysis of friction piles. These pile tip–settlement springs are replaced with fixed end conditions when analyzing bridge models with end-bearing piles. Figure 16.11 depicts the spring model of a steel pile used in this example.

16.5 ILLUSTRATED EXAMPLE OF A STEEL GIRDER BRIDGE IN 3D SOIL CONTINUUM FINITE ELEMENT MODEL

The same example used in Section 16.4 built with soil springs is also used here to demonstrate the soil continuum finite element model. A 3D nonlinear finite element model using ANSYS was built and listed in another PhD dissertation at the University of Maryland (Rasmi 2012). The nonlinearity is considered for the nonlinear effect of the material plasticity of steel piles. Due to symmetry and the complicity of the continuum modeling, only a quarter of the bridge was modeled.

The entire model was meshed using plane and hexahedral elements. The concrete slabs, piles, and girders were meshed using the 2D shell element

(SHELL181) of four nodes with six degrees of freedom per node. Cross bracings were modeled using the one-dimensional beam element (BEAM188). The concrete abutment and soil were modeled using a 3D solid element (SOLID185) of 20 nodes with three degrees of freedom per node. SOLID185 has plasticity, hyperelasticity, stress stiffening, creep, large deflection, and large strain capabilities and is used for 3D modeling of solid structures.

As described in Rasmi's work (Rasmi 2012; Rasmi et al. 2013), the geometry of the quarter model and the boundary conditions are shown in Figure 16.14. Symmetry boundary conditions are applied on the symmetry planes: $z = 0$ on symmetry surface 1 and $x = 0$ on symmetry surface 2. The bottom of the soil is fixed in the y- and z-directions to simulate the end-bearing type pile. The soil thickness in the positive z-direction (backfill soil thickness) is assumed to be 0.9 m (3′), and its thickness in the negative z-direction behind the piles is assumed to be 3 m (10′). Assuming that these soil layers are thick enough, the free surfaces of the soil are assumed to be stationary in the z-direction as the piles move. Therefore, the displacements perpendicular to these free areas (displacement in the z-direction) are assigned zero value as the boundary condition. Gravity is applied in the y-direction (Figure 16.14). The supports are provided in the y-direction underneath the slab at 15.2-m (50′) distances. The y-displacement at these constraints is zero.

As for material properties, steel material used in piles, girders, and cross bracings are modeled as elastic–plastic material with multilinear plastic behavior using a MISO command in ANSYS and only the deviatoric stress is assumed to cause yielding. Concrete, where it is only used for slab and abutment, is assumed to behave only elastically. For soil, the material cannot stand

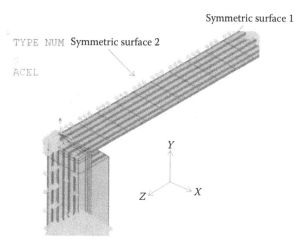

Figure 16.14 Geometry of the quarter model of the integral abutment bridge. (Data from Rasmi, J., "Thermo-Mechanical Fatigue of Steel Piles in Integral Abutment Bridges," PhD Dissertation, Civil and Environmental Engineering, University of Maryland, College Park, MD, 2012.)

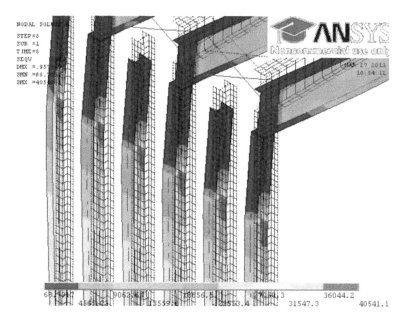

Figure 16.15 Model displacement due to temperature. (Data from Rasmi, J., "Thermo-Mechanical Fatigue of Steel Piles in Integral Abutment Bridges," PhD Dissertation, Civil and Environmental Engineering, University of Maryland, College Park, MD, 2012.)

tension and can support only compressive forces where their strength and yield are pressure dependent. For this type of material usually the Drucker–Prager (DP) model is used (Rasmi 2012). To define the DP model, a flow potential and yield function are required. Several different types of functions are available in ANSYS (linear, power low, and hyperbolic). For this analysis a linear yield function and a linear flow potential are used. Figure 16.15 shows the analysis results of model displacement due to temperature.

In this chapter, two examples, one using 3D soil spring finite element model and another using 3D soil continuum finite element model, are illustrated. The soil spring finite element model is more commonly used in research as well as in design. The benefit of using this model is the simplification of assigning $p-y$ curves to their respective soil springs, and only local soil has to be concerned. The soil continuum finite element model usually involves a more calculation-intensive modeling of the surrounding soil. In this case, soil–structure interaction and the artificial soil boundary to reflect the wave are of major concerns of the modeling process. The third modeling technique associated with boundary elements, which usually needs special programs or finite element library to solve the problem, is not discussed in this example. When modeling using the third method, soil elements should be carefully selected to simulate the behavior between the soil and the piled structure.

Chapter 17

Dynamic/earthquake analysis

17.1 BASICS OF DYNAMIC ANALYSIS

Structures may subject to both static and dynamic loading. Unlike static analysis, in which only static structural displacement is considered, acceleration and velocity are introduced as well in dynamic analysis. For a system that has only one degree of freedom (DOF), as shown in Figure 17.1, the forces resisting the applied loading are considered as the following:

1. A force proportional to displacement (the stiffness), which can be expressed as ky
2. A force proportional to velocity (the damping), which can be considered as $c\dot{y}$
3. A force proportional to acceleration (the inertia), which can be expressed as $m\ddot{y}$

So, as shown in Figure 17.1, the fundamental dynamic equilibrium equation is

$$m\ddot{y}(t) + c\dot{y}(t) + ky(t) = f(t) \tag{17.1a}$$

where y, $\dot{y}$, and $\ddot{y}$ are displacement, velocity, and acceleration, respectively.

For a system that has multiple DOFs, the equation corresponding to 17.1a can be rewritten as

$$M\ddot{a}(t) + C\dot{a}(t) + Ka(t) = f(t) \tag{17.1b}$$

where:
  M is the global mass matrix
  C is the global damping matrix
  K is the global stiffness matrix
  $a(t)$ is the displacement vector
  $f(t)$ is the external load vector

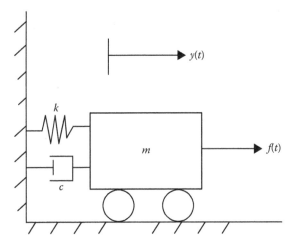

Figure 17.1 Dynamic forcing system.

In comparison with the static equation 3.3, forces due to acceleration and damping are introduced in dynamic analysis.

The damping ratio, or damping coefficient, ξ, is defined as $c/c_r = c/2\sqrt{km}$ where steel bridges normally have a low damping coefficient $\xi \leq 0.02$. Most commonly used experimental method to determine the damping in a structure is the half-power (bandwidth) method by two frequencies shown in Figure 17.2 and can be calculated by Equation 17.2 as

$$\xi = \frac{f_2 - f_1}{f_2 + f_1} \tag{17.2}$$

where two frequency points f_1 and f_2 (in cycle/sec) are on either side of the curve in Figure 17.2.

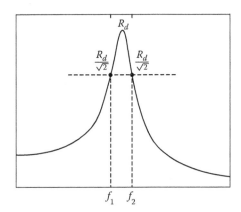

Figure 17.2 Half-power method to estimate damping by experiment.

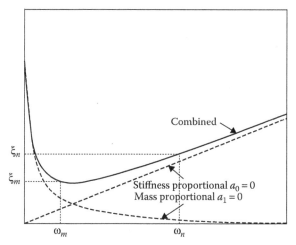

Figure 17.3 Relationship between damping ratio and frequency for Rayleigh damping.

Modal analysis is the most popular and efficient method for solving engineering dynamic problems. In order to apply modal analysis of damped systems, it is common to assume proportional damping. Mathematically the most common and easy way is to use Rayleigh damping method, with a linear combination of the mass and the stiffness matrices as

$$c = a_0 m - a_1 k \qquad (17.3)$$

where:

    c, m, and k are the damping, the mass, and the stiffness matrix, respectively

    a_0 and a_1 are proportional constants

The relationship between damping ratio and frequency for Rayleigh damping is shown in Figure 17.3. By simplification, this relationship leads to the next equation:

$$\left\{ \begin{matrix} \xi_n \\ \xi_m \end{matrix} \right\} = \frac{1}{2} \begin{bmatrix} \dfrac{1}{\omega_n} & \omega_n \\ \dfrac{1}{\omega_m} & \omega_m \end{bmatrix} \left\{ \begin{matrix} a_0 \\ a_1 \end{matrix} \right\} \qquad (17.4)$$

where ω_n and ω_m are the damping ratios (ξ_n and ξ_m) associated with two specific angular frequencies (ω_n and ω_m in radian/second) are known, the two Rayleigh damping factors (a_0 and a_1) can be calculated by Equation 17.4.

17.2 PRINCIPLE OF BRIDGE DYNAMIC ANALYSIS

In this section, five types of bridge dynamic analysis will be briefly discussed. The first type is the dynamic interaction between vehicle and bridge. The second type is the pedestrian bridge dynamics between pedestrian and bridge, which gained more attention recently. The third type is associated with the dynamic methods for analyzing bridge structures, including soil–foundation–structure interaction, when subjected to earthquake loads. The fourth type is the blast analysis, and the fifth type is the analysis of long-span bridge responses to wind. These five different types of analysis, though all based on linear or nonlinear analyses, have different emphases and thus different modeling techniques, which will be discussed in Sections 17.2.1 through 17.2.5.

17.2.1 Vehicle–bridge interaction

The aim of this subject is to analyze the effects of highway vehicle- or train-induced vibrations for impact analysis or fatigue analysis. The vibration-induced stresses could lead to fatigue or other types of failure, such as deck cracking. In the modeling process, only the superstructure is of a concern to be included in a beam (Figure 17.4), grid, or more sophisticated shell model. Although the structural analysis model is linear, the interaction between bridge and moving vehicle or train is often considered as a nonlinear dynamic problem in the aspect of time domain. To simulate the dynamic interaction,

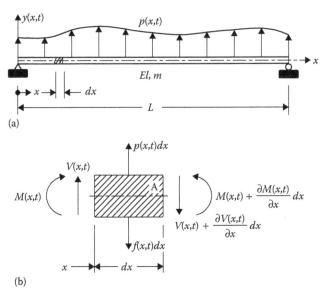

Figure 17.4 Basic beam subjected to dynamic loading. (a) Beam properties and coordinates. (b) Resultant forces acting on the differential element.

theoretically the bridge and the vehicle or train could be modeled into two elastic structures connected with contact force due to their relative movement. This contact force interacting with two structures is time dependent and nonlinear as the contact force might move from time to time. All vehicles possess the suspension system, either in air suspensions or steel-leaf suspensions. Air suspensions use hydraulic shock absorbers for damping, whereas steel-leaf suspensions use steel strips to provide damping through Coulomb friction between steel strips. Meanwhile, the bridge is also an elastic body subject to the dynamic loading due to moving vehicles (Figure 17.5). A two-dimensional nonlinear vehicle simulation program (NLVSP) was developed by Cole and Cebon (1992) to predict the tire forces of articulated vehicles with well-damped suspension modes under typical speed and road roughness. The steel-leaf-spring suspension model used in the NLVSP is simulated by nonlinear suspension elements. For an air-suspended vehicle, air-spring elements with parallel viscous dampers are then used. BridgeMoment, developed by Green and Cebon (1994), Varadarajan (1996), and Xie (1999), predicts the bending moments in a bridge due to the passing of a heavy transport vehicle. The bridge displacement is determined by the general equation of motion for a bridge and the convolution integral. For this bridge structure, the infinite number of DOFs can be discretized to a multi degree-of-freedom (MDOF) structure where the general equation of motion for the vertical vibration in a two-dimensional beam can be expressed as

$$m(x)\frac{\partial^2 y(x,t)}{\partial t^2} + c\frac{\partial y(x,t)}{\partial t} + ky(x,t) = f(x,t) \qquad (17.5)$$

where:
   k is the self-adjoint linear differential operator with respect to the spatial variables
   m is the distributed mass of the bridge

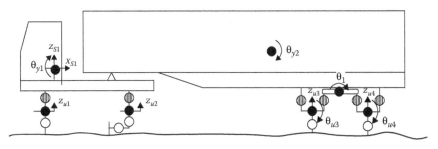

   $\bigcirc$ = linear spring and damper
   $\oplus$ = nonlinear leaf-spring element

Figure 17.5 Tractor and trailer vehicle model. (Data from Cole, D.J. and Cebon, D., "Validation of Articulated Vehicle Simulation," Vehicle System Dynamics, 21, 197–223, 1992.)

c is the viscous damping operator with respect to the spatial variables

$y(x,t)$ is the vertical deflection of the bridge along longitudinal x-direction at time t

$f(x, t)$ is the force exerted by the vehicle on the bridge

To obtain a unique solution, the boundary conditions and the initial displacement $y(x,0)$ and velocity $\dot{y}(x,0)$ must be defined. The eigenvalues and eigenvectors (modal shapes), all in the vertical direction, of Equation 17.5 can be easily handled by close-form solution or through mathematical modeling. Based on Green and Cebon (1994), Equation 17.5 can be solved with the convolution integral of

$$y(x,t) = \int_{\infty}^{-\infty} h(x,x_f,t - \tau)f(x_f,\tau)d\tau \qquad (17.6)$$

where $h(x,x_f,t - \tau) = $ impulse response function at position x for an impulse applied at position x_f, which is related to the mode shapes. Therefore, the bridge response is determined by the mode shapes and the forcing function.

The main factors affecting vehicle-induced bridge dynamics are bridge surface roughness, speed, frequency matching, and vehicle suspension type (Cantieni and Heywood 1997). Solving the problem can be described in the following steps (MacDougall et al. 2006):

Step 1: Simulate the vehicle within a routine to solve for the vehicle's natural frequency and the wheel static load. This routine is used to predict the tire forces of articulated vehicles where, for example, Figure 17.5 shows an 11-DOF vehicle model used by Cole and Cebon (1992).

Step 2: Apply the vehicular loads on the bridge model to calculate the force $f(x,t)$ exerted on the bridge at certain location x and time t due to the moving vehicle.

Step 3: Use the calculated force $f(x,t)$ from step (2) and the bridge's impulse response function $h(x,x_f,t - \tau)$ to determine the bridge's deflection $y(x,t)$.

Step 4: Based on the calculated bridge's deflection $y(x,t)$, the equivalent external loading applied on the bridge is equal to the sum of the bridge's self-weight and the bridge's inertia force:

$$F_{applied} = mg + ma = mg + m\ddot{y}(x,t) = mg + f(x,t) - c\dot{y} - ky \qquad (17.7)$$

In the case of highway bridges, moving vehicles on bridge are arranged randomly in terms of speeds, loads, direction, and location; however, for railway bridges, train vehicles generally provide uniformly distributed load and can be treated as a sequence of moving masses. Also, railway traffic provides inherent frequencies due to repetitive characters of wheel or bogie loads; a more significant resonance might be produced and affects the bridge durability. Recently, in order to accurately simulate the moving vehicle–bridge interaction, LS-DYNA (1998) with FEA was used.

17.2.2 Pedestrian bridge vibrations

Resonance has been ignored in the design of pedestrian bridges until recently. Pedestrian bridges, especially light bridges supported by cables, should be checked for vibration serviceability due to human activities. Unless the bridge is supported by flexible substructure or soil condition, only the superstructure is simplified in the modeling process as a beam linear dynamic analysis model. Modal analysis is the first step for the pedestrian bridge dynamic analysis for determining the natural frequencies and mode shapes of a structure, as well as the responses of individual modes to a given excitation. Vibration of the pedestrian bridge can be due to two sources, vertical and lateral vibrations. Lateral vibration is assuming synchronous lateral excitation. This occurs when a large enough group of pedestrians senses a lateral movement and subconsciously tries to counteract that movement by shifting their weight in opposition to the perceived movement, in effect creating a steady driving force. On the other hand, footsteps are the source of vertical vibration where the force $f(t)$ in Equation 17.1 can be represented by

$$\sum f(t) = P\left(1 + \sum \alpha_i \cos\left[2\pi i f_{\text{step}} t + \varphi_i\right]\right) \tag{17.8}$$

where:
P is the person's weight
α_i is the dynamic coefficient for the harmonic force
i is the harmonic multiple (1, 2, 3,...)
f_{step} is the step frequency of activity
t is the time
φ_i is the phase angle for the harmonic

In the assumption, f_{step} is commonly assumed at 2 Hz, or 2 steps per second. Values for alpha are typically taken at 0.5, 0.2, 0.1, and 0.05 for the first four harmonics of walking. It is when f_{step} matches the frequency of any of the modes of vibration of the structure that resonance will occur.

Figure 17.6 shows recommended peak acceleration for human comfort for vibrations due to human activities (Allen and Murray 1993; Murray et al. 1997). As shown in the figure, the tolerance limits for vibration frequencies between 4 and 8 Hz are lower, whereas outside this frequency range, people accept higher vibration accelerations. Two sources provide design-limiting values for bridges:

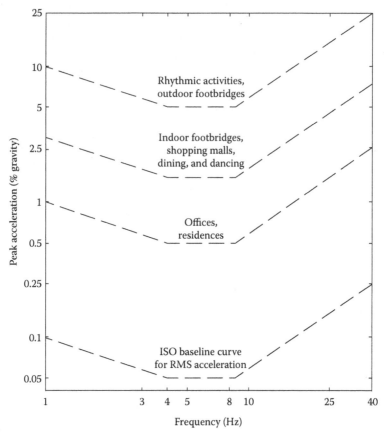

Figure 17.6 Recommended peak acceleration for human comfort for vibrations due to human activities. (Data from Allen, D.E. and Murray, T.M., *Engineering Journal*, 4th Qtr, AISC, 117–129, 1993; Murray, T.M., Allen, D.E., and Ungar, E.E., "Floor Vibrations due to Human Activity," *AISC Steel Design Guide #11*, Chicago, IL, 1997. https://www.aisc.org/store/p-1556-design-guide-11-floor-vibrations-due-to-human-activity-see.aspx.)

- AASHTO (2009)
  - $f \geq 3.0$ Hz and $f \geq 2.85 \ln\left(\dfrac{180}{w}\right)$
  - $W \geq 180 e^{-0.35f}$
  - Special cases: $f \geq 5.0$ Hz
- British Code (*1978 BS 5400*)/Ontario Bridge Code (1991)
  - $f_o \geq 5.0$ Hz
  - $a_{max} = 4\pi^2 f_o^2 y_s K\psi \leq 0.5(f_o)^{1/2}$ m/s^2
  - $F = 180 \sin(2\pi f_o T)$ N
  - $v_t = 0.9 f_o$ m/s (≥ 2.5 m/s per Ontario Code)

where f is the fundamental frequency of the pedestrian bridge where it can be manually calculated by assuming a single-DOF (SDOF) system or found from the computer model (shown in detail in the next section), K is a configuration factor varied from 0.6 to 1.0, and ψ is the dynamic response factor depending on the span length l and decay of vibration δ based on bridge composition.

The serviceability of a pedestrian bridge is important for obvious reasons. In design, the overriding factors for serviceability are the structure's dynamic characteristics—stiffness and its ability to avoid resonance.

17.2.3 Bridge earthquake analysis

AASHTO guide specifications in LRFD (2012), differing from the early practices, is adopting displacement-based design procedures instead of the traditional force-based "R-factor" method. It is widely recognized that the traditional force-based design (FBD) approach cannot provide the appropriate means for implementing concepts of performance-based design. Performance levels as shown in Table 17.2 are described in terms of displacements where damage is in closer correlation with displacements rather than forces. As a consequence, new design approaches, based on displacements, have been recently implemented. The former force approach was based on generating design-level earthquake demands by reducing ultimate elastic response spectra forces by a reduction factor (R-factor). The reduction factor was selected based on structure geometry, anticipated ductility, and acceptable risk. The newly adopted displacement approach is based on comparing the elastic displacement demand to the inelastic displacement capacity of the primary structural components while ensuring a minimum level of inelastic capacity at all potential plastic hinge locations.

Based on their requirements, four seismic design categories (SDCs) are established in AASHTO guide specifications (2012): SDC A (for simple-span bridges), B, C, and D. Three global seismic design strategies are allowed: type 1—ductile substructure/elastic superstructure, type 2—elastic substructure/ductile (steel) superstructure, and type 3—elastic superstructure/elastic substructure/fusing mechanism (seismic isolation or energy dissipation) in between.

Based on Equation 17.1, differential equation governing the response of a structure to horizontal earthquake ground motion $\ddot{u}_g(t)$ is converted to

$$\mathbf{m}\ddot{u} + \mathbf{c}\dot{u} + \mathbf{k}u = -\mathbf{m}\mathbf{l}\ddot{u}_g(t) \tag{17.9}$$

where:

  $\mathbf{u}$ is the vector of N lateral floor displacements relative to the ground
  $\mathbf{m}$, $\mathbf{c}$, and $\mathbf{k}$ are the mass, classical damping, and lateral stiff matrices of the system; each element of the influence vector $\mathbf{l}$ is equal to unity

By using modal response history analysis (RHA), the modal coordinate $q_n(t)$ is governed by

$$\ddot{q}_n + 2\zeta_n \omega_n \dot{q}_n + \omega_n^2 q_n = -\Gamma_n \ddot{u}_g(t) \tag{17.10}$$

In which ω_n is the natural vibration frequency and ζ_n is the damping ratio for the nth mode. The solution $q_n(t)$ can readily be obtained by comparing Equation 17.10 to the equation of motion for the nth-mode elastic SDOF system, an SDOF system with vibration properties—natural frequency ω_n and damping ration ζ_n—of the nth mode of the MDOF system, subjected to $\ddot{u}_g(t)$.

Besides RHA, modal response spectrum analysis (RSA) was also adopted for linear seismic analysis where the peak modal response can be combined by the conservative absolute sum (ABSSUM) modal combination rule:

$$r_n \leq \sum_{n=1}^{N} |r_{n0}| \tag{17.11}$$

or by the more reasonable square-root-of-sum-of-square (SRSS) rule:

$$r_n \cong \left\langle \sum_{n=1}^{N} r_{n0}^2 \right\rangle^{\frac{1}{2}} \tag{17.12}$$

or by the complete quadratic combination (CQC) rule to a system with closely spaced natural frequencies:

$$r_n \cong \left\langle \sum_{j=1}^{N} \sum_{n=1}^{N} \rho_{in} r_{i0} r_{n0} \right\rangle^{\frac{1}{2}} \tag{17.13}$$

17.2.3.1 Linear and nonlinear seismic analyses

Four distinct analytical procedures, as shown in Figure 17.7, can be used in systematic rehabilitation of structures (FEMA-273 1997): linear static, linear dynamic, nonlinear static (pushover), and nonlinear dynamic procedures (NDPs). Linearly elastic procedures (linear static and linear dynamic) are the most common procedures in seismic analysis and design of structures due to their simplicity. On the other hand, adjustments to overall deformations and material acceptance criteria can be incorporated to consider the inelastic response. Based on their importance, bridges can be classified as either ordinary or important bridges where ordinary bridges can be further defined as standard and nonstandard ordinary structures. In their Caltran study, Aviram et al. (2008) described bridge seismic analysis types based on the bridge classifications, which are also listed in Table 17.1.

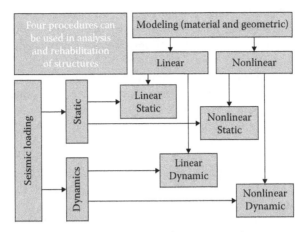

Figure 17.7 Four distinct analytical procedures for seismic analysis.

Table 17.1 Bridge seismic analysis types recommended by Caltrans

Bridge classification	Nonlinear static		Dynamic		
	Equivalent static analysis	Incremental static analysis (Pushover)	Response spectrum analysis—linear	Time-history analysis—direct integration	
				Linear	Nonlinear
Ordinary standard	A	R	A	A	A
Ordinary nonstandard	N	R	A	A	R
Important	N	R	A	A	R

A: Acceptable analysis type
N: Not acceptable analysis type
R: Acceptable and strongly recommended analysis type, not necessarily comprehensive

A large number of bridges were designed and constructed at a time when bridge codes had no seismic design provisions or when these provisions were insufficient according to current standards. Many of these bridges may suffer severe damage when struck by earthquakes, as evidenced by recent moderate earthquakes. Linear elastic procedures are sufficient as long as the structure behaves within elastic limits. If the structure responds beyond the elastic limits, linear analyses may indicate the location of first yielding but cannot predict failure mechanisms and account for redistribution of forces during progressive yielding. This fact makes the elastic procedures insufficient to perform assessments and retrofitting evaluations for those bridges in particular and structures in general. Nonlinear (static and dynamic) procedures are the solutions that can overcome this problem and show the performance level of the structures under any loading level. Nonlinear procedures can also help demonstrate how structures really work by identifying modes of failure and the potential for progressive collapse. Nonlinear procedures will help engineers to understand how a structure will behave when it is subjected to major earthquakes, assuming that the structure will respond beyond the elastic limits, and this will resolve some of the uncertainties associated with codes and elastic procedures. The performance approach, which was shown in AASHTO guide specifications (2012), is considered, as shown in Table 17.2.

Performance-based engineering, with their performance levels shown in Figure 17.8, is set to select design structural criteria such that at specified level ground motion, the structure will not be damaged beyond certain limiting states.

In this section, conventional dynamic analysis (nonlinear dynamic in Figure 17.7) and modal pushover analysis procedures (nonlinear static in Figure 17.7) to determine seismic demands for inelastic structures are presented.

Table 17.2 Performance approach

Probability of exceedance for design earthquake ground motions		Performance level	
		Life safety	Operational
Rare earthquake (MCE)	Service	Significant disruption	Immediate
3% in 75 years	Damage	Significant	Minimal
Frequency of expected earthquake	Service	Immediate	Immediate
50% in 75 years	Damage	Minimal	Minimal to none

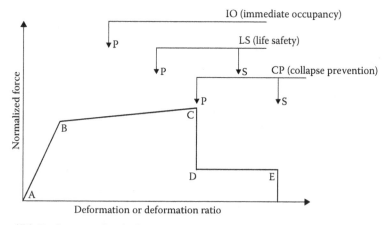

Figure 17.8 Performance level of structures.

17.2.3.2 Nonlinear time-history analysis

Time-history analysis (THA) is a step-by-step analysis of the dynamical response of a structure to a specified loading that may vary with time through a process of numerical integration of the equations of motion. It involves the development of a complete mathematical model of the bridge wherein an effort is made to model nonlinear forms of behavior in a highly localized (rather than global) manner. The mathematical model is formulated in such a way that the stiffness and even connectivity of the elements can be directly modified based on the deformation state of the structure. This permits the effects of element yielding, buckling, and other nonlinear behavior on structural response to be directly accounted for in the analysis. The model is then subjected to time histories of earthquake ground acceleration that may be in either historical records or design spectrum compatible records. In either case, an attempt is made to capture the full time history of the nonlinear structural response.

The use of multiple records in the analyses allows observation of the difference in response resulting from differences in record characteristics. As a minimum, suites of ground motions include at least three different records (FEMA-450 2003).

Different from linear THA, the differential equations of motion for nonlinear THA (NL-THA) cannot be considered as smooth functions. It is due to the nonlinear hysteresis of most bridge structural materials, friction forces developed between contacting surfaces, and buckling of elements. Therefore, only step-by-step methods are recommended for the solution of the nonlinear time history of bridge structures. The step-by-step solution methods attempt to satisfy dynamic equilibrium at discrete time steps and may require iteration, especially when nonlinear behavior is developed in the structure and the stiffness of the complete structural system must be recalculated due to degradation of strength and redistribution of forces (Aviram et al. 2008).

Unlike linear time history, the nonlinear case can take a significant amount of time to solve structural systems with just a few hundred DOFs. Engineers must be careful in the interpretation of the results and check the results using the applicable acceptance criteria. An example of a bridge case analyzed by linear THA and NL-THA methods plotted on the same graph is shown in Figure 17.9.

17.2.4 Blast loading analysis

Blast loads are considered as most extreme loads, and even a small amount of blast can produce a serious damage to the structure. The blast wave produced by explosion travels even faster than the speed of sound. When it arrives at a location, it causes a sudden rise in the normal pressure. The increase in atmospheric pressure over normal values is referred to as overpressure, and the simultaneous pressure created by the blast winds is called dynamic pressure. Blast pressure can create loads on structure that are

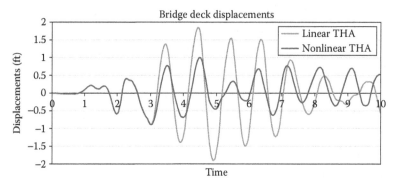

Figure 17.9 Linear versus nonlinear time-history analysis for a nine-span bridge model.

many times greater than the normal design loads, and blast winds can be much more severe than hurricanes.

Blast waves are produced whenever an explosion takes place. These waves propagate in the form of spherical waves, resulting in discontinuities in the structures. Some of these waves transfer across the structures while remaining are reflected back. During this wave propagation, high pressure and high temperature are generated, which travel across the least resistance path of the structure. This entire process of the wave generation and propagation last for a few milliseconds.

The initial step in blast design or analysis is the determination of the blast loads. The factors that consider attention are energy absorption, load combinations, critical elements, and structural redundancy to prevent progressive collapse of the structure.

If an explosion occurs on the top of the bridge, bridge deck will experience the downward thrust of the overpressure, which will be transmitted to other bridge components such as hangers, cables, and towers. Foundation will experience blast-induced vertical and overturning forces. If the blast load is applied at the bottom of the bridge, deck slab and the supporting girders will experience an upward pressure for which they are generally not designed. When they are subjected to vertical upward forces, the bottom of the deck member is subjected to compression and the top is subjected to tension, for which they are not normally designed for. Towers and foundations are also subjected to vertically upward lateral forces and overturning moments. Failure of the system is obvious unless otherwise they are designed for the vertical upward forces.

Several structural analysis options are available for blast-resistant design:

1. Equivalent static analysis (ESA). This method is generally for simple system to determine the equivalent static design load conservatively and neglects the inertial effects of members in motion.

2. SDOF linear/nonlinear dynamic analysis. This method is considered the current state-of-practice method that ignores higher-order failure, allowing for the analysis of a large number of load cases, bridge types, and structural configurations.

3. MDOF, uncoupled/coupled, nonlinear dynamic analysis. This method includes the finite element method (FEM) analysis. A coupled analysis accounts for coupled effects of structural response with fluid dynamics behavior of an explosion load, considering time and spatial coupling while uncoupled analysis does not.

The most common and simplified blast dynamic analysis method used in practice is an SDOF or MDOF, uncoupled, nonlinear dynamic analysis. The loads acting on a structure are usually determined using a shock-wave propagation program. Once the loads have been determined, the structural response can be analyzed using a dynamic structural analysis, accounting for the full plastic capacity of the members. In an uncoupled analysis, the blast load calculations are separated from the structural response. A coupled analysis, which is more refined, performs the blast load calculations and structural response simultaneously. This technique accounts for the motion and response of structural members as the blast wave proceeds around (or through) them and will mostly provide a more accurate prediction of the structural response. Several techniques exist for performing a coupled analysis, all of which involve time–space discretization. Uncoupled analyses will usually provide conservative yet reasonable results with much less effort and are best suited for typical design cases. The LS-DYNA (1998) with FEA-coupled analysis as mentioned in Section 17.2.1 can also be used here to simulate the blast load–bridge interaction.

The dynamic response of bridge structures under a blast load is quite complex due to the highly nonlinear nature of shock wave lasting around several milliseconds. It is hard to analyze accurate deformation or crack conditions of bridges subjected to blast wave. Nonlinear static analysis can be used to analyze the bridge structures with blast loading. Therefore, the blast pressures must be converted to equivalent static loads. In 1990, the U.S. Department of Defense published the TM 5-1300 Manual, *Structures to Resist the Effects of Accidental Explosions*. The manual contains an empirical formula to find the scaled distance (Z) of a blast wave.

$$Z = \frac{R}{W^{1/3}} \tag{17.14}$$

In Equation 17.14, R is the standoff distance of an object from the blast centroid, measured in feet, and W is the charge weight of TNT in pounds. The TM 5-1300 Manual (1990) contains a chart using this empirical formula. A typical pressure time-history curve in free field is shown in Figure 17.10. The positive phase is usually idealized to an equivalent triangular blast load having the same peak pressure and an idealized duration (t_d). The amplitude

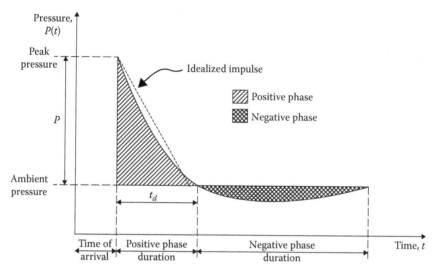

Figure 17.10 Pressure time history for free field blast. (Data from TM 5-1300, *Structures to Resist the Effects of Accidental Explosions*, Department of Army, Washington, DC, 1990.)

of the negative phase is much lower than that of the positive phase, and usually the negative phase is neglected in the design. Only for light structures does the negative phase have a significant effect (Winget et al. 2005).

Although blast load is a dynamic load, equivalent static loads due to explosion are usually used in assessing the structural performance because they impact the structure for a very short duration. If dynamic effect is considered, the transient overpressure loads used on the right-hand side of Equation 17.1 can be estimated where the decay of the reflected overpressure is assumed to obey the modified Friedlander exponential decay equation, which can be written as

$$p(t) = p_m[1 - t/t_p]e^{-\alpha t/t_p} \tag{17.15}$$

where:

  p is pressure
  p_m is peak pressure
  t_p is positive phase duration
  α is the waveform parameter

Since the structure behavior after sudden impact is localized, care must be exercised when performing dynamic analysis; in particular, all high modes of vibration should be included when using modal superposition or Ritz vector analysis methods. Direct step-by-step integration methods are preferable, since such algorithms account for all possible vibration modes associated with the given finite element mesh and analysis time step. Also considered in dynamic analysis are nonlinear dynamic loads and thus nonlinear behavior.

As mentioned above, many used ESA in their studies. Two examples, one for PC and one for steel girder bridges, are illustrated in Chapter 15 as part of the redundancy analysis. Bridge model, attack scenarios, and structural responses were discussed in Sections 15.4 and 15.5.

17.2.5 Wind analysis

Wind induces two typical aerodynamic phenomena in long-span bridges: fluttering and buffeting. The former is an aerodynamic instability that may cause failure of the bridge, and the latter is an aerodynamic random vibration that may lead to fatigue damage, excessive vibration, and large displacements. The wind velocities at which the bridge starts to flutter are called flutter velocities. Aerodynamic design must ensure that the critical flutter velocity is higher than the maximum wind velocity at the site and that the bridge does not vibrate excessively under gusty winds. Flutter may occur in both laminar and turbulent flows. Buffeting is a random response of structures to turbulent flow.

Natural winds, which are turbulent in nature, cause both flutter and buffeting problems (Cai et al. 1999).

Aerodynamic loading is commonly separated into self-excited and buffeting forces. The self-excited forces acting on a unit deck length are expressed as a function of the so-called flutter derivatives (Scanlan 1978a), which can be expressed as

$$\{F_{se}\} = \begin{Bmatrix} L_{se} \\ D_{se} \\ M_{se} \end{Bmatrix} = U^2[F_d]\{q\} + U^2[F_v]\{q\} \tag{17.16}$$

Similarly, the buffeting forces (Scanlan 1978b) are expressed as

$$\{F_b\} = \begin{Bmatrix} L_b \\ D_b \\ M_b \end{Bmatrix} = \bar{U}^2[C_b]\{\eta\} \tag{17.17}$$

where:

L_{se}, D_{se}, and M_{se} are the self-excited lift force, drag force, and torsional moment, respectively

$[F_d]$ and $[F_v]$ are the flutter derivative matrices corresponding to displacement and velocity, respectively

$[C_b]$ is the static coefficient matrix

$\{\eta\}$ is the vector of turbulent wind components normalized by mean wind velocity

U, which is distinguished from the mean value $\bar{U}$ in the previous expression, will be interpreted as the mean or instantaneous wind velocity in different cases

Equations 17.16 and 17.17 can be used to replace the forcing function shown on the right-hand side of Equation 17.1. Special technique and specialized program have to be adopted for the analysis. For details, please refer to Scanlan (1978a, 1978b) and Cai et al. (1999) for FEM formulation.

For wind analysis, many used ESA in their studies. To demonstrate blast analysis, two examples are illustrated in Chapter 15 as part of the redundancy analysis.

Long-span bridge design should follow special guidance for aerodynamic issues. Wind tunnel testing may be unavoidable for the design of long-span bridges. The aerodynamic stability issue is not covered in this chapter while wind load can be considered as a static wind load pressure. Its application is discussed in Chapter 11 for cable-stayed bridges and illustrated in Section 11.5 for the Sutong Bridge, China.

17.3 MODELING OF BRIDGE FOR DYNAMIC ANALYSIS

As introduced in the last section, bridge dynamic analyses can be categorized into five different types: (1) dynamic interaction between vehicle and bridge, (2) pedestrian bridge dynamics, (3) bridge earthquake analysis, (4) blast analysis, and (5) long-span bridge wind analysis. The first two types of bridge dynamic analysis, except few special cases, can be modeled with superstructure only where the substructure and foundation have little contribution on the dynamic behavior. Modeling for the other three types of analysis will include the whole system, super- and substructures, where the earthquake analysis even includes the foundation. The first, second, and fifth types can be handled by linear dynamic analysis, whereas the third and fourth types may involve nonlinear dynamic analysis.

Due to its uniqueness in analysis and popularity in usage, only the modeling technique of bridge earthquake analysis is discussed in detail here. The earthquake-resistant system (ERS) for bridges may be modeled with the entire super- and substructures (the *global* model) or an individual bent or column (the *local* model). Individual bridge components (the local model) shall have displacement capacities greater than the displacement demand from the global model to satisfy the performance requirement.

17.3.1 Linear elastic dynamic analysis

Linear elastic dynamic analysis (EDA) will be a minimum requirement for the global response analysis. The global analytical model should include the stiffness and mass distributions of the bridge. Commonly a three-dimensional (3D) model is used where it, as shown in Figure 17.11 (NHI 1996), can be a spine model, a grillage model, and a 3D FEM model where the spine and grillage models are the popular kinds. Because elastic analysis

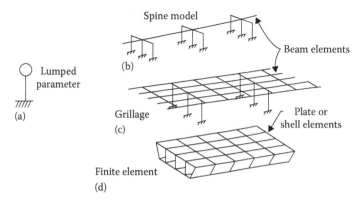

Figure 17.11 (a–d) Types of analytical models.

assumes linear relationship between stiffness and strength, effective section properties should be determined for seismic analysis of reinforced concrete structures with the consideration of concrete crack and steel yielding. One important note of the bridge modeling is that to catch all the essential modes, a minimum of three elements per flexible column and four elements per span should be used in the linear elastic model (AASHTO 2012).

The superstructure is idealized using equivalent linear elastic beam–column elements. For either spine or grillage model of concrete structures, the effective bending stiffness and thus the moment of inertia I_{eff} can be taken as

$$E_c I_{eff} = \frac{M_y}{\varphi_y} \tag{17.18}$$

And the shear stiffness parameter $(GA)_{eff}$ for pier walls in the strong direction may be determined as

$$(GA)_{eff} = G_c A_{cw} \frac{I_{eff}}{I_g} \tag{17.19}$$

And the effective torsional moment of inertia J_{eff} is determined by

$$J_{eff} = 0.2 J_g \tag{17.20}$$

where:
M_y is the moment capacity
φ_y is the curvature of section at first yield of the reinforcing steel
E_c is the modulus of elasticity
G_c is the shear modulus of concrete
I_g is the gross moment of inertia about the weak axis
A_{cw} is the cross-sectional area of pier walls
J_g is the gross torsional moment of inertia of the reinforced concrete
   section

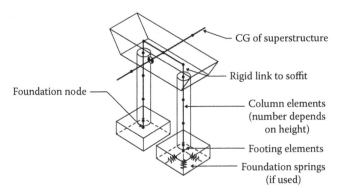

Figure 17.12 Illustration of a spine model.

In either spine or grillage model, elements are defined with superstructure simplified, substructure with ends released, fixed or directly modeled through soil spring elements, and proper connectivity between super- and substructures. A close-view illustration of such model is shown in Figure 17.12. The superstructure is represented by a single line (spine model) or multilines (grillage model) of 3D frame elements, which pass through the center of gravity (CG) of the superstructure. Each of the columns and the cap beam are represented by 3D frame elements, which pass through the geometric centers and midheight, respectively. Rigid end zone can be used to account for the offset between the centerline of the cap beam and the soffit of the superstructure.

17.3.2 Soil stiffness

Abutment may provide longitudinal stiffness K_{eff} due to passive soil pressure uniformly distributed over the height (H_w) and width (W_w) of the backwall or diaphragm.

$$P_p = p_p H_w W_w \tag{17.21}$$

For integral- or diaphragm-type abutments, equivalent linear secant stiffness K_{eff} is

$$K_{eff} = \frac{P_p}{(F_w H_w)} \tag{17.22}$$

where F_w is a factor taken between 0.01 and 0.05 for soils ranging from dense sand to compact clays.

The foundation modeling methods (FMMs) adopted are depending on their SDC where FMM I is for SDCs B and C while FMM II is for SDC D (Table 17.3). There are two ways to determine the foundation stiffness

Table 17.3 Foundation modeling methods

Foundation type	Modeling method I	Modeling method II
Spread footing	Rigid	Foundation spring required if footing flexibility contributes more than 20% to pier displacement
Pile footing with pile cap	Rigid	Foundation spring required if footing flexibility contributes more than 20% to pier displacement
Pile bent/drilled shaft	Estimate depth to fixity	Estimate depth to fixity or soil springs based on p–y curves

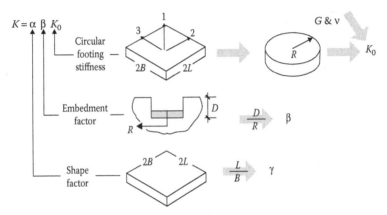

Figure 17.13 Half-spaced method for spread footings. (Data from NHI Course No. 13063 "Seismic Bridge Design Applications," April 25, Publication No. FHWA-SA-97-017 [Part One] and -018 [Part Two], 1996.)

(NHI 1996). One is elastic foundation method, and another is elastic half-space method (Figure 17.13).

In elastic foundation method, k_s for vertical stiffness (or subgrade reaction coefficient) and k_r for rotation stiffness can be determined by

$$k_s = \frac{P}{(\text{Area})(\text{Deflection})} \quad (\text{in kip/ft}^3 \text{ or kN/m}^3) \tag{17.23}$$

$$k_r = k_s \left(\frac{L^3 B}{16} \right) \left(\text{in} \frac{\text{kip} - \text{ft}}{\text{rad}} \text{ or } \frac{\text{kN} - \text{m}}{\text{rad}} \right) \tag{17.24}$$

where:
 P is the vertical load on the mat foundation
 L and B are the half sizes of the mat in their respective longitudinal and transverse directions

In elastic half-space method, footing is bonded to elastic half-space medium where shape (α factor) and embedment (β factor) are considered in the formulation (Figure 17.11):

$$k = \alpha\beta k_0 \qquad\qquad (17.25)$$

where unfactored stiffness k_0 of circular surface footing is listed in Table 17.4. Shape (α) and embedment (β) factors can be found in Figures 17.14 and 17.15, respectively.

As defined in FMM II for SDC D (Table 17.3), soil flexibility is modeled. Three types of foundation are illustrated in Figure 17.16, which are (1) spread footing, (2) piles/drilled shafts, and (3) seat or integral abutments. Discussion of types 2 and 3 are covered in Chapter 16.

Table 17.4 Stiffness of circular surface footing

Degree of freedom	Equivalent radius R	Stiffness K_0
Vertical translation	$R_0 = \sqrt{4BL/\pi}$	$4GR/(1-v)$
Lateral translation (both)	$R_0 = \sqrt{4BL/\pi}$	$8GR/(2-v)$
Torsion rotation	$R_1 = (4BL[4B^2 + 4L^2]/6\pi)^{1/4}$	$16GR^3/3$
Rocking about 2	$R_2 = ([2B]^3[2L]/3\pi)^{1/4}$	$8GR^3/3(1-v)$
Rocking about 3	$R_3 = ([2B][2L]^3/3\pi)^{1/4}$	$8GR^3/3(1-v)$

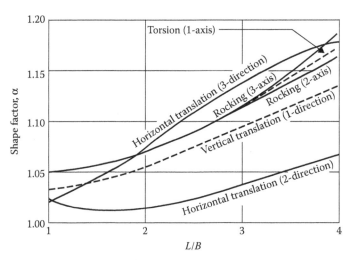

Figure 17.14 Shape factor (α) for rectangular footing. (Data from NHI Course No. 13063 "Seismic Bridge Design Applications," April 25, Publication No. FHWA-SA-97-017 [Part One] and -018 [Part Two], 1996.)

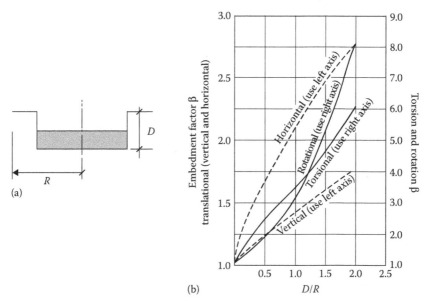

Figure 17.15 Embedment factor (β). (a) Embedment dimensions. (b) Embedment factor.

17.3.3 Nonlinear analysis

Nonlinear dynamic analysis typically involves the development of a complex bridge mathematical model with highly localized (rather than global) nonlinear behavior. The interior expansion joints and the abutment joints are modeled using zero-length elastoplastic gap-hook elements. Based on the report by Aviram et al. (2008), Table 17.5 summarizes the recommended linear and inelastic modeling of the primary components of an ordinary standard bridge structure. The behavior of the plastic hinge can be categorized by a yield surface and a moment–rotation relation. The yield surface defined the interaction between axial force, weak and strong bending moments, and even torque. However, it should be aware that nonlinear dynamic analysis is problematic for routine application with reasonable nonlinear components, sensibility to the details of the model, and intensive output interpretation (Fu and Ahmed 2012). However, for bridges of importance (those categorized as other than ordinary), an inelastic static analysis should be performed.

17.3.3.1 Nonlinear static—Standard pushover analysis

AASHTO guide specifications (2012) also recommend pseudostatic "pushover analysis" be used for the displacement-based performance design method. This procedure examined the nonlinear response of a structure as

Foundation type	Conceptual model	Analytical model

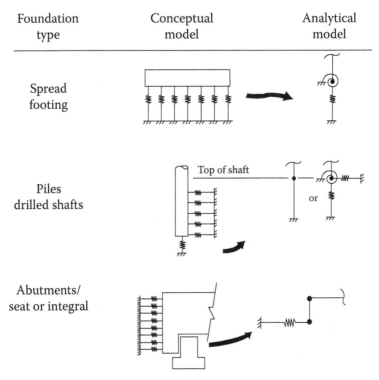

Spread footing

Piles drilled shafts

Top of shaft

or

Abutments/ seat or integral

Figure 17.16 Modeling soil flexibility.

Table 17.5 Linear and nonlinear component modeling

Component	Linear elastic	Nonlinear
Superstructure	X	
Column–plastic hinge zone		X
Column–outside plastic hinge zone	X	
Cap beam	X	
Abutment–transverse		X
Abutment–longitudinal		X
Abutment–overturning		X
Abutment–gap		X
Expansion joints		X
Foundation springs	X	
Soil–structure interaction	X	

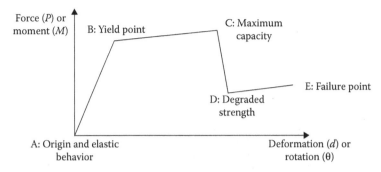

Figure 17.17 Pushover force–deformation (*P–d*) or moment rotation (*M–θ*) curve.

its members yield sequentially under increasing loads. The pushover analysis is stopped when the bridge reaches either a predefined displacement limit or the ultimate capacity limit. The ultimate capacity may correspond to either a localized failure (i.e., a plastic hinge reaching its curvature capacity) or the development of global collapse mechanism (i.e., sufficient plastic hinges developed to cause structure instability). The pushover curve (force vs. displacement) of the bridge, such as the one shown in Figure 17.17, allows identifying any softening behavior of the entire structure due to material strength degradation or *P*–Δ effects. The pushover analysis of the bridge is conducted as a displacement-controlled method to a specified limiting displacement value to capture the softening behavior of the structure by monitoring the displacement at a point of reference, such as one of the column's top nodes or the center of the superstructure span (Aviram et al. 2008). An illustrated example is shown in the Section 17.4.

17.3.3.2 Nonlinear static alternate—Modal pushover analysis

The modal pushover analysis (MPA) method has been presented by Chopra and Goel (2002) for complex building structures, which accounts for higher mode effects on the behavior of structures. Due to the nature of bridges, which extend horizontally, rather than buildings that extend vertically, some considerations and modifications should be taken into account to render the MPA applicable for bridges. Key elements of applying the MPA procedure for the case of bridges are the following:

- Definition of the control node. The control node is used to monitor displacement of the structure. Its displacement versus the base shear forms the capacity (pushover) curve of the structure.
- Development of the pushover curve and transformation of it into a capacity curve.

- Use of the capacity spectrum for defining the earthquake demand for each mode.
- The number of modes that should be considered.

Step-by-step extended MPA procedure for bridges was proposed and presented in detail in the works of Ahmed (2010) and Ahmed and Fu (2012).

17.4 3D ILLUSTRATED EXAMPLE OF EARTHQUAKE ANALYSIS BY SPA, MPA, AND NL-THA—FHWA BRIDGE NO. 4

This example is used to illustrate the MPA and its comparison with standard pushover analysis (SPA) and NL-THA. This bridge is one of the FHWA examples series (Mast et al.) and was modified for nonlinear analyses. It consists of three spans. The total length is 97.5 m (320′) with span lengths of 30.5–36.6–30.5 m (100′–120′–100′), respectively. All substructure elements are oriented at a 30° skew from a line perpendicular to a straight bridge centerline alignment. Figure 17.18 shows plan and

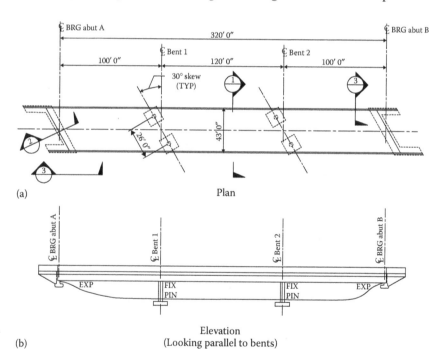

Figure 17.18 (a) Plan and (b) elevation views of illustration example 1. (Data from Mast, R., Marsh, L., Spry, C., Johnson, S., Grieenow, R., Guarre, J., and Wilson, W., Seismic Design of Bridges—Design Examples 1–7 [FHWA-SA-97-006 thru 012], USDOT/FHWA, September 1996.)

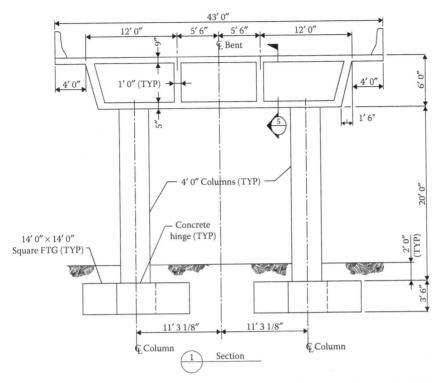

Figure 17.19 Cross-sectional view of illustration example 1. (Data from Mast et al. 1996.)

elevation views of illustration example 1. The superstructure is a cast-in-place concrete box girder with two interior webs. The intermediate bents have a crossbeam integral with the box girder and two round columns that are pinned at the top of spread footing foundations. Figure 17.19 shows the cross-sectional view.

17.4.1 Foundation stiffness

The intermediate bent foundations were modeled with equivalent spring stiffness for the spread footing. For this bridge, all of the intermediate bent footings used the same foundation springs. Values of stiffness were developed for the local bent supports and transformed to global support when input to SAP2000 (2007) program so as to have compatible results for the MPA analysis and the NL-THA. Values of stiffness for foundation springs provided by (Mast et al. 1996) are used in this study. The abutments were modeled with a combination of full restraints (vertical translation and superstructure torsional rotation) and equivalent spring stiffness (transverse translation); other DOFs were all released.

17.4.2 Finite element model and analyses

Figure 17.20 depicts the finite element model with their section properties shown in Table 17.6. The superstructure was modeled with four elements per span, and the elements axes are located along the centroid of the superstructure. The total mass of the structure was lumped to the nodes of the superstructure. The bents were modeled with 3D frame elements that represent the cap beams and individual columns. Since columns are pinned to the column bases, two elements were used to model each column between the top of footing and the soffit of the box girder superstructure; the upper element represents the plastic hinge, whereas the lower one represents the rest of column. A rigid link was used to model the connection in between. The first element from the bottom is a plastic hinge element, which represents the inelastic behavior of the column. The length of the plastic hinge was calculated using the following formula in English units (Priestly et al. 1996):

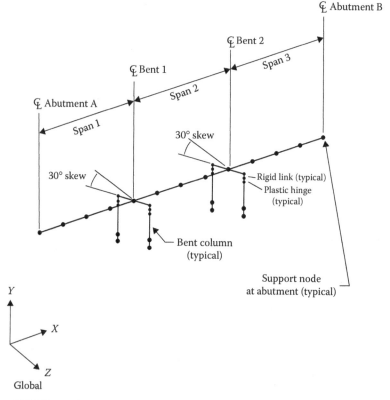

Figure 17.20 Finite element model of illustration example I. (Data from Mast et al. 1996.)

Table 17.6 Section properties for the bridge model

Element properties	CIP box superstructure	Bent cap beam	Bent column
Area in ft² (m²)	72.74 (6.76)	27.00 (2.51)	12.57 (1.17)
I_x–Torsion in ft⁴ (m⁴)[a]	1177 (10.16)	100,000[a]	25.13 (0.22)
I_y in ft⁴ (m⁴)[b]	401 (3.46)	100,000[b]	9.00 (0.08)
I_z in ft⁴ (m⁴)[c]	9697 (83.69)	100,000[c]	9.00 (0.08)

[a] This value has been increased for force distribution to bent columns. Actual value is $I_x = 139$ ft⁴ (1.20 m⁴).

[b] This value has been increased for force distribution to bent columns. Actual value is $I_y = 90$ ft⁴ (0.78 m⁴).

[c] This value has been increased for force distribution to bent columns. Actual value is $I_z = 63$ ft⁴ (0.54 m⁴).

$$L_p = 0.08L + 0.15f_{ye}d_{bl} \geq 0.3f_{ye}d_b \qquad (17.26)$$

where:

d_{bl} is the diameter of the longitudinal reinforcement (ft)
f_{ye} is the effective yield strength of steel reinforcement (ksi)
L is the distance from the critical section of the plastic hinge to the point of contraflexure (ft)

In this example, L is the clear height of the column since the column base is pinned. The second element is the actual column element. The third element represents the varying section between the column section and the column head, which is modeled by the fourth element. The moments of inertia for the column and the plastic hinge elements are based on a cracked section calculated using the moment–curvature and moment–rotation curves.

NL-THA was performed to the three bridges to compare its results with the SPA and MPA results. Three actual acceleration histories were implemented in this example; which were adjusted to match the design response spectrum for each analysis case. Those actual acceleration time histories are as follows:

- El Centro 1940
- Northridge 1994, Century City LACC North
- Santa Monica 1994, City Hall Grounds

Maximum seismic demand displacement of monitoring point is predicted using the SPA, MPA (without inelastic behavior correction for demand displacement), and the modified MPA (using modified control point displacement u'_{cn}) and then compared with the average demand displacement of the same node obtained from the NL-THA using three different ground acceleration histories closely matching the demand spectrum.

Results of the modal analysis, modal periods and frequencies, modal participation factors, and modal participating mass ratios are shown in Table 17.7a–c, respectively.

Table 17.7 Modal analysis results

(a) Modal periods and frequencies

OutputCase	StepType	StepNum	Period sec	Frequency cyc/sec	CircFreq rad/sec	Eigenvalue rad²/sec²
Modal	Mode	1.000000	0.966007	1.0352E+00	6.5043E+00	4.2306E+01
Modal	Mode	2.000000	0.526100	1.9008E+00	1.1943E+01	1.4263E+02
Modal	Mode	3.000000	0.210878	4.7421E+00	2.9795E+01	8.8777E+02
Modal	Mode	4.000000	0.125163	7.9896E+00	5.0200E+01	2.5200E+03
Modal	Mode	5.000000	0.081535	1.2265E+01	7.7061E+01	5.9385E+03
Modal	Mode	6.000000	0.068764	1.4543E+01	9.1373E+01	8.3491E+03
Modal	Mode	7.000000	0.048488	2.0624E+01	1.2958E+02	1.6792E+04
Modal	Mode	8.000000	0.034272	2.9178E+01	1.8333E+02	3.3610E+04
Modal	Mode	9.000000	0.030670	3.2605E+01	2.0486E+02	4.1969E+04
Modal	Mode	10.000000	0.024273	4.1198E+01	2.5885E+02	6.7004E+04
Modal	Mode	11.000000	0.022045	4.5361E+01	2.8501E+02	8.1233E+04
Modal	Mode	12.000000	0.018333	5.4547E+01	3.4273E+02	1.1747E+05

(b) Modal participation factors

OutputCase	StepType	StepNum	Period sec	UX kip-s²	UY kip-s²	UZ kip-s²
Modal	Mode	1.000000	0.966007	11.908197	0.254998	0.000000
Modal	Mode	2.000000	0.526100	0.273393	-11.131288	0.000000
Modal	Mode	3.000000	0.210878	-1.907E-12	-9.051E-14	0.000000
Modal	Mode	4.000000	0.125163	-0.001474	-4.179216	0.000000

Modal	Mode	5.000000	0.081535	1.396E-11	2.655E-13	0.000000
Modal	Mode	6.000000	0.068764	-3.028E-11	-8.692E-13	0.000000
Modal	Mode	7.000000	0.048488	0.000825	0.654170	0.000000
Modal	Mode	8.000000	0.034272	0.008127	-0.000365	0.000000
Modal	Mode	9.000000	0.030670	7.547E-11	1.627E-12	0.000000
Modal	Mode	10.000000	0.024273	5.445E-10	1.165E-11	0.000000
Modal	Mode	11.000000	0.022045	-0.000359	0.122636	0.000000
Modal	Mode	12.000000	0.018333	0.004605	-0.000223	0.000000

(c) Modal participating mass ratios

OutputCase	StepType	StepNum	Period sec	UX	UY	UZ	SumUX	SumUY
Modal	Mode	1.000000	0.966007	0.99947	0.00046	0.00000	0.99947	0.00046
Modal	Mode	2.000000	0.526100	0.00053	0.87331	0.00000	1.00000	0.87377
Modal	Mode	3.000000	0.210878	0.00000	0.00000	0.00000	1.00000	0.87377
Modal	Mode	4.000000	0.125163	1.531E-08	0.12310	0.00000	1.00000	0.99687
Modal	Mode	5.000000	0.081535	0.00000	0.00000	0.00000	1.00000	0.99687
Modal	Mode	6.000000	0.068764	0.00000	0.00000	0.00000	1.00000	0.99687
Modal	Mode	7.000000	0.048488	4.794E-09	0.00302	0.00000	1.00000	0.99989
Modal	Mode	8.000000	0.034272	4.65E-07	9.366E-10	0.00000	1.00000	0.99989
Modal	Mode	9.000000	0.030670	0.00000	0.00000	0.00000	1.00000	0.99989
Modal	Mode	10.000000	0.024273	0.00000	0.00000	0.00000	1.00000	0.99989
Modal	Mode	11.000000	0.022045	9.086E-10	0.00011	0.00000	1.00000	1.00000
Modal	Mode	12.000000	0.018333	1.494E-07	3.512E-10	0.00000	1.00000	1.00000

Pushover curve (Figure 17.21) uses mode 2 as the lateral load (mode shape as shown in Figure 17.22 multiplied by the mass). NL-THA was performed using three different acceleration histories, and average response was compared with those from the modal pushover analysis.

Peak ground accelerations (PGAs) of 0.30g and 0.45g were considered. Comparison is performed for the maximum demand displacement in the transverse direction, total base shear, and rotations of plastic hinges. Results of the standard and modal pushover approaches were evaluated by comparing them with those from the NL-THA; the latter is considered to be the

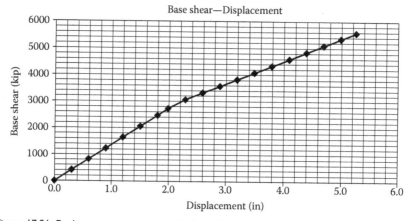

Figure 17.21 Pushover curve using mode 2 as the lateral load.

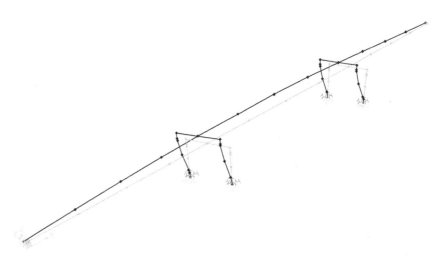

Figure 17.22 Deformed shape of mode 2 ($T_2 = 0.5621$s).

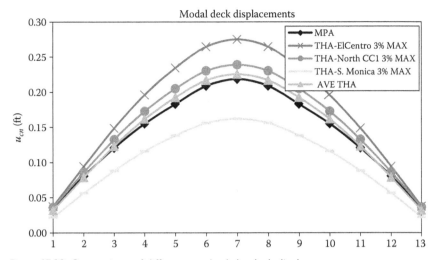

Figure 17.23 Comparison of different methods by deck displacement.

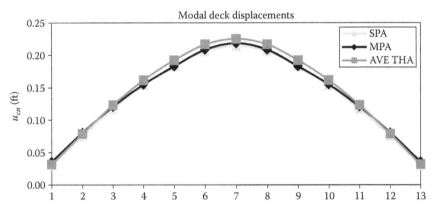

Figure 17.24 Comparison of different methods by deck displacement (PGA = 0.30g).

most rigorous procedure to compute seismic demands. To this effect, a set of three real-time acceleration records compatible with the design spectrum was used in the NL-THA analyses. The deck displacements determined from each of the SPA and MPA analyses with respect to the control point of the most critical pier were compared with those from NL-THA for increasing levels of earthquake excitation, as shown in Figure 17.22, for multiple earthquake with their average, and Figures 17.23 and 17.24 compared SPA, MPA, and NL-THA, both for PGA = 0.30g.

17.5 3D ILLUSTRATED EXAMPLE OF A HIGH-PIER BRIDGE SUBJECTED TO OBLIQUE INCIDENCE SEISMIC WAVES—PINGTANG BRIDGE, PEOPLE'S REPUBLIC OF CHINA

The time-lag effect in seismic wave propagation has an influence on large-span structure. This case study demonstrates the effect of the topography and the angle of the oblique incidence waves. 3D finite element analysis with equivalent artificial boundary is used to simulate the radial damping of continuous medium within a finite domain (Gu et al. 2014). This equivalent artificial boundary can be represented by viscoelastic artificial boundary elements to simulate the spring and dash system around the soil outside boundary.

The numerical model of a continuous rigid bridge with the total length of 560 m and its spans of 100 m + 180 m + 180 m + 100 m was built. The pier is a RC double wall of 2 m thickness, 9 m depth, and about 30 height. The diameter of the circular pile is 2.8 m. The bridge layout is shown in Figure 17.25(a). The numerical model of the bridge with special topographic shape is shown in Figure 17.25(b). The parameters of the ANSYS model elements (ANSYS 2012) and soil are shown in Tables 17.8 and 17.9, respectively.

Anza Earthquake record of a short duration was selected for this study. The results of the internal forces are compared by inputting the seismic wave in vertical direction and on an oblique incidence. The bottom and the top of the pier are the most vulnerable locations in this rigid continuous bridge. The force at the bottom is larger than that at the top in this case. So the internal forces at the bottom of piers are studied. Locations of the numbered piers are shown in Figure 17.25(a). The amplitudes of the internal

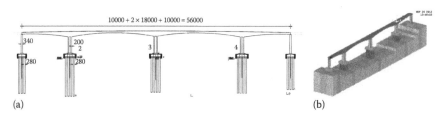

Figure 17.25 Layout of Pingtang bridge and FEA model: (a) elevation (in cm); (b) soil-bridge finite element model to simulate the spring and dash system.

Table 17.8 3D FEA model and elements

Element	BEAM188	MASS21	SOLID45	BEAM4
structure	Girder, pier, pile	Lump mass	Soil	Rigid

Table 17.9 Material parameters of soil

Velocity of P wave V_S (m/s)	Velocity of P wave V_P (m/s)	Poisson ratio μ	Soil density ρ (kg/m³)	Thickness of soil (m)
500	866	0.25	2200	30

forces of piers subjected to P waves of different input angles of the oblique incidence wave are shown in Figure 17.26. The results showed that the shear force of the middle pier is smaller than the other two piers. The shear force and moment of the piers with seismic wave in the vertical direction are smaller than the force and moment with inputting wave at a 30° angle. Conversely, the axial force of the piers is larger in the vertical direction.

The amplitudes of the internal forces of piers with SV waves of different input angles of the oblique incidence waves are shown in Figure 17.27. The shear force of the middle pier is smaller than those of the other two piers as well. But the results of the SV waves showed a reverse trend from the P waves. The shear force and moment of the piers with seismic wave in the vertical direction is larger than the force and moment with inputting wave at a 30° angle. Conversely, the axial force of the piers is smaller in the vertical direction.

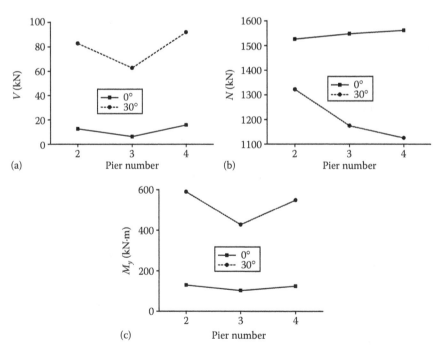

Figure 17.26 Maximum amplitude of internal forces at bottom of piers under oblique incidence P waves: (a) shear force (V); (b) axial force (N); (c) Moment (M_y).

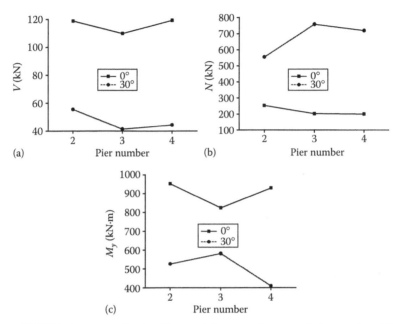

Figure 17.27 Maximum amplitude of internal forces at bottom of piers under oblique incidence SV waves: (a) shear force (V); (b) axial force (N); (c) moment (M_y).

Table 17.10 Indices of nonuniform effects for piers under oblique incidence waves at 30°

Internal forces	η_P			η_{SV}		
	Pier 2	Pier 3	Pier 4	Pier 2	Pier 3	Pier 4
V	552%	874%	477%	−53.28%	−62.39%	−62.80%
N	−13%	−24%	−27%	118.86%	274.26%	259.41%
M_y	359%	319%	344%	−44.76%	−29.41%	−56.02%

Considering ratio of the values computed from vertical and oblique incidence input, an index can be calculated as follows:

$$\eta = \frac{\max|F| - \max|F_0|}{\max|F_0|} \tag{17.27}$$

where:

F_0 is the internal force of the piers including shear forces, axial forces, and moment computed by vertical input

F is the internal force computed by oblique incidence wave

The indices of nonuniform effects for piers in valley under oblique incidence waves at 30° are shown in Table 17.10. It is shown that the oblique incidence waves have great effect on the piers of bridges in valley.

Chapter 18

Bridge geometry

18.1 INTRODUCTION

Bridges are counted as a part of road facilities to serve the purpose of transportation. Most bridges are designed and built to satisfy their roads' requirements. To satisfy requirements of a road alignment, the axis of a bridge may have to be curved in both horizontal and vertical directions, and the finished grade of bridge deck must comply with the transverse slopes set forth in road geometries. Also, a bridge has to be designed under certain engineering aesthetic guidelines, which may force a bridge axis, profiles, or its components in curve or complex shapes. In other words, both the bridge axis and a bridge component are more complex than what they are usually described in mathematics and mechanics models.

A few questions may arise when building an analysis model of a geometrically complicated bridge, for example, how a curved girder axis is calculated and meshed into small elements and how a haunched girder profile is defined and simulated. In this chapter, bridge geometry-related principles and practical methods will be introduced.

18.2 ROADWAY CURVES

The design of a roadway curve is usually separated into horizontal and vertical curves. The horizontal curve, the projection of a roadway on plane, defines the transition from one tangent to another allowing a vehicle to turn in a graduate horizontal rate; the vertical curve, the projection of a roadway on elevation, defines the transition from one slope to another allowing a vehicle to change grade in a graduate vertical rate. In addition to the constraints of sight distances and drainages, the design of both curves must provide a roadway with graduate changes of curvatures or grade, rather than a sharp change. Due to the different requirements of horizontal turns and vertical grade changes, characteristics of horizontal curves and vertical curves are different.

For the geometry modeling purpose, a term of *mainline* is introduced. Roadway mainline is just the centerline of the roadway, the geometry at the center of the roadway in terms of roadway design, not necessarily the middle of the roadway. Not only many roadway geometric characteristics but also their bridge components depend on the mainline geometry. Figures 18.1 through 18.3 show the relationship between the mainline of a roadway and bridge girders. In general, the girder geometry should follow the mainline in both horizontal and vertical curves as shown in Figures 18.1 and 18.2, respectively (Wang and Fu 2013).

The location of the mainline or its alignment in transverse and vertical curves is critical and should be unique. For the purpose of roadway geometry design, mainline is always aligned with the control point in the transverse direction on top of the roadway surface, which is not necessarily the center of a roadway. The thick line in the middle of Figure 18.3 shows the mainline and its location.

18.2.1 Types of horizontal curves

When a vehicle runs on a curve, the horizontal centrifugal force is proportional to the reciprocal of curve radius or curvature of the roadway. To provide an acceptable riding smoothness and to meet physical requirements,

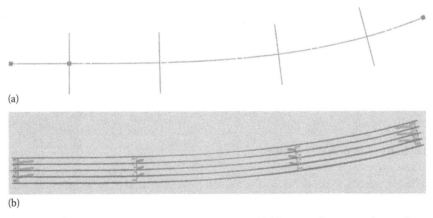

(a)

(b)

Figure 18.1 Roadway and girder horizontal curves. (a) Horizontal curves of a roadway centerline and span layout. (b) Girder axes follow roadway horizontal curves.

Figure 18.2 Girder axes follow roadway vertical curves, and curves in girder profiles.

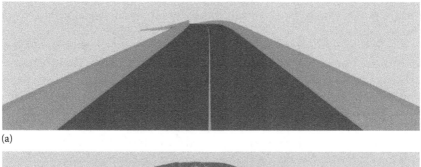

(a)

(b)

Figure 18.3 Roadway mainline. (a) Mainline is the roadway centerline, but not necessarily in the center of roadway. (b) Mainline is aligned on the top of the deck.

when roadway transits from one tangent to another the rate of curvature changes and the maximal of curvature has certain limitation. This requirement of roadway makes arcs, spirals, and their combinations suitable and very common for horizontal transitions.

An arc has a constant curvature. When used as a part or the whole transition as shown in Figures 18.1 and 18.4, the absolute value of its curvature should meet the maximal curvature requirement and the changes to the previous segment or to the next segment cannot be too sharp. For example,

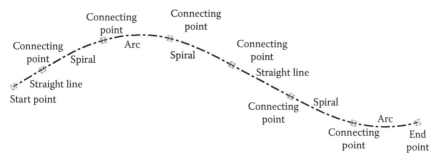

Figure 18.4 An example of a plane curve—components of a compound curve.

in situations where only one arc segment is used to connect two tangents, the arc radius cannot be too small as the curvature change is from zero to the arc's curvature.

A spiral can be a perfect fit for a transition from one curvature to another as it is so defined that the curvature change is proportional to the curve distance. As shown in Figure 18.4, a spiral is used to connect two segments that have different radius. In addition to the requirements of maximal curvatures at two end segments, the spiral length, which controls the rate of the curvature change, cannot be too short.

Parabolic curve, which is exclusively used for vertical and transverse curves, fails to possess the advantages of arcs or spirals as horizontal curve transitions (Hickerson 1959). In most horizontal transition situations, tangents (straight lines), arcs, spirals, and their combinations are commonly used.

18.2.2 Types of vertical curves

Vertical curves are used to make a transition from one slope to another. Parabolic curve is the only type of curve used in vertical curves. As shown in Figure 18.5, the parabolic segment is called sag vertical curve when the transition of slopes is from negative to positive and crest vertical curve vice versa.

18.2.3 Types of transverse curves

A roadway is usually required to have certain crowns in the middle and cross slopes on sides to help water draining from roadway laterally. Thus, the cross section or profile at any station of a roadway contains two tangents and one transition curve. As shown in Figures 18.6 and 18.7, the parabolic curve is widely used in transverse curves.

18.2.4 Superelevation and superwidening

When traveling along curve transition segment, vehicles will overcome centrifugal forces by mainly lateral tire friction to maintain movement in circular. In cases where either roadway curvature is big or design speed is high, the transverse slope on the outer side of a roadway should be raised up to flat or even positive toward inner side so that extra horizontal forces

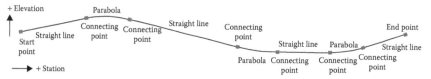

Figure 18.5 An example of a vertical curve—components of a vertical curve.

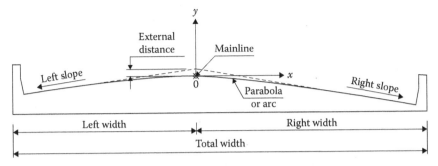

Figure 18.6 Transverse curve and local coordinate system.

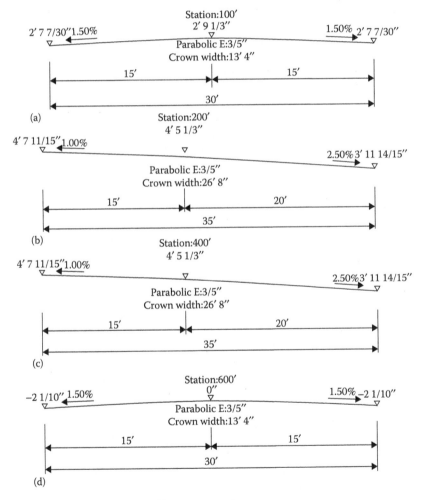

Figure 18.7 (a–d) An example of key cross sections.

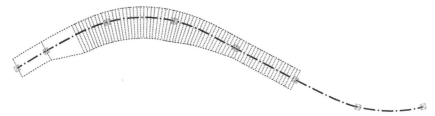

Figure 18.8 Plane view of a transverse curve transition example.

can be gained from vehicle gravity to help balancing centrifugal forces. The change of transverse slope from a normal rate to accommodate turning is called superelevation of a roadway. When designing superelevations, the superelevation change rate, the final transverse slope, and the runout/run-off length should meet certain requirements. However, the transition rate of superelevation or the change of transverse slope within the runout/runoff length is usually linear as shown in Figure 18.7.

When turning along a curve segment, vehicle wheels are easily off-tracking, hence curve roadway widening is needed to ensure safety and to protect shoulders from impacting. Similarly, transitions from normal-width segment to widened segment, as shown in Figures 18.7 and 18.8, should be designed. As superelevation transitions, superwidening transitions are usually linear, that is, the change of the width is linear in terms of transition length.

18.2.5 Bridge curves

As part of a roadway, most bridge structures have to comply with geometries set forth by a roadway globally. Therefore, the geometry of a bridge axis or mainline is the same as a roadway curve in both horizontal and vertical directions (Figures 18.1 through 18.3). Deck curves in transverse direction, including superelevation and superwidening, must also meet requirements from road transverse curves.

For most girder bridges, geometry of a girder axis follows the geometry of a bridge axis, or mainline, so as to form the deck plan accordingly. For example, a girder in a multiple-girder bridge is parallel to the bridge main-line in horizontal curve and has a vertical curve as defined by the vertical curve of the bridge mainline and transverse curve of the roadway. However, cases where deck curves in both horizontal and transverse directions are made up of deck components themselves are common too. For example, straight girders are often used in small-curvature bridges, especially those in simply supported multiple-span bridges.

In addition to bridge mainline geometries, girders in a continuous bridge may be haunched longitudinally to incorporate changes of internal forces. Haunches can happen in both concrete and steel bridges. Figure 18.1 shows the curved girders in horizontal view, and Figure 18.2 shows the elevation view of these haunch girders.

18.3 CURVE CALCULATIONS

Given the most commonly used curve types, such as straight lines, arcs, spirals, and parabolas, calculations needed to obtain a point on curve are simple and straightforward mathematically. Challenges, however, arise from the engineering depiction of a curve in a way of easy representing actual roadway curves in three-dimensional (3D) space and accurately controlling geometries of any bridge component. To model a bridge in 3D based on a spatial curve preset by roadway, defining an appropriate curve model is fundamental. Procedures for sampling points along a 3D curve or road surface can then be established.

18.3.1 Bridge mainline curve model

The bridge mainline, or the deck centerline, is the reference line of modeling a bridge in 3D. Geometries and locations of most bridge components can be derived or located by referring to the bridge mainline. Often the mainline of a bridge can be the same as the centerline of the roadway. Vertically, it is aligned on top of the deck. Figure 18.3 shows an example of a bridge mainline. As a spatial curve shown in Figure 18.3, geometries of a bridge mainline contain plane curves, or the horizontal curves, and vertical curves (Wang and Fu 2013).

Following the practices of roadway design and route locations (Hickerson 1959), spatial curves can be described in horizontal (plane) and vertical curves separately. A pure mathematical description of a roadway curve in spatial is not practical at all in road engineering. Therefore, a bridge mainline can be described separately by its (1) plane curves and (2) vertical curves.

Plane curves are compound curves in general, which may contain straight lines, arcs, and spirals with smooth connections from one to another. Figure 18.4 shows a plane curve as an example. Smooth connection from one component to another means tangents at connecting points are continuous at least. In most cases, as the example shown in Figure 18.4, smooth connection can further mean that curvatures at connecting points are continuous. Under such a restriction, a spiral segment is needed to connect a straight line and an arc. The connecting of straight line and arc, especially those with small radii, which causes the curvature changes from zero to a constant, is discouraged. Having parameters of each connecting components including the type and length of curve segment and radius of an arc or ending radius of a spiral, plus the starting point and tangent defined, geometric properties, such as location, tangent, and curvature at any given curve length ordinate, can be obtained by a simple calculation procedures. Section 18.3.4, for instance, provides principles and steps to calculate a spiral segment.

Vertical curves are compound curves too but contain only straight lines and parabola. Figure 18.5 shows an example of vertical curves. Unlike plane curves, connections in vertical curves are simple. When grade transition is needed, parabola is always used to connect one straight line to another one. As slopes or tangents in the vertical curve of both connecting grades are known, only external distance is needed to define a parabola filleting two straight lines. The parabola used in vertical curves can be called as vertical parabola. Similarly for spirals used in a plane curve, in which the change of curvature is proportional to the curve length, the grade change of a vertical parabola is proportional to curve horizontal length. Curve tangents at connecting points, as shown in Figure 18.5, are continuous.

Having the earlier definitions on both plane and vertical curves, the mainline of a bridge, or the roadway centerline, can be described separately. When defining the vertical part, the horizontal ordinate is the unfolded curve length of the corresponding plane curve, that is, the stations of roadway centerline; the vertical ordinate is the elevations (Figure 18.5). The following list provides examples of definitions of bridge mainline curves:

Plane curves. (1) A straight line with a length of 61 m (200′); (2) a spiral with a length of 152 m (500′) connecting the straight line to the next arc segment with a radius of 244 m (800′), curve goes clockwise; (3) an arc segment with a length of 122 m (400′) and a radius of 244 m (800′); (4) another spiral segment with a length of 152 m (500′) connecting the arc to the next straight line; (5) a straight line with a length of 122 m (400′); (6) a spiral with a length of 122 m (400′) connecting the line and next arc segment with a radius of 274 m (900′), curve goes counterclockwise; and (7) last arc segment with a length of 152 m (500′) and a radius of 274 m (900′); starting tangent is 120° to latitude axis and location is (0 longitude, 0 latitude).

Vertical curves. (1) Control point at station 0: altitude = 0; (2) control point at station 274 m (900′): altitude = +20′ (6.1 m), parabola fillet with an external distance of 1.2 m (4′); (3) control point at station 610 m (2000′): altitude = −15′ (4.6 m), parabola fillet with an external distance of 0.3 m (1′); (4) control point at station 762 m (2500′): altitude = −15′ (4.6 m), parabola fillet with an external distance of 0.5 m (1.5′); and (5) last control point at station of 884 m (2900′): altitude = 0.

18.3.2 Roadway transverse curve model

Transverse curve model defines the roadway transverse slopes, crowns, superelevations, and superwidening. As shown in Figure 18.6, the transverse curve at a roadway cross section can be defined by (1) left width, the horizontal distance from the mainline to road edge on the left; (2) right width, the horizontal distance from the mainline to road edge on the right;

(3) left and right slopes; and (4) external distance of the crown. Although most roadway crowns are parabolic, an arc crown can be simply included in this definition by using a signed value of the external distance. For example, a negative external distance indicates a parabolic crown and an arc crown if otherwise.

Considering the vertical curve model that contains only parabolic fillets, a generic parabolic/arc vertical curve model can be shared among vertical and transverse curves. When this model is applied to vertical curves, only parabolic fillets are applicable.

The local coordinates system, which is used to describe transverse curve at any cross section, is important in roadway surface calculations. Figure 18.6 shows the transverse curve coordinate system, whose origin is aligned with roadway mainline. From the definitions of plane and vertical curves of mainline, once geometric parameters (e.g., longitude, latitude, altitude, and tangent of the plane curve) of a given point on the mainline are known, any point on the roadway surface along a cross section will be known. For design purposes, these separated representations of roadway cross sections are practical and accurate enough. For the purpose of digital visualization, triangular surface meshes can be easily established, given two consecutive roadway cross sections.

18.3.3 Transitions of transverse curves

As discussed in the previous sections, transverse curves may vary in curve segments as superelevation and superwidening are required. Key transverse curves can be explicitly specified at certain known locations along curve segments. Transverse curve properties of cross sections in between consecutive key locations, such as widths, slopes, and external distances, can be interpolated by linear, circular, or parabolic methods. When linear method is used to interpolate a geometry property, only two key cross sections are required. When the circular or parabolic method is used, three consecutive key cross sections are required. The following list provides examples of transverse curve transition definitions: (1) cross section at station 30 m (100′) is symmetric with a total width of 9.1 m (30′), a slope of 1.5% and a parabolic crown with an external distance of 15 mm (3/5″); (2) cross section at station 61 m (200′) has a 1.5 m (5′) superwidening on the right side, superelevation on the left side causes the slopes to +1% and –2.5% on the right side, crown maintains the same; (3) cross section at station 122 m (400′) remains the same as that at 61 m (200′); and (4) cross section at station 600′ changes back to that at station 30 m (100′) (Figure 18.7).

As a generic example, Figure 18.8 shows the plane view of transverse curve transitions. Dot lines are roadway edges; radial lines are cross sections of interested. Figure 18.9 shows the perspective view of mainline and interested cross sections.

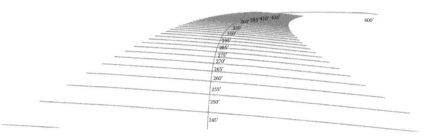

Figure 18.9 Perspective view mainline and cross sections.

18.3.4 Spiral calculation

Spirals used in roadway plane curves are for making a curvature transition. A spiral for this purpose is simply defined as a curve whose curvature is proportional to curve length, or the curvature change-to-curve length ratio is constant:

$$c(l) = c_s + \frac{c_e - c_s}{L} l \tag{18.1}$$

where:

  c denotes curvature
  subscripts s and e denote starting and ending of spiral, respectively
  L is the total length of spiral
  l is the curve length ordinate

From the definition of curvature, reciprocal of curve radius, the differential of sweeping angle θ is

$$d\theta = c(l)dl \tag{18.2}$$

Integrating Equation 18.2 with the substitution of 18.1 and considering zero initial sweeping angle, sweeping angle at any given curve length ordinate can be obtained as

$$\theta(l) = c_s l + \frac{c_e - c_s}{2L} l^2 \tag{18.3}$$

Taking a local coordinate system as shown in Figure 18.10, the differentials of ordinates x and y can be written as

$$dx = \cos\left(c_s l + \frac{c_e - c_s}{2L} l^2\right) dl; \, dy = \sin\left(c_s l + \frac{c_e - c_s}{2L} l^2\right) dl \tag{18.4}$$

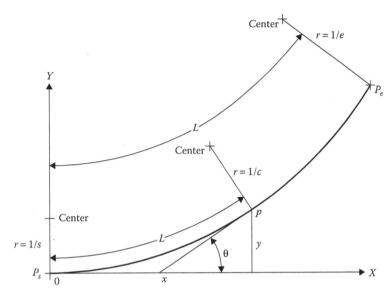

Figure 18.10 Spiral curve and its local coordinate system.

Therefore, local coordinates x and y at any given curve length are the integration forms of Equation 18.4:

$$x = \int_0^l \cos\left(c_s l + \frac{c_e - c_s}{2L} l^2 \right) dl, \quad y = \int_0^l \sin\left(c_s l + \frac{c_e - c_s}{2L} l^2 \right) dl \qquad (18.5)$$

Given a curve length ordinate l, point on a spiral and curve properties can be computed by Equations 18.1, 18.3, and 18.5. When computing coordinates by Equation 18.5, Simpson's Rule can be used as a generic numerical integration method.

As a spiral is a part of compound plane curve as usual, local coordinates and tangent at any point on a spiral as shown in Figure 18.10 have to be transformed to global coordinate system by a simple rotation and a translation.

18.3.5 Vertical parabola calculation

Parabolas used in roadway vertical curves are for making a grade transition. Similar to spiral, the grade of a vertical parabola is proportional to horizontal curve length, or the grade change to horizontal length is constant:

$$g(x) = \frac{dy}{dx} = g_s + \frac{g_e - g_s}{X} x \qquad (18.6)$$

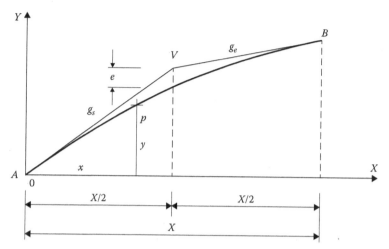

Figure 18.11 Vertical parabola and its local coordinate system.

where:

    g denotes grade

    subscripts s and e denote starting and ending of parabola, respectively

    X is the total horizontal length of parabola

    x is the curve horizontal ordinate

Taking the starting point as the origin of the local coordinate system as shown in Figure 18.11, by integrating Equation 18.6, the ordinate y of point on curve at x can be obtained as

$$y = g_s x + \frac{g_e - g_s}{2X} x^2 \tag{18.7}$$

In addition to connecting grades, g_s and g_e, the horizontal length for the transition X is a critical characteristic of a vertical parabola as it controls the rate of grade change. Considering that the vertical line at $x = (1/2)X$ passing through point V as shown in Figure 18.11, the relationship between the external distance e and X can be written as

$$X = \frac{8e}{g_e - g_s} \tag{18.8}$$

18.4 CURVE AND SURFACE TESSELLATION

When showing a 3D road curve that contains spirals/arcs in plane and parabolas in vertical curve on screen, curve has to be subdivided into small straight lines or arc segments as computer graphics technologies cannot reproduce and render such a 3D curve. This subdivision process is the so-called curve tessellation. As tessellations are for visualization purpose only, the minimum length of subdivision can be 1 m or a couple of feet. When tessellating a curve, several different types of points on curve have to be subdivided. These points, the so-called ensured points include (1) geometry control points where either plane curve or vertical curve changes, (2) girder or beam section change points, (3) point of interests, (4) support points, (5) diaphragm points, (6) roadway cross-sectional points, and (7) control points for superelevation and superwidening. Once the ensured points are obtained according to geometry theories, points in between any two consecutive ensured points will be inserted according to a minimum tessellation segment length.

When the tessellation is for the purpose of plane view, tessellated segments may contain arcs and straight lines, for arcs can be rendered by general computer graphics technologies. When the tessellation is for 3D view, the tessellated segments can only be straight lines.

When showing roadway surface or deck in 3D, a similar tessellation process is needed to produce triangle planes in space so that the surface can be shown as 3D views. In addition to the longitudinal tessellation on 3D roadway centerline, cross sections at each longitudinal tessellation point will be further evaluated, as discussed in Section 18.3. Each cross-section curve will be tessellated transversely. The ensured points on transverse curve include (1) geometry control points such as where a parabola starts or ends; (2) locations of all girder centerlines; and (3) locations of mainline, road edges, curbs, or medians. Given two transverse segments on two consecutive cross sections, as shown in Figure 18.12, two triangle planes can be produced for 3D rendering. Figure 18.12 shows a roadway in 3D with the wireframe mode so that the tessellated triangles can be illustrated, whereas Figure 18.14 shows the roadway with the solid rendering mode.

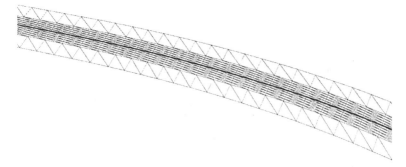

Figure 18.12 Tessellations of roadway surface or deck.

18.5 BRIDGE DECK POINT CALCULATIONS

During the construction of a bridge, certain points on the bridge deck may need to be verified to control bridge geometry, specifically to control the finished grade on the deck. In general, any point on the deck can be designated as a control point. However, intersections of girders, curbs, or medians and diaphragms are usually the default control points for deck grade verifications.

As discussed in Section 18.3, having the separated representations in plane and vertical curves of a 3D roadway mainline and transverse curve definitions of each cross section, any point on the roadway surface or deck can be evaluated. When reporting elevations of these control points, preset camber values on these points should be separated from derived elevations that are obtained by roadway geometry definitions. Cambers are usually required for a bridge to counter vertical structural displacements due to dead loads and/or part of live loads.

Figures 18.13 and 18.14 show an example of bridge deck control points. In Figure 18.13, each triangle mark indicates the location of a control point

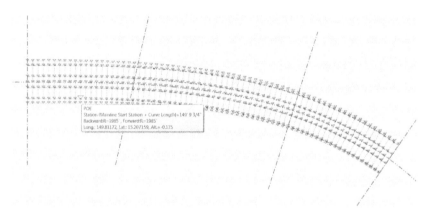

Figure 18.13 Calculated deck points on plane view.

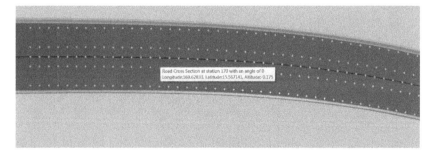

Figure 18.14 Deck points in 3D view.

in plane. In Figure 18.14, cross symbols show such locations in 3D. With modern computer graphics techniques adopted, detailed information about each point such as plane locations and elevations can also be shown when hovering over a symbol, as rectangle boxes shown in both Figures 18.13 and 18.14.

18.6 PRECAST SEGMENTAL BRIDGE GEOMETRY CONTROL

Precasting concrete girder segments in yard while substructure is being built and assembling in place later is a popular construction method for concrete box girder bridges. Many advantages such as eliminating time to curing concrete and reducing concrete creep and shrinkage at earlier ages made segmentally precasting method widely adopted in concrete box girder bridges. As girder segments are casted in yard and assembled later in place, how to ensure the finished bridge curve in close agreement with theoretical bridge curves in both horizontal and vertical directions becomes a critical issue in this type of construction method. Geometry control of girder segments during casting in yard so that errors in the finished bridge curve are under control is a common and very important topic of precast segmental bridge. In this section, key concepts and principles in precast segmental bridge geometry control will be introduced.

18.6.1 Basics

18.6.1.1 Long-line casting and short-line casting

When girder segments are casted in yard, there are two different types of casting: (1) long-line casting and (2) short-line casting (Baker 1980). In long-line casting system, all segments of a cantilever or a span are casted in their correct relative position on a continuous soffit of sufficient length. When one segment is cast, the forms will be moved to the next segment position along the soffit. Figure 18.15 shows the schematic of a long-line casting system. Geometry control in long-line system is established by adjusting forms and soffit before pouring concrete. In the perspective of geometry control, the long-line casting system is easy to set up. The disadvantage of long-line system is obvious that substantial space is required.

In short-line casting system, only one girder segment is casted at one time on the casting bed, and cured segments are moved to the storage yard. Figure 18.16 shows the schematic of a short-line casting system. The pros and cons of this type of precasting are obvious too. As there is only one segment to be casted in one time, the length of casting bed is limited. Forms can be used repeatedly for other segments. The most important advantage

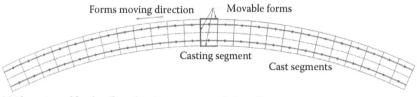

(a) Plane view of fixed soffit and casting sequence of a long-line system

(b) Elevation view of fixed soffit and casting sequence of a long-line system

Figure 18.15 (a) Plane and (b) elevation view of a long-line casting system.

is that the forms can be built as machinery so as to be easily unfolded and folded, for both quantities and size permit to do so. Therefore, higher quality of precast can be achieved. The disadvantage of short-line system is the geometry control during the casting of each segment. Imaging a curved box being sliced into many short segments, the deliberated geometry control measurements have to ensure different segments are casted in their right shapes so that the theoretical girder can be reproduced when all are resembled in place. This section will mainly discuss on the geometry control during precasting in short-line system.

18.6.1.2 Final curve and theoretical casting curve

As casting segments are laid on casting bed or supported, conditions are different from when they are assembled where structural displacements due to dead loads and/or poststressing happened. There are two types of girder curves involved during precasting and assembling of a precast segmental bridge. The first one is called the *final curve*, which is what engineers designed and expected after a bridge is built. For a segmental constructed bridge, there will be many permanent load applications after a segment is assembled that cause the girder curve change from the initial condition. Examples of these loads include structural weight of a girder segment, prestressing, concrete creep or shrinkage, and superimposed dead loads. The second curve is called the *theoretical casting curve*, which is what geometry control is aiming at. It can be imagined that the theoretical casting curve is what all segments should form after assembled without any load application, as there is no load applied on segments while casting in yard. The theoretical casting curve can be obtained by backward analyses, in which each applied permanent load is removed one by one from closure stage. From the perspective of precasting geometry control, it can be simply taken that a theoretical casting curve, which is different from the final curve, should

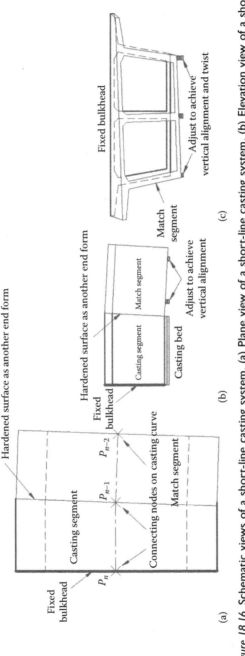

Figure 18.16 Schematic views of a short-line casting system. (a) Plane view of a short-line casting system. (b) Elevation view of a short-line casting system. (c) Side view of a short-line casting system.

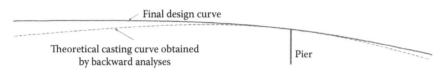

Figure 18.17 Difference between the final curve and the curve to control casting.

be achieved during the precasting of segments. Figure 18.17 illustrates the difference of these two curves. Note that the theoretical casting curve may be above the final curve depending on the quantity of prestressing tendons and other loads.

18.6.1.3 Casting segment and match cast segment

In short-line casting system, only one segment is casted at one time, which is called casting or wet segment. Comparing the short length of each precast segment and the length of the entire span, it can be easily understood that the reproduction of theoretical casting curve is solely controlled by the connections between segments. The accuracy of the shape of each individual segment does not control the geometry as a whole. Therefore, the geometry control of precasting segments is really the control of the connection face for any two consecutive segments. This is achieved by casting a segment against the segment it connects that is already casted. The segment is called match cast segment, that is, the segment used to be matched for a new casting segment. Figure 18.18 illustrates these two segments. Figure 18.16 shows an actual short-line casting system.

18.6.2 Casting and matching

As shown in Figures 18.16 and 18.18, the formworks for the segment to be casted are laid on the fixed casting bed. One end form is the bulkhead, which is fixed as well. On the opposite of the bulkhead, the connection face of the matching segment is used as another end form directly. Before casting, debonder is applied on the connection surface to prevent bonding of the concrete. This match casting against the hardened surface of its connection segment leaves an almost invisible joint when segments are assembled.

As the match cast segment sits on top of the supporting soffit, both vertical and horizontal alignments can be reached by adjusting screw jacks beneath or the horizontal locations of the soffit. In case the superelevation exists, the casting segment may be twisted relatively as shown in Figure 18.16c. This can also be achieved by adjusting the support of match cast segment.

Figure 18.19 shows a 3D rendering of a segmental bridge being assembled. Coordinate system Long–Alt–Lat as shown in Figure 18.19 is the global coordinate system in which both final and theoretical casting curves

Figure 18.18 Adjustment of match segment and formworks for casting segment. (Courtesy of Ninive™ CASSEFORME, http://www.ninive.it/bridge-formwork/ segmental-box-girder-forms/.)

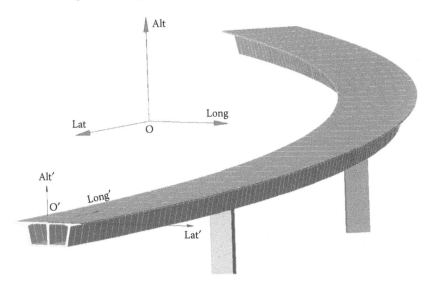

Figure 18.19 Segments assembled and global/local coordinate systems.

are established. Coordinate system Long′–Alt′–Lat′ is the casting coordinate system or the local coordinate system. By transforming the global coordinate system to casting coordinate system as shown in Figures 18.19 and 18.20, the relative locations of the match segment to the casting segment in both vertical and horizontal directions can be obtained. Therefore,

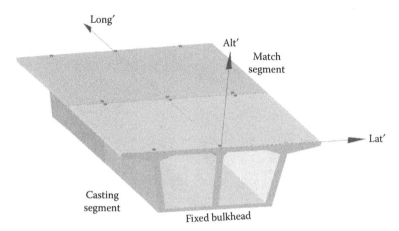

Figure 18.20 Local (casting yard) coordinate system and control points for alignment.

locating of the match segment to ensure a perfect (theoretical casting) vertical/ horizontal curve and superelevation, if applicable, can be achieved.

18.6.3 Control points and transformation

As shown in Figure 18.20, adjustments of match segment are controlled by measurements of certain control points on concrete segments. Mathematically, three points on a rigid body are enough to determine its location in space. For the purpose of easy practice and the need to control the shape of casting segment, six points on top of a segment are used as control points: two points on each web centerline and two on the theoretical centerline. Longitudinally, the control points are located at the segment edge as close as possible, 51 mm (2″) offset from the edge, for example, so as to produce enough control of the vertical alignment.

Measurements of control points are done in casting yard under a local coordinate system (Long′–Alt′–Lat′), whereas theoretical curve values are established in global coordinate system. Transforming of a 3D point from the global system to the casting system is essential. As described in detail in Section 18.6.7, given a defined local system, transforming between global and local is simple.

18.6.4 Procedures of casting and control

The first segment to be casted is different from successive segments as both ends are against bulkhead, rather than the normal casting as shown in Figure 18.16. After the first segment is cured and before moving to match cast position, all control points' coordinates in the casting system are surveyed and are taken as as-cast values.

The first segment is moved to match cast position, and the casting bed is ready for the second segment. A new local system aligned in face of the second segment is established. Of the 12 points, six points for the first segment (match cast segment), which is already casted, and six for the second segment (casting segment) are transformed from the global system to this new local system. Transformed values for the match cast segment are used to guide the adjustment of supporting jacks so as to ensure the end form of the casting segment is in correct position and orientation. Values for the casting segment are used to guide the punching of six marking bolts on top of casting segment (through the connection to rebars beneath). This step is a main part of geometry assurance and is called *setup*. Control values in this setup are also called setup values (LoBuono 2005).

Again, after the casting segment is cured and before moving to match cast position, as-cast values are surveyed again. Theoretically, as-cast values are the same as setup values. The earlier process will be repeated till all segments are casted.

18.6.5 Error finding and correction

However, as-cast values are not exactly the same as setup values in reality. Their difference from setup values indicates the existence of geometry error. If these errors are simply ignored, they will be accumulated along the casting, and thus the final curve after assembling will be out of control. To make sure the as-cast curve is in close agreement with the theoretical curve, the error must be detected and corrected during casting.

It should be noted first that the as-cast values of a segment are measured in the segment's own local system. For example, as-cast values of the second segment are under the local system of the second segment and as-cast values of the third segment are under the third's local system. Transforming these local as-cast values from each individual local system, as-cast curve in the global system can be obtained. Finding casting error can be done by further comparing as-cast curve with the theoretical curve. Figure 18.21 shows the elevation errors between as-cast curve and theoretical cast curve. Errors on plane and twist along the longitudinal axis are in the same manner.

As shown in Figure 18.21, corrections can be done using adjusted setup values for match cast segment when casting segment is set up. The adjusted setup values are obtained by transforming the as-cast points of match cast segment, instead to the local system of casting segment. Chances are that accumulated as-cast errors are too big to be fully corrected in the immediately followed casting segment, for too big a kink in vertical, planar, or longitudinal twist not satisfying the smooth geometry requirements. In that case, as shown in Figure 18.21, the error can be partially corrected in the next

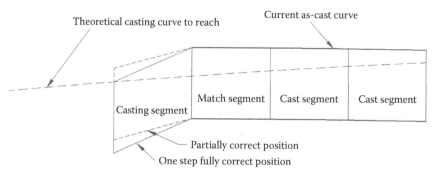

Theoretical casting curve to reach

Current as-cast curve

Match segment

Cast segment

Cast segment

Casting segment

Partially correct position

One step fully correct position

Figure 18.21 Error of as-cast curve and correction.

casting segment and the remaining error can be carried to the next segment. This technique is called split error correction.

18.6.6 Evolution of geometry control in precast segmental bridge

The precast segmental bridge construction method was first developed in 1930s by French engineer Eugene Freyssinet. It was first practiced in 1973 in the United States. Precasting and geometry control methods had long been developed ever since. However, advancing of computer and survey technologies impacts the evolution of geometry control techniques. For example, the modern computer software technologies have enabled controlling of more sophisticated 3D curves and the replacement of traditional optical theodolite by modern total station has greatly improved the survey accuracy and field efficiency. Real-time survey and control technologies have also enabled the automation of the whole process of measuring, calculating, and adjusting (Kumar et al. 2008).

18.6.7 Geometry transformation

As the underlying process of geometry control, geometry transformation is the basis. The goal of geometry transformation is to find the ordinate representation of a fixed point in another coordinate system, given that another coordinate system is defined under one coordinate system. Geometry transformation is also a primary process in computer graphics; its principle can be widely found in computer graphics books. One point that should be noted is what is used of geometry transformation in match cast geometry control is much simpler than what computer graphics may be used of. In computer graphics, transformations are most aimed at object transforming, in which a point is either translated/rotated along an axis or scaled. In this section,

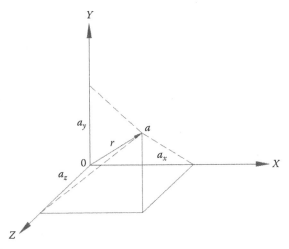

Figure 18.22 Direction cosines.

transformation of a fixed point from one coordinate system to another is briefly summarized.

18.6.7.1 Direction cosines

Given a direction a with a length of r and coordinate components of a_x, a_y, a_z as shown in Figure 18.22, the cosines of its angles to three axes $a_x/r, a_y/r, a_z/r$, respectively, are called direction cosines. Specifically when a is a unit direction, it can be represented by its direction cosines as

$$a = (a_x, a_y, a_z) \tag{18.9}$$

Having a direction's cosines as shown in Equation 18.9, the point on direction a with a length ordinate of l_a can be represented by $l_a \cdot (a_x, a_y, a_z)$.

18.6.7.2 Direction cosines matrix of a local coordinate system

Similar to Equation 18.9, direction cosines of three axes of a defined local system X', Y', Z' can be found as

$$X' = (x'_x, x'_y, x'_z); \; Y' = (y'_x, y'_y, y'_z); \; Z' = (z'_x, z'_y, z'_z) \tag{18.10}$$

Given a point P, which has coordinates of (x', y', z') in a local coordinate system defined as Equation 18.10, its representation in the global coordinate system where the local coordinate system is defined as

$$(x, y, z) = (x', y', z')\lambda \tag{18.11}$$

where λ is called direction cosine matrix of local coordinate system and is defined as

$$\lambda = \begin{bmatrix} x'_x & x'_y & x'_z \\ y'_x & y'_y & y'_z \\ z'_x & z'_y & z'_z \end{bmatrix} \tag{18.12}$$

18.6.7.3 Transformation between two coordinate systems

Once a local coordinate system is defined, λ is known. A point represented by a local coordinate system (x', y', z') can be transformed to the global system by Equation 18.11 so its coordinates in the global system (x, y, z) can be obtained. Similar to Equation 18.11, its reverse transformation can be found as

$$(x', y', z') = (x, y, z)\lambda^{-1} \tag{18.13}$$

By using Equation 18.13, a point represented by its global coordinate system (x, y, z) can be transformed to a local system so its local coordinate system (x', y', z') can be obtained.

Transforming Equation 18.11 to 18.13 is based on that the two coordinate systems have the same origin as shown in Figure 18.22. When applying to transformation between the global and local systems as shown in Figure 18.19, the origin of the local coordinate system has to be translated to the same origin with the global system before the transformation. The transformed coordinates will then be translated back to the true origin of the local system.

18.6.7.4 Definition of the casting system in global system

The connecting nodes between segments on the theoretical casting curve are known, as shown in Figure 18.16. Therefore, the origin and the longitudinal axis of the local system (Long') for the current casting segment as shown in Figure 18.20 can be established. As the vertical axis of the local system (Alt') cannot be generally assumed being parallel to the global vertical axis (Alt) due to the existence of superelevation, the transverse axis of the local system (Lat') has to be defined instead.

Because the transverse axis (Lat') is always parallel to the bulkhead, Lat' can be known as long as a point along the bulkhead is known. This can be obtained by the control point shown in positive Lat' axis in Figure 18.20. By constructing a line on the plane constructed by points p_n, p_{n-1} and the control point that is perpendicular to line p_n, p_{n-1}, the axis of Lat′ as a line can be known. Further, the direction of positive Lat′ can be determined by referring the control point.

Having Long′ and Lat′ defined, Alt′ can be simply derived from a cross-product operation of Lat′ × Long′. Thus, the direction cosine matrix, as defined by Equation 18.12 for the transformation between the current casting system and the global system, is established. Once the λ matrix is obtained for the current casting segment, transforming of control points between the global coordinate system and the casting system can be performed further.

When implementing geometry control program for precast segmental bridges, geometry transformation can be simply called as underlying functionalities. Other procedures such as calculating setup values, collecting as-cast values, and detecting errors can follow the discussions in the earlier sections. Together with regular tabular data reports, a 3D rendering, as shown in Figure 18.20, truly reflecting segment geometries, setup values, and as-cast measurements, will be greatly helpful for both designers and field engineers.

18.6.8 An example of short-line match casting geometry control

To demonstrate the geometry control of precast segmental bridges in casting yard, an example of a single-span curve bridge of radius 183 m (600′) is presented. As the plane curve shown in Figure 18.23, the example span contains 16 segments with a total length of 37.6 m (123.5′). Figure 18.23 contains two centerlines: (1) The theoretical curve and (2) as-cast curve, or the obtained curve (these two curves, however, are overlapped on each other due to minor discrepancies). However, they may not show clearly because the obtained curve is very close to the theoretical curve. Figure 18.24 shows a prediction of setups for segment No. 10 so that casting of segment No. 11 would be in the correct position. This prediction is based on the survey of all previous segments. Values shown in Figure 18.24 indicate that the dry segment should be positioned by jacking its support so that the segment to be casted will be at the right position after assembled. Figure 18.25 shows the survey points and values after segment No. 13 is casted, which will affect the prediction of the setup of this segment after it is moved to match position.

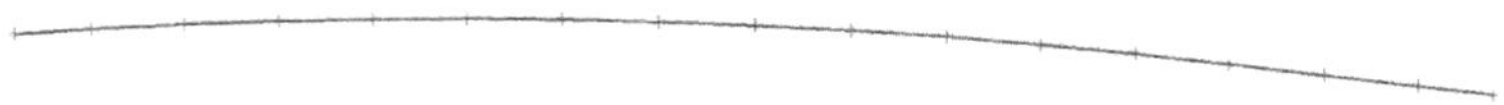

Figure 18.23 The plane curve of a segmental bridge with 16 precast segments.

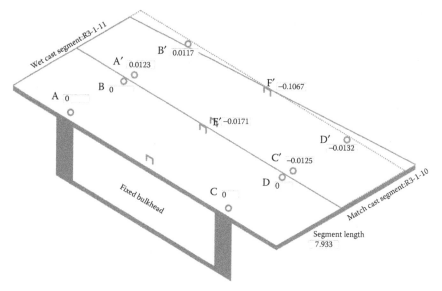

Figure 18.24 Prediction of setup values for a match cast segment.

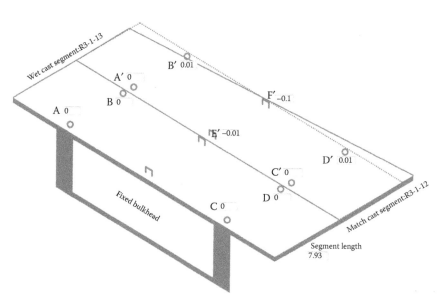

Figure 18.25 Survey values after cast of a segment.

18.7 TREND OF BRIDGE COMPUTER MODELING AND VISUALIZATION

As a close issue to bridge geometry, bridge computer modeling and visualization have long been the focus of computer technology applications in bridge engineering. Comparing with applications in other fields, the advancing of bridge computer modeling and visualization, however, does not match what modern computer graphics technologies promise and what bridge engineers expect. Most bridge analysis and design software available nowadays are still based on mathematics or mechanics model of a bridge, rather than the engineering model of a true project. What the current bridge software provide a typical process of bridge analysis and design to engineers still is (1) to establish and analyze a bridge's mechanical model, (2) to check design code for each component based on the analysis results, and (3) to resize components or adjust structural dimensions and repeat the previous process if necessary. The benefit of fast technology advancement in both computer hardware and software improves only the performance of each step; challenges such as abstracting mechanical model from engineering model and representing analysis results in the way engineers used to are still governing the whole process of analysis and design of bridge structures. Another aspect that shows great potential for advancing in computer application is visualization. Showing only 2D or 3D frame lines of mechanics model or bridge schematics cannot meet the demanding of bridge analysis and design nowadays.

The modern computer graphics technologies are now well capable of processing virtual 3D bridge models in great detail. The key to take the advantage of it is to establish a bridge engineering model, rather than a simplified and abstracted bridge mechanical model. The complexity of bridge engineering model can be greatly simplified so as to be feasible when a particular, commonly used bridge type is focused. Figure 18.26 illustrates some modeling and visualization features as trends envisioned in bridge analysis, design, and rating applications. Roadway and bridge geometries are the first part to describe a bridge project. Detailed bridge component dimensions and materials can be defined further. While a bridge engineering model is being edited, its true 3D rendering will be reflected in real time so engineers can get visual feedback instantly. A key feature that makes 3D rendering more useful, not merely a visual confirmation, is to allow identification of any component on-the-fly and to bring up its detailed design parameters for editing. For example, when the highlighted stiffener as shown in Figure 18.26 is clicked, the stiffener's definition will be showing up on screen so as to be edited instantly in place. Because the engineering model is established, the mechanical model can be automatically created and analyzed. The tedious error-prone process of converting the structural analysis model and analyzing can be automated. 3D bridge components

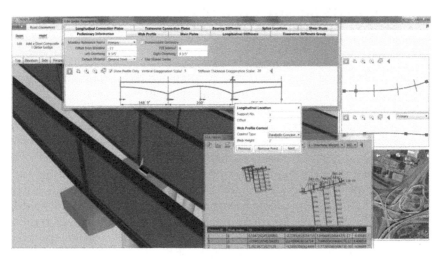

Figure 18.26 Trend of computer modeling and visualization.

can then be rendered by color codes (not shown here) reflecting the analysis or rating results. When existing bridges are to be rated, geographic information system and centralized database system can be used as the underlying support technologies.

In short, the great demand of structural rating due to the deterioration of bridge structures and changing of traffic loading patterns in large geographic scales, and the availability of highly advanced modern computer hardware and software technologies are enabling the development of new-generational bridge software applications toward automation, visualization, and virtualization.

References

CHAPTER 1

AASHTO, *AASHTO LRFD Bridge Design Specifications*, 6th Edition, American Association of State Highway and Transportation Officials, Washington, DC, 2012 with 2013 Interim.

EN 1991-2, "Eurocode 1: Actions on Structures—Part 2: Traffic Loads on Bridges," https://ia601600.us.archive.org/30/items/en.1991.2.2003/en.1991.2.2003.pdf.

Technical Committee CEN/TC 250, "Structural Eurocodes," *General Code for Design of Highway Bridges and Culverts (JTG D60-2004)*, Ministry of Communications; China Communications Press, Beijing, People's Republic of China, 2004.

Jaramilla, B. and Huo, S., "Looking to Load and Resistance Factor Rating," Public Road, July/August 2005, Vol. 69, No.1, https://www.fhwa.dot.gov/publications/publicroads/05jul/09.cfm.

O'Connor, C. and Shaw, P.A., *Bridge Loads: An International Perspective*, Spon Press, New York, 2000.

OHBDC, Ontario Highway Bridge Design Code, 1991 and Commentary, Ministry of Transportation and Communications, Toronto, Ontario, Canada, 1991.

Tonias, D.E., *Bridge Engineering: Design, Rehabilitation, and Maintenance of Modern Highway Bridges*, McGraw-Hill, New York, 1994.

CHAPTER 2

AASHTO, *AASHTO LRFD Bridge Design Specifications*, 6th Edition, American Association of State Highway and Transportation Officials, Washington, DC, 2012 with 2013 Interim.

Bakht, B. and Jaeger, L.G., *Bridge Analysis Simplified*, McGraw-Hill, New York, 1985.

Coletti, D. and Puckett, J.P., "Steel Bridge Design Handbook: Structural Analysis," Publication No. FHWA-IF-12-052—Vol. 8, November 2012, http://www.fhwa.dot.gov/bridge/steel/pubs/if12052/volume08.pdf.

FHWA, *Curved Girder Workshop Notebook*, University of Maryland, College Park, MD, 1990.

Fu, C.C., "Bridge Rating using Influence Surface for Curved Bridge Structures," Presented to the *First Congress on Computing in Civil Engineering*, June 20–22, 1994.

Fu, C.C., "Load and Resistance Factor Rating (LRFR) for Steel Bridges," *2013 IBC/FHWA Workshop on Application of Software in Bridge Load Rating*, Pittsburgh, PA, June 5, 2013.

Fu, C.C. and Hsu, Y.T., "The Development of an Improved Curvilinear Thin-Walled Vlasov Element," *Computers & Structures*, 54(1), 147–159, 1995.

Fu, C.C. and Schelling, D.R., "TRAP Theoretical and User's Manual," Federal Highway Administration, Publication No. FHWA-RT-89-054, February 1989.

Hambly, E.C., *Bridge Deck Behavior*, 2nd Edition, E & FN Spon, New York, 1991.

Heins, C.P. and Hall, D.H., *Designer's Guide to Steel Box Girder Bridges*, Bethlehem Steel Corporation, Bethlehem, PA, 1981.

Hsu, Y.T., Fu, C.C., and Schelling, D.R., "An Improved Horizontally Curved Beam Element," *Computers & Structures*, 34(2), 313–316, 1990.

Hwang, H., Yoon, H., Joh, C., and Kim, B.S., "Punching and fatigue behavior of long-span prestressed concrete deck slabs," *Engineering Structures*, 32, 2861–2872, 2010.

Jategaonkar, R, Jaeger, L.G., and Cheung, M.S., *Bridge Analysis using Finite Elements*, Canadian Society for Civil Engineering, Montréal, Québec, Canada, 1985.

Kumarasena, T., Scanlan, R.H., and Morris G.R., "Deer Isle Bridge: Efficacy of Stiffening Systems," *Journal of Structural Engineering*, 115(9), 2313–2328, 1989.

OHBDC, *Ontario Highway Bridge Design Code, 1991 and Commentary*, Ministry of Transportation and Communications, Toronto, Ontario, Canada, 1991.

Richart, F.E., Hall, J.R., and Woods, R.D., *Vibrations of Soils and Foundations*, Prentice Hall, Englewood Cliffs, NJ, 1970.

Troitsky, M.S., *Cable-Stayed Bridges—An Approach of Modern Bridge Design*, 2nd Edition, BSP Professional Books, London, 1988.

CHAPTER 3

Bažant, Z.P., Hubler, M.H., and Yu, Q., "Excessive Creep Deflections: An Awakening," *Concrete International*, August 2011.

Fan, L., *Prestressed Concrete Continuous Bridges*, People's Transportation Press, Beijing, People's Republic of China, 1998.

Hambly, E.C., *Bridge Deck Behaviour*, 2nd Edition, E & FN Spon, New York, 1991.

Shi, D. et al., *Bridge Structure Computations*, Tongji University Press, Shanghai, People's Republic of China, 1987.

Wang, S., "Automatic Incremental Creep Analysis and Its Implementation," *Journal of Tongji University—Natural Science*, 28(2), 138–142, 2000.

Zhu, B., *The Finite Element Method Theory and Applications*, 2nd Edition, China Water and Power Press, Beijing, People's Republic of China, 1998.

Zienkiewicz, O.C. and Taylor, R.C., *The Finite Element Method*, 3rd Edition, McGraw-Hill, London, 1977.

CHAPTER 4

AASHTO, *AASHTO LRFD Bridge Design Specifications*, 6th Edition, American Association of State Highway and Transportation Officials, Washington, DC, 2013a.

AASHTO, *Manual for Condition Evaluation and Load and Resistance Factor Rating of Highway Bridges*, American Association of State Highway and Transportation Officials, Washington, DC, 2013b.

ANSYS®, *ANSYS Mechanical User Guide*, ANSYS Inc., Canonsburg, PA, 2005.

Bakht, B., Jaeger, L.G., and Cheung, M.S. "Cellular and Voided Slab Bridges," *Journal of the Structural Division*, 107(9), 1797–1813, 1981.

CSiBridge®, "Integrated 3D Bridge Design Software," Computers and Structures, Inc., Berkeley, CA, 2010, http://www.csiamerica.com/products/csibridge.

Darwin, D., "Reinforced Concrete," in *Finite Element Analysis of Reinforced Concrete Structures II*, Isenberg, J., Ed., American Society of Civil Engineers, New York, 1993, pp. 203–232.

Elsaigh, W., Kearsley, E., and Robberts, J. "Modeling the Behavior of Steel-Fiber Reinforced Concrete Ground Slabs. II: Development of Slab Model," *Journal of Transportation Engineering*, 137(12), 889–896, 2011a.

Elsaigh, W., Robberts, J., and Kearsley, E. "Modeling the Behavior of Steel-Fiber Reinforced Concrete Ground Slabs. I: Development of Material Model," *Journal of Transportation Engineering*, 137(12), 882–888, 2011b.

Fu, C.C., "Merlin-DASH® User's Manual," the Bridge Engineering Software and Technology (BEST) Center, University of Maryland, College Park, MD, 2012, http://best.umd.edu/software/merlin-dash/.

Fu, C.C., Briner, T.L, and Getaneh, T., "Theoretical and Field Experimental Evaluation of Skewed Modular Slab Bridges," Report No. MD-12- SP109B4N, Maryland State Highway Administration, Baltimore, MD, 2012.

Fu, C.C and Graybeal, B., "Shrinkage and Creep Study of Ultra High Performance Concrete Girders," (11-2229) *The Proceedings of Transportation Research Board*, January 23–27, Washington, DC, 2011.

Fu, C.C., Pan, Z.F., and Ahmed, M.S., "Transverse Post-tensioning Design of Adjacent Precast Solid Multi-beam Bridges," *Journal of Performance for Constructed Facilities*, 25(3), 223–230, 2011.

Gao, D.Y., "Stress-Strain Curves of Steel Fiber Concrete Under Axial Compression," *Hydraulic Journal* (in Chinese), 10, 43–47, 1991.

Hambly, E.C., *Bridge Deck Behavior*, E & FN Spon, London, 1976.

Hognestad, E., *A Study on Combined Bending and Axial Load in Reinforced Concrete Members*. University of Illinois Engineering Experiment Station, Urbana-Champaign, IL, 1951, pp. 43–46.

Kachlakev, D.I., "Strengthening Bridges Using Composite Materials," FHWA Report OR-RD-98-08, FHWA, Corvallis, OR, 1998.

Kent, D.C., and Park, R., "Flexural Members with Confined Concrete," *Journal of the Structural Division, Proceedings of the American Society of Civil Engineers*, 97(ST7), 1969–1990, 1971.

Mander, J.B., Priestley, M.J.N., and Park, R., "Observed Stress-Strain Behavior of Confined Concrete," *Journal of Structural Engineering*, 114(8), 1827–1849, 1988a.

Mander, J.B., Priestley, M.J.N., and Park, R., "Theoretical Stress-Strain Model of Confined Concrete," *Journal of Structural Engineering*, 114(8), 1804–1826, 1988b.

Menassa, C., Mabsout, M., Tarhini, K., and Frederick, G., "Influence of Skew Angle on Reinforce Concrete Slab Bridges," *Journal of Bridge Engineering*, 12(2), 205–214, 2007.

O'Brien, E.J. and Keogh, D., *Bridge Deck Analysis*, E & FN Spon, London, 1999.

Park, S.H., *Bridge Inspection and Structural Analysis*, 2nd Edition, Trenton, NJ, 2000.

Rajagopalan, N., *Bridge Superstructure*, Alpha Science International, Oxford, October 12, 2006.

SAP2000®, "Integrated Software for Structural Analysis & Design," Computers and Structures Inc., Berkeley, CA, 2007, http://www.csiamerica.com/products/sap2000.

Sen, R., Issa, M., Sun, X., and Gergess, A., "Finite Element Modeling of Continuous Posttensioned Voided Slab Bridges," *Journal of Structural Engineering*, 120, 2, 1994.

Timoshenko, S. and Woinowsky-Krieger, S. *Theory of Plates and Shells*. McGraw-Hill, London, 1959.

CHAPTER 5

AASHTO, *AASHTO LRFD Bridge Design Specifications*, 6th Edition, American Association of State Highway and Transportation Officials, Washington, DC, 2013 Interim.

ACI-209, *Prediction of Creep, Shrinkage and Temperature Effects in Concrete Structures, Designing for Creep and Shrinkage in Concrete Structure*, ACI Publication SP-76, American Concrete Institute, Detroit, MI, 1982.

Bakht, B. and Jaeger, L.G., *Bridge Analysis Simplified*, McGraw-Hill, New York, 1985.

FLDOT/Corven Engineering, Inc., *New Directions for Florida Post-Tensioned Bridges*, Volume 1 of 10: Post-Tensioning in Florida Bridges, Florida Department of Transportation, Tallahassee, FL, February 2002.

Fu, C.C., "Merlin-DASH/PBEAM® User's Manual," BEST Center, University of Maryland, College Park, MD, 2012.

Fu, C.C. and Wang, S., "Prestressed Concrete Girder Bridges from 2D to 3D Modeling," (02-3102) Transportation Research Board Practical Papers, *2002 Catalog of Practical Papers for State Departments of Transportation*, Washington, DC, 2002.

JTG D62-85, *Code for Design of Highway Reinforce Concrete and Prestressed Concrete Bridges and Culverts* (in Chinese), People's Transportation Press, Beijing, People's Republic of China, 1985.

Ketchum, M.A. and Scordelis, A.C., *Redistribution of Stresses in Segmentally Erected Prestressed Concrete Bridges*, University of California, Berkeley, CA, Report No. UCB/SESM-86-07, 1986.

LUSAS®, "LUSAS Bridge/Bridge Plus Bridge Engineering Analysis," 2012, http://www.lusas.com/products/information/eurocode_pedestrian_loading.html.

McDonald, D., "Comparison of Design Practices of Prestressed Concrete Beam Bridge in the U.S.," Unpublished M.S. scholarly paper supervised by C.C. Fu, December 2005.

MIDAS®, "MIDAS Civil Integrated Solution System for Bridge and Civil Engineering," 2007–2014, http://en.midasuser.com/products/products.asp?nCat=352&idx=29134.

Pan, Z.F., Fu, C.C., and Lü, Z., "Impact of Construction Technology on Long-Term Deformation of Long-Span Prestressed Concrete Bridges," *The Proceedings of the 5th International Conference on Bridge Maintenance, Safety and Management 2010*, July 11–14, Philadelphia, PA, 2010.

Precast/Prestressed Concrete Institute (PCI), *Precast Prestressed Concrete Bridge Design Manual*, 3rd Edition, PCI, Chicago, IL, 2011.

Schellenberg, K., Vogel, T., Fu, C., and Wang, S., "Comparison of European and U.S. Practices Concerning Creep and Shrinkage," *The Proceedings of Fib Symposium Structural Concrete and Time*, La Plata, Argentina, September 28–30, 2005.

Wang, S.Q. and Fu, C.C., "*VBDS®*: Visual Bridge Design System Version 1.0," 2005, www.best.umd.edu/program/VBDS_UsersManual.pdf.

CHAPTER 6

CalTran, "Structural Modeling and Analysis," LRFD Bridge Design Practice, August 2012, http://www.dot.ca.gov/hq/esc/techpubs/manual/bridgemanuals/bridge-design-practice/pdf/bdp_4.pdf.

Fu, C.C. and Tang, Y., "Torsional Analysis for Prestressed Concrete Multiple Cell Box," *Journal of Engineering Mechanics*, 127(1), 45–51, 2001.

Fu, C.C. and Yang, D., "Design of Concrete Bridges with Multiple Box Cells due to Torsion Using Softened Truss Model," *ACI Structural Journal*, 93(6), 696–702, 1996.

Hsu, T.T.C., "ACI Shear and Torsion Provision for Prestressed Hollow Girders," *ACI Structural Journal*, Technical Paper, Title no. 94-S72, 1994.

Hsu, T.T.C., *Unified Theory of Reinforced Concrete*, CRC Press, Boca Raton, FL, 1993.

Mast, R., Marsh, L., Spry, C., Johnson, S., Grieenow, R., Guarre, J., and Wilson, W., *Seismic Design of Bridges—Design Examples 1–7 (FHWA-SA-97-006 thru 012)*, USDOT/FHWA, September 1996.

Nutt, R. and Valentine, O., "NCHRP Report 620—Development of Design Specifications and Commentary for Horizontally Curved Concrete Box-Girder Bridges," Transportation Research Board, Washington, DC, 2008.

Priestley, M.J.N., Seible, F., and Calvi, G.M., *Seismic Design and Retrofit of Bridges*, Wiley, New York, 1996.

Sennah, K.M. and Kennedy, J.B., "Literature Review in Analysis of Curved Box-Girder Bridges," *Journal of Bridge Engineering*, 7(2), 134–143, 2002.

Wang, S.Q. and Fu, C.C., "*VBDS®*: Visual Bridge Design System Version 1.0," 2005. www.best.umd.edu/program/VBDS_UsersManual.pdf.

CHAPTER 7

AASHTO, *AASHTO LRFD Bridge Design Specifications*, 6th Edition, American Association of State Highway and Transportation Officials, Washington, DC, with 2013 Interim.

AASHTO/NSBA Steel Bridge Collaboration Task Group 13, *Guidelines for Steel Girder Bridge Analysis*, Document G13.1, 1st Edition, American Association of State Highway and Transportation Officials, Washington, DC, p. 155, 2011.

ABAQUS, I., *ABAQUS/Standard User's Manual*, Dassault Systèmes, 2007, http://www.3ds.com/support/documentation/users-guide/.

ACI 209R-92, *Prediction of Creep, Shrinkage, and Temperature Effects in Concrete Structures*, ACI, Detroit, MI, 2008.

AISC, *Design Guide 9: Torsional Analysis of Structural Steel Members*, AISC, Chicago, IL, 2003.

Barr, P.J., Eberhard, M.O., and Stanton, J.F. "Live-Load Distribution Factors in Prestressed Concrete Girder Bridges," *Journal of Bridge Engineering*, 6(5), 298–306, 2001.

Baskar, K., Shanmugam, N.E., and Thevendran, V., "Finite-Element Analysis of Steel-Concrete Composite Plate Girder," *Journal of Structural Engineering*, 128(9), 1158–1168, 2002.

Chen, S.S., Aref, A.J., Ahn, I.-S., Chiewanichakorn, M., Carpenter, J.A., Nottis, A., and Kalpakidis, I., *NCHRP Report 543—Effective Slab Width for Composite Steel Bridge Members*, National Cooperative Highway Research Program, Transportation Research Board, Washington, DC, 2005.

Chen, Y., "Distribution of Vehicular Loads on Bridge Girders by the FEA Using ADINA: Modeling, Simulation, and Comparison," *Computers & Structures*, 72(1–3), 127–139, 1999.

Chung, W. and Sotelino, E.D., "Three-Dimensional Finite Element Modeling of Composite Girder Bridges," *Engineering Structures*, 28(1), 63–71, 2006.

CSiBridge, Computers and Structures, Inc., Berkeley, CA, 2011, http://www.csiamerica.com/products/csibridge.

DESCUS-I (Design and Analysis of Curved I-Girder Bridge Systems) Users' Manual, Production Software, Inc., August 2012, http://www.cee.umd.edu/best/Descus-I.pdf.

Eamon, C.D. and Nowak, A.S. "Effect of Secondary Elements on Bridge Structural System Reliability Considering Moment Capacity," *Structural Safety*, 26(1), 29–47, 2004.

Ebeido, T. and Kennedy, J.B., "Girder Moments in Simply Supported Skew Composite Bridges," *Canadian Journal of Civil Engineering*, 23(4), 904–916, 1996.

Elhelbawey, M.I. and Fu, C.C., "Effective Torsional Constant for Restrained Open Section," *Journal of Structural Engineering*, 124(11), November 1998.

FHWA/NSBA/HDR, "Steel Bridge Design Handbook FHWA-IF-12-052—Vol. 8: Structural Analysis," 1363–1365, Federal Highway Administration, USDOT, November 2012, http://www.fhwa.dot.gov/bridge/steel/pubs/if12052/volume08.pdf.

Fu, C.C. and Hsu, Y.T., "Bridge Diaphragm Elements with Partial Warping Restraint," *Journal of Structural Engineering*, 120(11), 3388–3395, November 1994.

Fu, C.C. and Hsu, Y.T., "The Development of an Improved Curvilinear Thin-Walled Vlasov Element," *Computers & Structures*, 54(1), 147–159, 1995.

Fu, K.-C. and Lu, F., "Nonlinear Finite-Element Analysis for Highway Bridge Superstructures," *Journal of Bridge Engineering*, 8(3), 173–179, 2003.

Hays Jr., C., Sessions, L.M., and Berry, A.J., "Further Studies on Lateral Load Distribution Using a Finite Element Method," *Transportation Research Record*, 6–14, 1986.

Hsu, Y.T., Fu, C.C., and Schelling, D.R., "An Improved Horizontally Curved Beam Element," *Computers & Structures*, 34(2), 313–316, 1990.

Issa, M.A., Yousif, A.A., and Issa, M.A. "Effect of Construction Loads and Vibrations on New Concrete Bridge Decks," *Journal of Bridge Engineering*, 5(3), 249–258, 2000.

Mabsout, M.E., Tarhini, K.M., Frederick, G.R., and Tayar, C., "Finite-Element Analysis of Steel Girder Highway Bridges," *Journal of Bridge Engineering*, 2(3), 83–87, 1997.

Nakai, H. and Yoo, C.H., *Analysis and Design of Curved Steel Bridges*, McGraw-Hill, New York, 1988.

Queiroz, F.D., Vellasco, P.C.G.S., and Nethercot, D.A. "Finite Element Modelling of Composite Beams with Full and Partial Shear Connection," *Journal of Constructional Steel Research*, 63(4), 505–521, 2007.

SAP2000®, "Integrated Software for Structural Analysis & Design," Computers and Structures, Inc., Berkeley, CA, 2007, http://www.csiamerica.com/products/sap2000.

Sebastian, W.M. and McConnel, R.E., "Nonlinear FE Analysis of Steel-Concrete Composite Structures," *Journal of Structural Engineering*, 126(6), 662–674, 2000.

Tabsh, S.W. and Tabatabai, M., "Live Load Distribution in Girder Bridges Subject to Oversized Trucks," *Journal of Bridge Engineering*, 6(1), 9–16, 2001.

Tarhini, K.M. and Frederick, G.R., "Wheel Load Distribution in I-Girder Highway Bridges," *Journal of Structural Engineering*, 118(5), 1285–1294, 1992.

White, D.W. et al., "Guidelines for Analysis Methods and Construction Engineering of Curved and Skewed Steel Girder Bridges," NCHRP Project 12-79 Report 725, TRB, Washington, DC, 2012.

CHAPTER 8

AASHTO, *Guide Specifications for the Design of Horizontally Curved Girder Bridges*, American Association of State Highway and Transportation Officials, Washington, DC, 2003.

AASHTO, *AASHTO LRFD Bridge Design Specifications*, 6th Edition, American Association of State Highway and Transportation Officials, Washington, DC, with 2013 Interim.

ANSYS Mechanical User Guide, ANSYS Inc., Canonsburg, PA, 2012.

Begum, Z., "Analysis and Behavior Investigations of Box Girder Bridges" (Advisor C. C. Fu) degree of Master of Science, University of Maryland, College Park, MD, 2010.

Fan, Z. and Helwig, T., "Distortional Loads and Brace Forces in Steel Box Girders," *Journal of Structural Engineering*, 128(6), 710–718, 2002.

FHWA/NSBA/HDR, "Steel Bridge Design Handbook FHWA-IF-12-052—Vol. 8: Structural Analysis," 1363–1365, Federal Highway Administration, USDOT, November 2012, http://www.fhwa.dot.gov/bridge/steel/pubs/if12052/volume08.pdf.

Fu, C.C. and Hsu, Y.T., "The Development of an Improved Curvilinear Thin-Walled Vlasov Element," *Computers & Structures*, 54(1), 147–159, 1995.

Heins, C.P., "Box Girder Bridge Design—State of the Art," American Institute of Steel Construction, *Engineering Journal*, 4th quarter, 15(4), 126–142, 1978.

Hsu Y.T., "The Development and Behaviour of Vlasov Elements for the Modeling of Horizontally Curved Composite Box Girder Bridge Superstructures," PhD dissertation, University of Maryland, College Park, MD, 1989.

Hsu, Y.T. and Fu, C.C., "Application of EBEF Method for the Distortional Analysis of Steel Box Girder Bridge Superstructures During Construction," *International Journal of Advances in Structural Engineering*, 5, 4, 211–222, November 2002.

Hsu, Y.T., Fu, C.C., and Schelling, D.R., "An Improved Horizontally Curved Beam Element," *Computers & Structures*, 34(2), 313–316, 1990.

Hsu, Y.T., Fu, C.C., and Schelling, D.R., "EBEF Method for Distortional Analysis of Steel Box Girder Bridges," *Journal of Structural Engineering*, 121(3), 557–566, 1995; 122(8), 1996.

Kollbrunner, C.F. and Basler, K., *Torsion*, Springer-Verlag, Berlin, 1966 (in German).

SCI-The Steel Construction Institute, Ascot, UK: ESDEP-European Steel Design Education Programme, CD ROM, ESDEP Society, 2000.

Vlasov, V.Z., *Thin-Walled Elastic Beams*, OTS61-11400, National Science Foundation, Washington, DC, 1965.

White, D.W. et al., "Guidelines for Analysis Methods and Construction Engineering of Curved and Skewed Steel Girder Bridges," NCHRP Project 12-79 Report 725, TRB, Washington, DC, 2012.

Wright, R.N., Abdel-Samad, S.R., and Robinson, A.R., "BED Analogy for Analysis of Box Girder," *Journal of the Structural Division*, 94, 1719–1744, 1968.

CHAPTER 9

Brown, D.J., *Bridges: Three Thousand Years of Defying Nature*, Octopus Publishing Group Ltd, London, 2005.

Ellis, L.J.H., "Critical Analysis of the Lupu Bridge in Shanghai," Department of Civil and Architectural Engineering, University of Bath, Somerset, 2007, http://www.bath.ac.uk/ace/uploads/StudentProjects/Bridgeconference2007/conference/mainpage/Ellis_Lupu.pdf.

Kawamura, T., Fujimoto, Y., and Palmer Jr., W.D., "Wrapping an Arch in Concrete," *Concrete International*, 12(11), 26–31, 1990.

Li, X.S., Sun, M., and Fu, C.C., "Fast assessment method of arch-girder composite bridges," *The Proceedings of 7th International Conference on Bridge Maintenance, Safety and Management*, July 7–11, 2014, Shanghai, People's Republic of China.

Pellegrino, C., Cupanis, G., and Modena, C., "The Effect of Fatigue on the Arrangement of Hangers in Tied Arch Bridges," *Engineering Structures*, 32(4), 1140–1147, 2010.

Wang, S.Q. and Fu, C.C., "*VBDS®*: Visual Bridge Design System Version 1.0," 2005, www.best.umd.edu/program/VBDS_UsersManual.pdf.

Yao, X., "Influenced Factors on Fatigue of Hangers of Tied Arch Bridges," *Highway Journal* (in Chinese) 2007(12), 37–44, December 2007.

CHAPTER 10

AASHTO LRFD Bridge Design Specifications, US unit 2012, American Association of State Highway and Transportation Officials, Washington, DC, with 2013 Interim.

Bentley, STAAD.Pro v8i Technical Reference Manual, 2012, https://communities
.bentley.com/cfs-file.ashx/__key/telligent-evolution-components-attachments/
13-275895-00-00-00-24-18-54/Technical_5F00_Reference_5F00_V8i.pdf.

Bergeron, K.A., "The Future is Now," Public Roads, May/June 2004, Vol. 67, No. 6,
http://www.fhwa.dot.gov/publications/publicroads/04may/06.cfm.

Federal Highway Administration, "Tied Arch Bridges: T 5140.4," September 1978,
http://www.fhwa.dot.gov/bridge/t514004.cfm.

FHWA/NSBA/HDR, "Steel Bridge Design Handbook FHWA-IF-12-052—Vol. 5: Selecting the Righ Bridge Type," Federal Highway Administration, USDOT, November
2012, http://www.fhwa.dot.gov/bridge/steel/pubs/if12052/volume05.pdf.

Fu, C.C., "TRAP (Truss Rating and Analysis Program) User's Manual," the BEST
Center, University of Maryland, College Park, MD, 2012, http://best.umd.edu/
software/trap/.

Fu, C.C. and Zhang, N., "Investigation of the Bridge Expansion Joint Failure using
Field Strain Measurement," *Journal of Performance for Constructed Facilities*,
25(4), 309–316, July/August 2011.

MIDAS®, "MIDAS Civil Integrated Solution System for *Bridge and Civil Engineering*,"
2007, http://en.midasuser.com/products/products.asp?nCat=352&idx=29134.

Kulicke, J.M., "Highway Truss Bridges," in *Bridge Engineering Handbook*, Chen,
W.-F. and Duan, L., Eds., CRC Press, Boca Raton, FL, 2000.

National Steel Bridge Alliance, "Selecting the Right Bridge Type," *Steel Bridge
Design Handbook*.

Wang, S.Q. and Fu, C.C., "*VBDS*®: Visual Bridge Design System Version 1.0," 2005,
www.best.umd.edu/program/VBDS_UsersManual.pdf.

CHAPTER 11

ANSYS, *ANSYS Mechanical User Guide*, ANSYS Inc., Canonsburg, PA, 2012.

Chen, C., Yan, D., and Dong, D., "Prediction of parameters error in method of construction control in cable-stayed bridge," *Advanced Materials Research*, Vols. 163–167,
2385–2389, Trans Tech Publications Inc., Zurich, Switzerland, 2011.

Chen, W.-F. and Duan, L., *Bridge Engineering Handbook*, CRC Press, Boca Raton,
FL, 1999.

Ernst, J. H., "Der E-Modul von Seilen unter berucksichtigung desDurchhanges." *Der
Bauingenieur*, 40(2), 52–55, 1965.

Hambly, E.C., *Bridge Deck Behaviour*, 2nd edn., E & FN SPON, London, 1991.

Li, Y., Li, X., and Yang, A., "The Prediction Method of Long-Span Cable-Stayed Bridge
Construction Control Based on BP Neural Network," *The Proceedings of the 9th
WSEAS International Conference on Mathematical and Computational Methods
in Science and Engineering*, Stevens Point, Wisconsin, November 5, 2007.

Lin, Y., "The Application of Kalman's Filtering Method to Cable-Stayed Bridge
Construction," *China Civil Engineering Journal*, 3, 8–15, 1983.

Ministry of Transport of China, *Wind-Resistant Design Specifications for Highway
Bridges (JTG/T D60-01-2004)*, People's Transportation Press, Beijing, People's
Republic of China, 2004.

Su, C., Chen, Z., and Chen, Z., "Reliability of Construction Control of Cable-Stayed
Bridges," *The Proceedings of the ICE—Bridge Engineering*, 164(1), 18–22, 2011.

Tabatabai, H., "NCHRP Synthesis 353—Inspection and Maintenance of Bridge Stay Cable Systems," Transportation Research Board, Washington, DC, 2005.

Wang, S. and Fu, C.C., "Static and Stability Analysis of Long-Span Cable-Stayed Steel Bridges" (03-2337), *The Proceedings of Transportation Research Board*, January 12–16, Washington, DC, 2003.

Wang, S. and Fu, C.C., "Structural Design and Analysis of Long Span Bridges," *The Proceedings of IABMAS*, Italy, July 8–12, 2012.

Wang, S.Q. and Fu, C.C., "*VBDS®*: Visual Bridge Design System Version 1.0," 2005, www.best.umd.edu/program/VBDS_UsersManual.pdf.

You, Q. et al., "Sutong Bridge—A Cable-Stayed Bridge with Main Span of 1088 Meters," IABSE Congress Report, *17th Congress of IABSE*, Chicago, IL, pp. 142–149(8), 2008.

CHAPTER 12

Chen, W.-F. and Duan, L., *Bridge Engineering Handbook*, CRC Press, Boca Raton, FL, 1999.

Ji, L. and Feng, Z., *Construction of Suspension Bridges across the Yangtze River in Jiangsu, China*, IABSE Workshop - Recent Major Bridges, May 11–20, 2009 Shanghai, People's Republic of China. Jiangsu Provincial Yangtze River Highway Bridge Construction Commanding Department, Taizhou, People's Republic of China.

Kawada, T., *History of the Modern Suspension Bridge*, American Society of Civil Engineers, 2010.

SAP2000®, "Integrated Software for Structural Analysis & Design," Computers and Structures Inc., Berkeley, CA, 2007, http://www.csiamerica.com/products/sap2000.

Wang, S. and Fu, C.C., "Structural Design and Analysis of Long Span Bridges," *The Proceedings of IABMAS*, Italy, 2012.

Wang, S.Q. and Fu, C.C., "*VBDS®*: Visual Bridge Design System Version 1.0," 2005, www.best.umd.edu/program/VBDS_UsersManual.pdf.

CHAPTER 13

AASHTO, *AASHTO LRFD Bridge Design Specifications*, 6th Edition, American Association of State Highway and Transportation Officials, Washington, DC, with 2013 Interim.

ACI Committee 318, *Building Code Requirements for Structural Concrete (ACI 318-02) and Commentary (ACI 318R-02)*, American Concrete Institute, Farmington Hills, MI, 2002.

Fu, C.C., "Study of Crane Beam Check by using the Strut-and-Tie Model," An internal study report to the Maryland Port Administration, University of Maryland, College Park, MD, 1994.

Fu, C.C., Sircar, M., and Robert, J., "Maryland Experience in using Strut-and-Tie Model in Infrastructure (05-0698)," *The Proceedings of Transportation Research Board*, January 9–13, Washington, DC, 2005.

Kuchma, D., "Strut-and-Tie Website," 2005, http://dankuchma.com/stm/index.htm.

MacGregor, J.G., Wight, J.K., and MacGregor, J., *Reinforced Concrete: Mechanics and Design*, 5th Edition, Prentice Hall, Englewood Cliffs, NJ, 2008.

Martin Jr., B.T., and Sanders, D.H., "Analysis and Design of Hammerhead Pier Using Strut and Tie Method," Final Report-Project 20-07_ Task 217, National Cooperative Highway Research Program, Transportation Research Board, Washington, DC, November, 2007.

SAP2000®, "Integrated Software for Structural Analysis & Design," Computers and Structures, Inc., Berkeley, CA, 2007, http://www.csiamerica.com/products/sap2000.

Schlaich, J. and Werschede, D., Detailing of Concrete Structures (in German), Bulletin d'Information 150, Comite Euro-International du Beton, Paris, France, March 1982, 163pp.

Scott, R.M., Mander, J.B., and Bracci, J.M., "Compatibility Strut-and-Tie Modeling: Part I—Formulation," *ACI Structural Journal*, 109(5), 635–644, 2012.

CHAPTER 14

AASHTO, *AASHTO LRFD Bridge Design Specifications*, 6th Edition, American Association of State Highway and Transportation Officials, Washington, DC, 2013 Interim.

ANSYS®, *ANSYS Mechanical User Guide*, ANSYS Inc., Canonsburg, PA, 2005.

Cook, R.D., Malkus, D.S., Plesha, M.E., and Witt, R.J., *Concepts and Applications of Finite Element Analysis*, 4th Edition, Wiley, New York, 2002.

Ermopoulos, J.C., Vlahinos, A.S., and Wang, Y.C. "Stability Analysis of Cable-Stayed Bridges," *Computers & Structures*, 44(5), 1083–1089, 1992.

Fu, C.C., "The Top Chord Evaluation of Welded Pony Type Truss Bridge," The Bridge Engineering Software and Technology (BEST) Center, Department of Civil Engineering, University of Maryland, College Park, MD, 2006.

Galambos., T.V., Ed., *Guide to Stability Design: Criteria for Metal Structures*, SSRC, Structural Stability Research Council, 5th Edition, John Wiley & Sons, Inc., Hoboken, NJ, 1998.

Li, G., *Stability and Vibration of Bridge Structures*, The Press of Rail Road Department, Beijing, People's Republic of China, 1996.

Murray, N.W., *Introduction to the Theory of Thin-Walled Structures*, Oxford University Press, Oxford, 1984.

Ren, W.X. "Ultimate Behavior of Long-Span Cable-Stayed Bridges," *Journal of Bridge Engineering*, 4(1), 30–37, 1999.

Rostovtsev, G.G., "Calculation of a Thin Plane Sheeting Supported by Ribs," Trudy Leningrad Institute, Inzhenerov Grazhdanskogo Vosdushnogo Flota, No. 20, 1940 (in Russian).

Ryall, M.J., Parke, G.A.R., and Harding, J.E., *Manual of Bridge Engineering*, Thomas Telford Publishing, London, 2000.

Tang, M.C. "Bulking of Cable-Stayed Bridges," *Journal of the Structural Division*, 102(ST7), 1675–1684, 1976.

Timoshenko, S., *Theory of Elastic Stability*, McGraw-Hill, New York, 1936.

Wang, S. and Fu, C.C., "Static and Stability Analysis of Long-Span Cable-Stayed Steel Bridges" (03-2337), *The Proceedings of Transportation Research Board*, January 12–16, Washington, DC, 2003.

Wang, S.Q. and Fu, C.C., "*VBDS®*: Visual Bridge Design System Version 1.0," 2005.

CHAPTER 15

AASHTO, *AASHTO LRFD Bridge Design Specifications*, 6th Edition, American Association of State Highway and Transportation Officials, Washington, DC, with 2013 Interim.

AASHTO, *The Manual for Bridge Evaluation*, 2nd Edition, American Association of State Highway and Transportation Officials, Washington, DC, 2010 with 2012 Interim.

ANSYS®, *ANSYS Mechanical User Guide*, ANSYS Inc., Canonsburg, PA, 2005.

Applied Research Associates, Inc. AT-Blast Version 2.2, 2004, http://www.ara.com/products/AT-blast.htm.

ASCE/SEI7-10, *Minimum Design Loads of Buildings and Other Structures*, American Society of Civil Engineers, Reston, VA, 2010

BEST Center, "TRAP (Truss Rating and Analysis Program) User's Manual," University of Maryland, College Park, MD, 2006.

Caltrans, "Memo to Designers 12-2: Guidelines for Identification of Steel Bridge Members," August 2004, http://www.dot.ca.gov/hq/esc/techpubs/manual/bridge-manuals/bridge-memo-to-designer/page/Section%2012/12-2m.pdf.

FEMA-310, *Handbook for the Seismic Evaluation of Buildings*, Federal Emergency Management Agency, Washington, DC, 1998.

FHWA/NSBA/HDR, "Steel Bridge Design Handbook FHWA-IF-12-052—Vol. 9: Redundancy," Federal Highway Administration, USDOT, November 2012, http://www.fhwa.dot.gov/bridge/steel/pubs/if12052/volume09.pdf.

Fu, C.C., "Report on the Determination of Redundancy of the U.S. Bridge Corporation Bridge 3000," College Park, MD, 2000.

Fu, C.C. and Schelling, D.R., "Report on the Determination of Load and Fatigue Capacity and Redundancy of the U.S. Bridge Corporation Bridge 2000/1000," College Park, MD, 1989.

Fu, C.C. and Schelling, D.R., "Report on the Determination of Load and Fatigue Capacity and Redundancy of the U.S. Bridge Corporation Bridge 3000," College Park, MD, 1994.

Imhof, D, Middleton, C.R., and Palmer, A.C., "Redundancy Quantification in the Safety Assessment of Existing Concrete Beam-and-Slab Bridges," *The 5th International PhD Symposium in Civil Engineering*, J. Walraven, J. Blaauwendraad, T. Scarpas & B. Snijder (Eds.), 2004, Taylor & Francis Group, London, 2004, pp. 373–381.

Mahoney, E.E., "Analyzing the Effects of Blast Loads on Bridges using Probability, Structural Analysis, and Performance Criteria," Master Thesis (Advised by Dr. C. C. Fu), University of Maryland, College Park, MD, August 2007.

NCHRP, "Report 403—A Redundancy in Highway Bridge Superstructures," NCHRP, Washington, DC, 1998.

NIST, "Report on Application of Seismic Rehabilitation," National Institute of Standards and Technology (NIST), September 2001.

Penn DOT. "Design Manual Part IV," Harrisburg, PA, 2000, ftp://ftp.dot.state.pa.us/public/PubsForms/Publications/PUB%2015M.pdf.

Razmi, J., Ladani, L., and Aggour, M.S., "Fatigue Crack Initiation and Propagation in Piles of Integral Abutment Bridges," *Computer-Aided Civil and Infrastructure Engineering*, 28(5), 389–402, May 2013.

SAP2000®, "Integrated software for structural analysis & design," Computers and Structures Inc., Berkeley, CA, 2007, http://www.csiamerica.com/products/sap2000.

CHAPTER 16

American Petroleum Institute. Recommended Practice for Planning, Designing, and Constructing Fixed Offshore Platforms – Working Stress Design. Report RP 2A-WSD, 20th Ed., 1993.

Arockiasamy, M., Butrieng, N., and Sivakumar, M., "State-of-the-Art of Integral Abutment Bridges: Design and Practice," *Journal of Bridge Engineering*, 9(5), 497–506, 2004.

Barker, R.M., Duncan, J.M., Rojiani, K.B., Ooi, P.S.K., Tan, C.K., and Kim, S.G., Eds., "Manuals for Design of Bridge Foundations," National Cooperative Highway Research Program (NCHRP) Report 343, Transportation Research Board, Washington, DC, 1991.

FHWA, *Steel Bridge Design Handbook:* Substructure Design, Publication No. FHWA-IF-12-052, Vol. 16, Washington, DC, November, 2012, http://www.fhwa.dot.gov/bridge/steel/pubs/if12052/volume16.pdf.

Greimann, L.F. "Rational Design Approach for Integral Abutment Bridge Piles," Transportation Research Record 1223, National Research Council, Washington, DC, 1989.

Greimann, L.F., and Wolde-Tinsae, A.M., "Design Model for Pile in Jointless Bridges," *Journal of Structural Engineering*, 114(6), 1354–1371, 1988.

Khodair, Y. and Hassiotis, S. "Numerical and Experimental Analyses of an Integral Bridge," *International Journal of Advanced Structural Engineering*, 2013, http://www.advancedstructeng.com/content/5/1/14.

Rasmi, J., "Thermo-Mechanical Fatigue of Steel Piles in Integral Abutment Bridges," PhD Dissertation, Civil and Environmental Engineering, University of Maryland, College Park, MD, 2012.

Shah, B.R., "3D Finite Element Analysis of Integral Abutment Bridges Subjected to Thermal Loading," M.S. Thesis, Kansas State University, Manhattan, KS, 2007.

Sisman, B. and Fu, C.C., "Use of Integral Piers to Enhance Aesthetic Appeal of Grade Separation Structures (04-4021)," *The Proceedings of Transportation Research Board*, January 11–15, Washington, DC, 2004.

Thanasattayawibul, N. "Curved Integral Abutment Bridges," PhD Dissertation, Civil and Environmental Engineering, University of Maryland, College Park, MD, pp. 72–81, 2006.

Wasserman, E.P. and Walker, J.H., "Integral Abutments for Steel Bridges," Virginia DOT, October 1996, http://www.virginiadot.org/business/resources/semi-integral-20.pdf.

CHAPTER 17

AASHTO, *AASHTO Guide Specifications for LRFD Seismic Bridge Design*, 2nd Edition, American Association of State Highway and Transportation Officials, Washington, DC, with 2012 Interim.

Ahmed, M.S., "Seismic Assessment of Curved Bridges Using Modal Pushover Analysis," PhD Dissertation, Department of Civil and Environmental Engineering, University of Maryland, College Park, MD, 2010.

Ahmed, M.S. and Fu, C.C., "Seismic Assessment of Long Curved Bridges Using Modal Pushover Analysis: A Case Study," *The Proceedings of the 6th International Conference on Bridge Maintenance, Safety and Management*, July 8–12, Como, Italy, 2012.

Allen, D.E. and Murray, T.M., "Design Criterion for Vibrations due to Walking," *Engineering Journal*, 4th quarter, AISC, 117–129, 1993.

ANSYS, *ANSYS Mechanical User Guide*, ANSYS Inc., Canonsburg, PA, 2012.

Aviram, A., Mackie, K.R., and Stojadinovic, B., "Guidelines for Nonlinear Analysis of Bridge Structures in California," Report No. UCB/PEER 2008/03, University of California, Berkeley, CA, 2008, http://peer.berkeley.edu/publications/peer_reports/reports_2008/web_PEER803_AVIRAM_etal.pdf.

BSI, "Steel, Concrete and Composite Bridges: Specification for Loads," British Standard BS 5400, Part 2, Appendix C, British Standards Institute, London, UK, 1978.

Cai, C.S., Albrecht, P., and Bosch, H., "Flutter and Buffeting Analysis. I: Finite-Element and RPE Solution," *Journal of Bridge Engineering*, 4(3), 174–180, 1999.

Cantieni, R. and Heywood, R., "OECD DIVINE Project: Dynamic Interaction between Vehicle and Infrastructure Experiment, Element 6, Bridge Research: Report on the Tests Performed in Switzerland and Australia," Draft EMPA Rep., EMPA, Dübendort, Switzerland, 1997.

Chopra, A.K. and Goel, R.K., "A Modal Pushover Analysis Procedure for Estimating Seismic Demands for Buildings," *Earthquake Engineering and Structural Dynamics*, 31(3), 561–582, 2002.

Cole, D.J., and Cebon, D., "Validation of Articulated Vehicle Simulation," *Vehicle System Dynamics*, 21, 197–223, 1992.

Federal Emergency Management Agency (FEMA), "NEHRP Guidelines for the Seismic Rehabilitation of Buildings," FEMA 273/October 1997, Applied Technology Council (ATC-33 Project), Redwood City, CA, 1997.

Federal Emergency Management Agency (FEMA), "NEHRP Recommended Provisions for Seismic Regulations for New Buildings and Other Structures," FEMA 450, Washington, DC, 2003.

Fu, C.C. and Ahmed, M.S., "Nonlinearity in Bridge Structural Analysis," in *Focus on Nonlinear Analysis Research*, G. Padovani and M. Occhino, editors, Mathematics Research Developments series, Nova Science Publishers, 2012.

Green, M.F. and Cebon, D., "Dynamic Response of Highway Bridges to Heavy Vehicle Loads: Theory and Experimental Validation," *Journal of Sound and Vibration*, 170(1), 51–78, 1994.

Gu, Y., Fu, C.C., and Aggour, M.S. "Topographic effect on Seismic response of high-pier Bridge subjected to Oblique incidence waves." *The Proceedings of iBridge Conference*, August 11–13, 2014, Istanbul, Turkey.

LRFD Guide Specifications for Design of Pedestrian Bridges, 2nd Edition, American Association of State Highway and Transportation Officials, Washington, DC, 2009.

LSTC, "LS-DYNA Theoretical Manual", Livermore Software Technology Corporation, Livermore, CA, 1998. http://www.lstc.com/.

MacDougall, C., Green, M.F., and Shillinglaw, S., "Fatigue Damage of Steel Bridges due to Dynamic Vehicle Loads," *Journal of Bridge Engineering*, 11(3), 2006.

Mast, R., Marsh, L., Spry, C., Johnson, S., Griebenow, R., Guarre, J., and Wilson, W., *Seismic Design of Bridges—Design Examples 1–7 (FHWA-SA-97-006 thru 012)*, USDOT/FHWA, September 1996.

Murray, T.M., Allen, D.E., and Ungar, E.E., "Floor Vibrations due to Human Activity," *AISC Steel Design Guide #11*, Chicago, IL, 1997, https://www.aisc.org/store/p-1556-design-guide-11-floor-vibrations-due-to-human-activity-see.aspx.

NHI Course No. 13063 "Seismic Bridge Design Applications," April 25, Publication No. FHWA-SA-97-017 (Part One) and -018 (Part Two), 1996.

OHBDC, Ontario Highway Bridge Design Code, 3rd edition, Highway Engineering Division, Ministry of Transportation and Communication, Downsview, Ontario, CA, 1991.

Priestly, M.J.N., Seible, F., and Calvi, G.M., *Seismic Design and Retrofit of Bridges*, Wiley, New York, 1996.

SAP2000®, "Integrated Software for Structural Analysis & Design," Computers and Structures Inc., Berkeley, CA, 2007, http://www.csiamerica.com/products/sap2000.

Scanlan, R. H., "The Action of Flexible Bridges under Wind. Part I: Flutter Theory," *Journal of Sound and Vibration*, 60(2), 187–199, 1978a.

Scanlan, R. H., "The Action of Flexible Bridges under Wind. Part II: Buffeting Theory," *Journal of Sound and Vibration*, 60(2), 202–211, 1978b.

TM 5-1300, *Structures to Resist the Effects of Accidental Explosions*, Department of Army, Washington, DC, November 1990.

Varadarajan, G. "An Assessment of 'Bridge-Friendliness' of Heavy Vehicles with Different Suspensions," M.S. Thesis, Queen's University, Kingston, Ontario, Canada, 1996.

Winget, D.G., Marchand, K.A., and Williamson, E.B., "Analysis and Design of Critical Bridges Subjected to Blast Loads," *Journal of Structural Engineering*, 131(8), 1243–1255, 2005.

Xie, H. "The Effects of Surface Roughness and Vehicle Suspension Type on Highway Bridge Dynamics," M.S. Thesis, Queen's University, Kingston, Ontario, Canada, 1999.

Yang, Y.B., Yau, J.D., and Wu, Y.S., *Vehicle-Bridge Interaction Dynamics with Applications to High-Speed Railways*, World Scientific Publishing, Singapore, 2004.

CHAPTER 18

Baker, J.M., "Construction Techniques for Segmental Concrete Bridges," *The Long Span Concrete Bridge Conference*, Hartford, CN, March 1980.

Hickerson, T.F., *Route Location and Design*, 5th Edition, McGraw-Hill, New York, 1959.

Kumar, K., Senthil, K.N., Koshy, V., and Ananthanarayanan, K., "Automated Geometry Control of Precast Segmental Bridges," *The 25th International Symposium on Automation and Robotics in Construction*, Vilnius, Lithuania, June 2008.

LoBuono, J.P., "MC3D—Evolution of Segmental Bridge Software, Engineering Professional," The official publication of the Wisconsin Society of Professional Engineers, Vol. 3, No.5, September/October 2005.

Wang, S.Q., and Fu, C.C., "Visual Bridge Geometry Modeling User's Manual," The Bridge Engineering Software and Technology (BEST) Center, University of Maryland, College Park, MD, 2013, http://best.umd.edu/software/.

Index